PROCEEDINGS OF THE 3RD INTERNATIONAL CONFERENCE ON MODELING AND COMPUTATION IN ENGINEERING (CMCE 2014), WUXI, CHINA, 28–29 JUNE 2014

Modeling and Computation in Engineering III

Editors

Lei Zhang

Shanghai Institute of Technical Physics of the Chinese Academy of Sciences, Shanghai, China

Liquan Xie

Department of Hydraulic Engineering, Tongji University, Shanghai, China

CRC Press
Taylor & Francis Group
Boca Raton London New York Leiden

CRC Press is an imprint of the
Taylor & Francis Group, an **informa** business
A BALKEMA BOOK

Published by:
CRC Press/Balkema
P.O. Box 447, 2300 AK Leiden, The Netherlands
e-mail: Pub.NL@taylorandfrancis.com
www.crcpress.com – www.taylorandfrancis.com

ISBN 13: 978-1-138-02680-3 (hbk)
ISBN 13: 978-1-315-74945-7 (ebk)

Typeset by V Publishing Solutions Pvt Ltd., Chennai, India

Modeling and Computation in Engineering III – Zhang & Xie (Eds)
© 2014 Taylor & Francis Group, London, ISBN 978-1-138-02680-3

Table of contents

Modeling and Computation in Engineering III – Zhang & Xie (Eds)
© 2014 Taylor & Francis Group, London, ISBN 978-1-138-02680-3

Preface

The demands of modelling and computation in engineering are rapidly growing as a multidisciplinary area with connections to the engineering, mathematics and computer science. The 3rd International Conference on Modelling and Computation in Engineering (CMCE 2014) will be held during 28–29 June 2014 in Wuxi, China and hopes to offer researchers an occasion to exchange their experiences of modelling and simulation technology, which include modelling technology, simulation technology and tools, computation methods and their engineering applications. The 2014 Hydraulic Engineering and Environment Workshop (HEEW 2014) is an important track of CMCE 2014 and focuses on the frontier research of hydraulic engineering and environmental technology.

More than 40 technical papers are published in the proceedings. Each of the papers has been peer reviewed by recognized specialists and revised prior to acceptance for publication. The topics of the contributed papers covered the fields as:

- civil engineering, e.g. long-span multiple bar steel tower, deep foundation excavation
- hydraulic structures, e.g. slope stability in high earth-rockfill dams, overtopping risk for Xiashan Reservoir in China
- hydropower and management, e.g. chaos ant colony optimization in cascade reservoir power generation
- coastal reclamation and environmental assessment, e.g. reclamation in Bohai Bay, a largescale reclamation project in the Oujiang River estuary
- flood control, irrigation and drainage, e.g. revised model AREM for forecasting precipitation, flood control and/or drainage hydraulic control structures in urban areas
- water resources and water treatment, e.g. newly developed slightly acidic electrolyzed water (pH6.25–6.53), water quality assessment in Yanjiaxiang and Xiyangsi Power Station (China), drought monitoring in Dongting Lake basin based on RS, cold water emission for CNOOC Tianjin floating LNG project in China
- environmental management and sustainability, e.g. Dianchi watershed in west China, case studies between the Rhine River, Thames, Biwa Lake and Qinhuai River, Taihu Lake in China
- waste management and environmental protection, e.g. water exchange for offshore artificial islands in semi-open bays, sediment movements under typhoon for Tianwan nuclear power plant in China
- pollution and control, e.g. GIS-based source pollution analysis on Xichong River Basin in west China
- modelling technology in hydraulic engineering, e.g. LU-SGS implicit algorithm proposed in DanDong port project, hydrodynamic-wave coupling models proposed for sediment transport in Oujiang River Estuary, hydrodynamics in Poyang Lake, hydrodynamics at XiangShan harbor, tidal modelling of Chengmai Bay at Hainan Island (China)
- geology and geography, e.g. GPS elevation fitting, 3D geological modelling, forest soil's physico-chemical properties by fractal theory
- mechanics in engineering, e.g. grit trajectory models in grinding engineering, mechanism of progressive cavity pumps (oil production)

- numerical software and applications
- other modelling and computation technology, e.g. climate modelling by permutation entropy, support vector machine for settling particle in petroleum engineering, inventory optimization within manufacturing company, GPU model for fingerprint matching, modelling in the optical camera imaging process.

Last but not least, we would like to express our deep gratitude to all authors, reviewers for their excellent work, and Léon Bijnsdorp, Lukas Goosen and other editors from Taylor and Francis Group for their wonderful work.

Modeling and Computation in Engineering III – Zhang & Xie (Eds)
© 2014 Taylor & Francis Group, London, ISBN 978-1-138-02680-3

Study on the performances different supported Ziegler-Natta catalysts in propylene polymerization

Xiudong Xu & Bingquan Mao
College of Materials Science and Engineering, Beijing University of Chemical Technology, Beijing, China
Beijing Research Institute of Chemical Industry, SINOPEC, Beijing, China

Xianzhi Xia & Zhong Tan
Beijing Research Institute of Chemical Industry, SINOPEC, Beijing, China

ABSTRACT: Four kinds of supported catalysts A, B, C and D were prepared and compared. The catalyst composition and microstructure was studied via surface properties measurement, IR and XPS. The polymerizing ability was evaluated and the corresponding polymers were characterized using isotacticity measurement, ^{13}C-NMR and DSC. As a result, the polymerization activities are in the order of A > B > D > C, which are in accordance with the order of surface area value of the four kind of catalysts. XPS data showed that the $Ti_{2P3/2}$ peaks of catalyst move to a lower energy range, and the value is in the order of Cat A > B > D = C. This indicates that the $TiCl_4 \cdot MgCl_2 \cdot$ DBP (di-butyl phthalate) binding energy of different catalyst are in the same order with that of XPS. Which lead to the isotactic index and the [mm] value of polymers of four catalysts are in the order of Cat A > B > D > C and the T_m of four catalysts are in the order of Cat A > B > C > D. Therefore, we show that the combination ability of complexe $TiCl_4 \cdot MgCl_2 \cdot$ DBP could influence the polymer isotacticity and the polypropylene melting point increases with the isotacticity.

1 INTRODUCTION

Ziegler-Natta (Z-N) catalyst plays an important role in olefin polymerization, and plenty of polypropylene is produced by using this catalyst (Lou, 2010, Hakim, 2008, Tan, 2011, Albrecht A, 2008, Anantawaraskul, 2009, Zhang, 2011 & Tan, 2013). Generally, the Z-N catalyst including titanium compounds and organo-aluminum compounds is used to obtain polymers in the olefin polymerization field. Some patents (Tan, 2011, Zhang, 2011 & Tan, 2013) disclose that a catalyst comprising a solid catalyst component mainly consisting of titanium, magnesium, chlorine and electron donor compounds, an organo-aluminum compound as a cocatalyst and an organo-silicon compound as an agent for improving stereoregularity, is used in the preparation of polypropylene to obtain propylene polymer. Recently, the researches for catalysts mainly focus on the following aspects, improving the polymerization activity of catalysts, improving the stereotegularity of polyolefins, improving the hydrogen response of catalysts, improving the particle morphology olefin polymers, reducing the residual chloride in the polymers and so on.

Titanium-based Z-N catalysts of the four generation for isospecific propylene polymerization contain $TiCl_4$, $MgCl_2$ or $Mg(OEt)_2$ as a support, and alkyl esters of aromatic diacids as modifying organic components of solid catalysts. To date, most Z-N catalysts in the fabrication of propylene use di-butyl phthalate (DBP) as internal electron donor (Tan, 2011, Albrecht, 2008, Anantawaraskul, 2009, Zhang, 2011 & Tan, 2013). Supported Z-N catalysts can replicate the morphology of carrier particle, and the calalyst morphology can be replicated by polymer. That is, the polymer particle morphology can be controlled through the carrier.

It has found some increasing applications for the characterization of polymers prepared using supported catalysts (Albrecht, 2008, Anantawaraskul, 2009 & Zhang, 2011).

Cat A, B, C and D are four kinds of supported Z-N catalysts. The composition and microstructure of these four kinds of catalysts were analyzed via surface area measurement, FTIR and XPS. The polymerization evaluation of the four catalysts was conducted, and the isotactcity and melting point of the polymer were characterized. Also, the kinetics behaviors of the four catalyst was compared when they were be used to polymerize propylene.

2 EXPERIMENTAL

2.1 *Materials*

Propylene of industrial grade, Isopropyl Alcohol, Ethanol, $MgCl_2$, toluene and $TiCl_4$ (Beijing Chemical Company, Beijing, China), triethylaluminium (TEA, 95% purity) and cyclohexanemethyldimethyloxyilane (CHMDMS) were used without further purification. Nitrogen of an extra pure grade (99.999%) was dried through a molecular sieves 4 Å column. Dibutyl phthalate (DNBP, Beijing chemical company, Beijing, China), hexane and decane (Tianjin Chemical Reagent Company) were used after dehydration with molecular sieves (4 Å).

2.2 *Catalyst preparation*

Cat A was prepared according to Pat. EP2617739 (Tan, 2013). All operations of A catalyst were carried out under dry nitrogen using standard Schlenk techniques. Catalyst preparation was carried out via the reaction of $Mg(EtO)_2$ with $TiCl_4$ in the presence of DNBP.

Cat B was prepared according to Pat. CN1,091,748 (Mao, 1994). The internal donor and Carrier were Diisobutyl phthalate (DIBP) and $MgCl_2{\cdot}nEtOH$, respectively.

Cat C was prepared according to Pat. WO03,082,930 (Evangelisti, 2003). The internal donor was DIBP, the Carrier was $MgCl_2{\cdot}nEtOH$.

Cat D was prepared according to Pat. CN101,054,424 (Li, 2007). The internal donor is DNBP, the Carrier was $Mg(EtO)_2$.

2.3 *Propylene polymerization*

In a 500 cm^3 reactor purged with nitrogen and propylene, 200 cm^3 of anhydrous n- decane was added and the system was charged with propylene. Then, 50 mg of catalyst, prescribed amount of $AlEt_3$ (Al/Ti = 100 molar ratio) and cyclohexanemethyl dimethyloxyilane (CHMDMS) (Al/Si = 20 molar ratio) were added when the polymerization temperature was 70°C. Polymerization was carried out under the pressure of 0.03 Mpa for 4h. The polymer was recovered through filtration and dried for certain time until it was dryness. The polymerization rate was determined by measuring the rate of monomer consumption from a mass flow meter.

2.4 *Characterization*

The amount of internal electron donors in a catalyst dissolved in a proper solvent was determined by Waters 600E high-performance liquid chromatographic (HLPC). Symptoms XPS was determined by Sigma Probe SYSTEM X-ray photoelectron spectrometer of UK VG company. The $Ti_{2p3/2}$ binding energy is determined by the $Au_{4f7/2}$ in 84.0 eV peak.Symptoms IR was determined by Magna-IR 760 infrared spectrometer of U.S. Nicolet company, solid sample was wrapped with liquid paraffin and ground into a paste in the glove box, and sandwiched KBr film to analyse.

The isotacticity index (II) of polymer was obtained by refluxed heptane extraction for 4h.

^{13}C-NMR spectra of isotactic PPs were run with a Bruker DMX 400 spectrometer operating at 100.6 MHz, using 10–20 mg/cm^3 solutions in deuterated dichlorobenzene at 110_C (Conditions: 5 mm probe; pulse repetition 4 s; number of scans, 20,000).

Polypropylene melting point was determinated by Perkin-Elmer DSC-7. Second temperature curve was used as the standard Determination of melting point.

3 RESULTS AND DISCUSSION

3.1 *Catalyst structure and characterization*

As shown in Table 1 the surface area of catalyst A was the largest. The surface area size of four kinds of catalyst was in the order Cat A > B > D > C. It could be speculated that polymerization activity of catalysts may be as this sequence Cat A > B > D > C.

Because The negativity of $MgCl_2$ is lower than the $TiCl_4$, and the electron can be transfer to Ti by the way Mg → Cl → Ti, which makes electron cloud density increase around activity center Ti. The literatures in ref (Fregonese, 2002, Mori, 1999 & Magni, 1996) reported the binding energy of $TiCl_4$ loaded $MgCl_2$ is 458.8 eV. When internal electron donor coordinated on an atom, it would increase the electron density on this atom and drive the atomic binding energy (Pan, in press) to lower energy band. When internal electron donor was added to catalyst system and coordinated on Ti and Mg atom, it can drive the $Ti_{2P3/2}$ peak to lower energy band similarly. As a result, the $Ti_{2P3/2}$ binding energy of coordination compound $TiCl_4 \cdot MgCl_2 \cdot$ DBP preparation should be lower than that of $TiCl_4 \cdot MgCl_2$. As shown in Table 1, the binding energy of Cat A and B is 458.6 eV, which are lower than that of $TiCl_4 \cdot MgCl_2$. In addition, the binding energy of Cat C and D are 458.5 eV and 458.7 eV respectively, which are also lower than 458.8 eV. This is because these catalysts provide electron to the atom, and the $Ti_{2P3/2}$ peak move to lower energy band. These $Ti_{2P3/2}$ peaks of the four catalysts move to lower energy band in the order Cat A > B > C = D. This indicates that the $TiCl_4 \cdot MgCl_2 \cdot$ DBP binding energy of catalyst A is the largest, and that of catalyst B is smaller, and that of Cat D and C are the smallest.

Infrared (IR) spectrum result proved the fact above. As shown in Figure 1, the ν(C = O) of DNBP appeared strong absorption peak at 1728.8 cm^{-1}, the ν(= C-O-C) appeared absorption peak at the 1284.4 cm^{-1}. The ν(C = O) of DIBP appeared strong absorption peaks at the 1726.2 cm^{-1}, the ν(= C-O-C) appeared absorption peak at the 1281.8 cm^{-1}. Because the internal donors interact with $MgCl_2$ and $TiCl_4$ in the catalyst, the ν(C = O) of DBP in catalyst would move to the low wave number area, while the ν(= C-O-C) would move to the high wave number area. As shown in Figure 2, the ν(C = O) of DBP in Cat A appeared absorption peak at 1698.6 cm^{-1}, and the ν(= C-O-C) appeared absorption peak at 1338.7 cm^{-1}; the ν(C = O) of DBP in Cat B appeared absorption peak at 1684.0 cm^{-1}, and the ν(= C-O-C) appeared absorption peak at 1311.5 cm^{-1}; the ν(C = O) of DBP in Cat C appeared absorption peak at 1693.5 cm^{-1}, and the ν(= C-O-C) appeared absorption peak at 1313.9 cm^{-1}; and the ν(C = O) of DBP in Cat D appeared absorption peak at 1698.1 cm^{-1}, and the ν(= C-O-C) appeared absorption peak at 1314.7 cm^{-1}. As far as the scope that the ν(= C-O-C) of DBP moved to high wavenumber in the four kinds of catalysts, the Cat A migration scope was the biggest, the other catalysts were close. As far as the scope that the ν(C = O) of DBP moved to low wavenumber in the four kinds of catalysts, the Cat B migration scope was biggest, the other catalysts were close. The migration scope of ν(C = O) was not as good as that of the ν(= C-O-C). Therefore, it can be infer that the combination ability of complexe $TiCl_4 \cdot MgCl_2 \cdot$ DBP was in the order Cat A > B > C = D.

Table 1. Catalyst characterization data.

Cat.	Ti %	DBP content %	Surface area m^3/g	$Ti_{2p3/2}$ ev
A	2.11	DNBP:10.45	3.350e + 02	458.1
B	2.3	DIBP:9.76	3.291e + 02	458.3
C	2.17	DIBP:10.81	2.318e + 02	458.6
D	2.19	DNBP:11.46	2.595e + 02	458.6

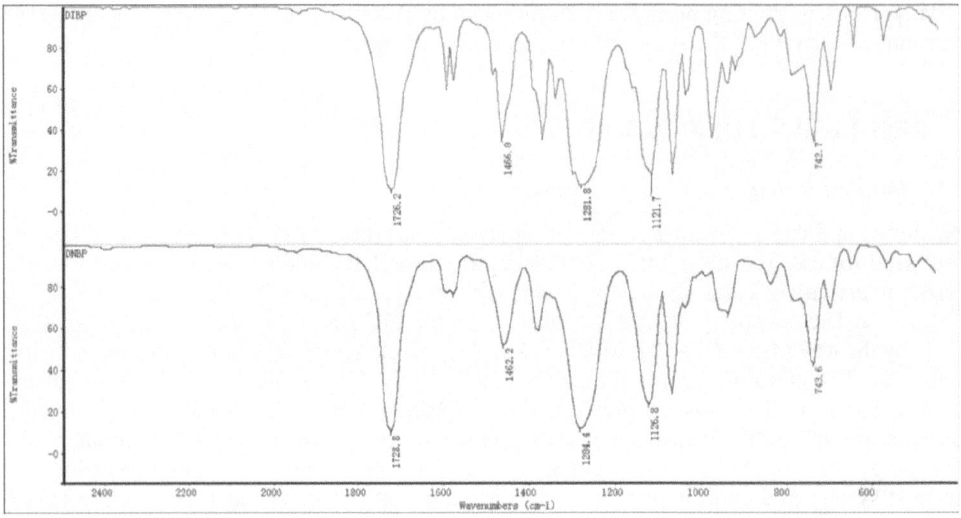

Figure 1. IR spectrum of DIBP and DNBP.

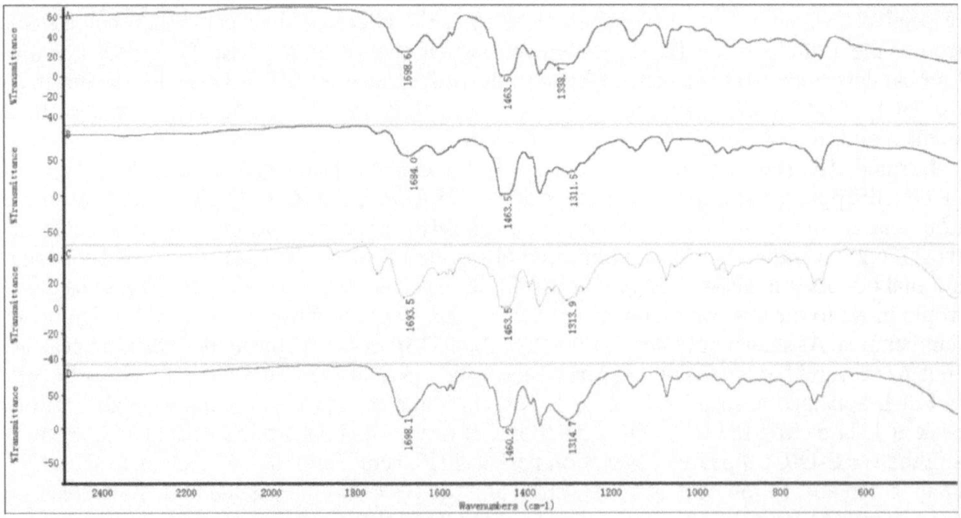

Figure 2. IR spectrum of Cat A, B, C and D.

3.2 *Reaction kinetics of propylene polymerization*

The influence of polymerization temperature on the polymerization of propylene can be considered from two perspectives (Gao, 2011 & Zhong, 2003). The first is that in a solvent propylene solubility may reduce with the temperature increasing, and the polymerization monomer concentration is reduced. The second is that the active center of catalyst would change due to temperature difference. On one hand, the structure of the active center changes. On the other hand, the rate of polymerization constant increases. When traditional Z-N catalysts were used in the production of polypropylene, the temperature of propylene polymerization is usually between 60~90°C. In this work, polymerization experiments conducted at 70°C for four different catalysts.

From Figure 3, it can be seen the polymerization activity decay rate of Cat A is similar to Cat C and Cat B, while the decay rate of Cat D is the highest in the four catalysts. The

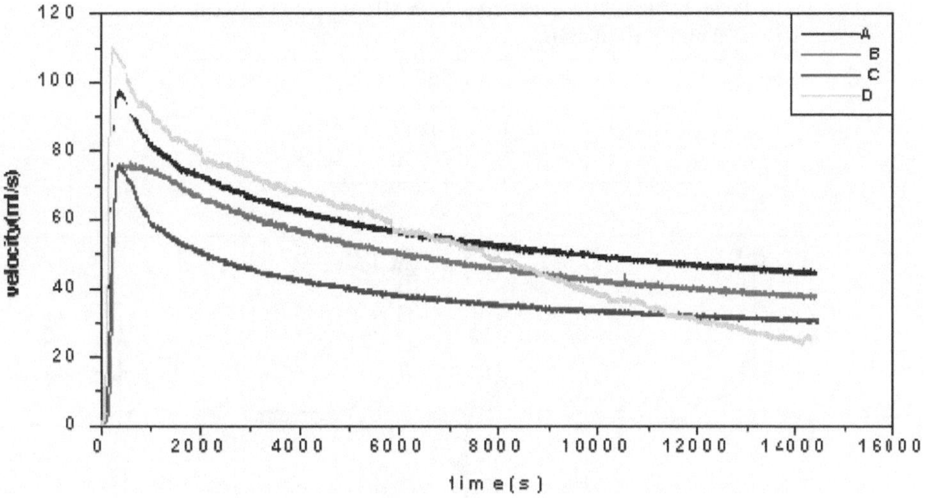

Figure 3. 70°C polymerization kinetics curve.

polymerization decay rate of different catalysts increased in the order Cat B < A ≈ C < D. The polymerization activity of different catalysts decreased in the order Cat A > B > D > C.

According to the above results, the decay rate of Cat A was similar to Cat C, Cat B was slightly faster than Cat A catalyst, Cat D was the fastest one in the four catalyst; the polymerization activity of Cat A was equal to Cat B catalyst, Cat C was the lowest.

Barino and Scordamaglia (Barino, 1998) indicated that the best electron donors can located with $MgCl_2$ strongly. The strong coordination between electron body and the $MgCl_2$ was believed to be the reason of energy stability of catalyst surface, which can prevent the migration of active species from the surface and lead to the raise of the catalyst activity. It was induced the coordination capacity between electron donor and $MgCl_2$ is in the order Cat A > B > D > C. And this was consistent with surface area and XPS order of catalysts in table 1.

3.3 *PPs isotacticity*

Polypropylene isotacticity plays an important role in controlling its quality and developing a new brand resin. isotactic index characterized the spatial structure arrangement of polypropylene resin. If isotactic index of PP was higher, the continuous regular repeat configuration unit of main chain would be more, and the degree of crystallinity would be higher. So high isotactic index is advantageous to the hardness, stiffness modulus, yield strength of PP resin. At the same, melting point, thermal stability and resistance to ageing of the PP were increased with high isotactity index, but the toughness, impact resistance, elongation performance were declined. Isotactic index was high, the resin processing requirements were higher (Li, 1992).

From Table 2 it can be seen the isotactic index of polymers with Cat A was the highest in the four kinds catalysts, and catalyst B was higher than the other two, and it was in the order Cat A > B > C > D. The results of ^{13}C-NMR proved the fact. It means that the PP of Cat A has more regular repeat configuration unit than the other three. This indicates that high isotacticity PP could be produced by Cat A and B.

Buscio et al analyzed the structure of polymer chain of catalytic system (Busico, 1991, Busico, 1999 & Talarico, 2002). They pointed out that there were three active centers, which existed in a dynamic change. The main difference was that the Ti atom ligands (such as Cl, Mg, O) were different, which made the chemical environment around Ti-C bond change.

As shown in Figure 4, two adjacent magnesium ion ligand L1, L2 (such as Cl, ester) with the active center were presented, active center as C2 or false C2 symmetry was high isotactic

Table 2. Isotactic index and ^{13}C-NMR of polypropylene used different catalysts.

Cat	A	B	C	D
Isotactic index/%	98.38	98.14	97.16	96.54
mm/%	97.51	96.76	96.28	95.87

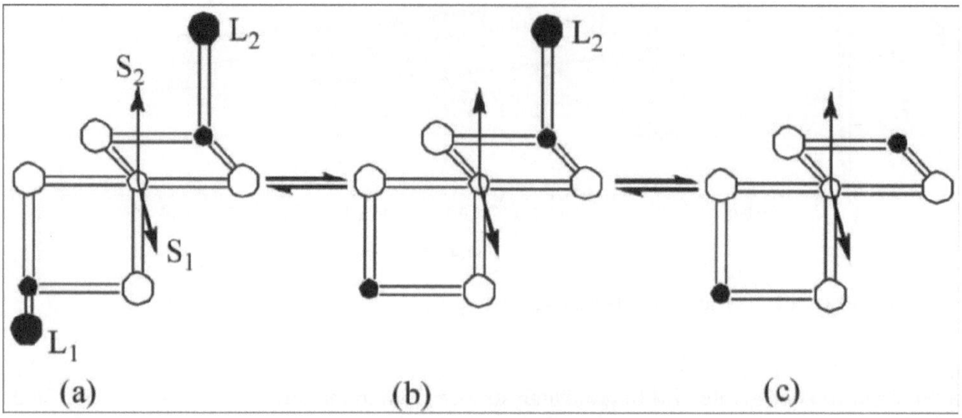

Figure 4. Model of Possible active species of highly isotactic (a), isotactoid (b), and syndiotactic (c) propagation. S1 and S2 are vacant positions 27 ○ = Ti; ● = Ti or Mg; ○ = Cl; ● = Cl or donor, S1 and S2 are vacant positions.

center. When there was only one adjacent magnesium ion ligand with the active center and the ligand volume was very large to prevent monomer coordinate on both sides of active center, this active center was also a high isotactic center. The polymerization product was high isotactic polypropylene.

As Figure 4 (b) shows when two adjacent magnesium ion ligands with the active center lost one, activity center as C1 symmetry was medium isotactic center. The polymerization product was medium isotactic polypropylene.

As Figure 4 (c) shows when two adjacent magnesium ion ligands with the active center both lost, activity center as chain end control was syndiotactic center, The polymerization product was low isotactic polypropylene.

Therefore the different method for preparing catalyst made the catalyst composition and active center distribution different, which could get different isotacticity polypropylene. Cat A had higher isotactic center so that the polypropylene showed higher isotactic index. This was consistent with IR spectrum and XPS results from Figure 2 and Table 1. The stronger combination ability of complex $TiCl_4 \cdot MgCl_2 \cdot$ DBP is, the higher isotactic index and [mm] value of polymer is. It showed that a stable donor coordination in the vicinity of active species gave higher steroregularity than a labile donor coordination in the polymer chain (Chadwick, 2001 & Chadwick, 2001).

3.4 *PPs melting point*

Polypropylene melting point(Tm) was an important parameter of resin grade. Several polypropylene melting points were determined below.

It can be seen from Table 3 that polymer of A catalyst has higher melting point than the other polymers in all polymerization temperature obviously. The T_m of four catalysts were in the order Cat A > B > C > D, This result was not consistent with IR and XPS result in Figure 2 and Table 1. Therefore the combination ability of complex $TiCl_4 \cdot MgCl_2 \cdot$ DBP could affect the melting point. It was also verified that polypropylene melting point increases with the isotacticity.

Table 3. Melting points of polypropylene used different catalysts.

Cat	A	B	C	D
T_m/°C	161.53	161.20	161.03	160.37

4 CONCLUSIONS

This paper selected four kinds of supported catalysts A, B, C and D, and analyzed catalyst composition, surface area, IR, XPS. Moreover, we studied the peformances in propylene polymerization with four catalysts at polymerization temperature 70°C. According to the results of the catalysts, we summarize the corresponding rule.

1. The surface area size of four kinds of catalyst was in the order Cat C < D < B < A. These $Ti_{2P3/2}$ peaks of catalyst move to lower energy range in the order Cat D = C < B < A. This indicates that the $TiCl_4 \cdot MgCl_2 \cdot$ DBP binding energy of Cat A is largest, and that of Cat B is smaller, and that of Cat D and C are smallest. IR spectrum result proved the fact.
2. To the four catalysts, the decay rate of Cat A was similar to Cat C, Cat B was slightly faster than Cat A, Cat D was the fastest one in the four catalyst; the polymerization activity of Cat A was equal to Cat B, Cat C was the lowest. And this was consistent with surface area and XPS sequences of catalysts in Table 2. So this showed that Cat A was suitable for long period running in propylene polymerization and copolymerization device, such as gas phase polymerization apparatus of multi-reactor, combined polymerization device of gas phase with bulk unit.
3. The isotactic index and the [mm] value of polymers with four catalysts were in the order Cat A > B > D > C. This showed the PP of high isotacticity could be produced by Cat A and B. This was consistent with IR spectrum and XPS results of catalyst from Figure 2 and Table 1. The stronger combination ability of complexe $TiCl_4 \cdot MgCl_2 \cdot$ DBP is, the higher isotactic index of polymer would be.

The T_m of four catalysts were in the order Cat A > B > C > D, This result was not consistent with analysis result with IR and XPS in Figure 2 and Table 1. Therefore the combination ability of complexe $TiCl_4 \cdot MgCl_2 \cdot$ DBP could affect the melting point. It was also verified by polypropylene melting point increasing with the isotacticity.

REFERENCES

Albrecht, A., Brull, R., Macko, T., Sinha, P. & Pasch, H. 2008. Analysing the chemical composition distribution of ethylene-acrylate copolymers: Comparison of HT-HPLC, CRYSTAF and TREF, Macromol Chem Phys, 209, 1909–1919.

Anantawaraskul, S., Somnukguandee, P., Soares, J.B.P. & Limtrakul, J. 2009. Application of a Crystallization Kinetics Model to Simulate the Effect of Operation Conditions on Crystaf Profiles and Calibration Curves, *J Polym Sci Part B-Polym Phys*, 47, 866–876.

Barino, L., Scordamaglia, R. 1998. Modeling of isospecific Ti sites in $MgCl_2$ supported heterogeneous Ziegler-Natta catalysts, *Macromol Theory Simul*, 7, 407–419.

Busico, V., Cipullo, R., Monaco, G., Talarico, G., Vacatello, M., Chadwick, J.C., Segre, A.L. & Sudmeijer, O. 1999. High-resolution ^{13}C NMR configurational analysis of polypropylene made with $MgCl_2$-supported Ziegler-Natta catalysts. 1 the "model" system $MgCl_2/TiCl_4$ 2,6-dimethylpyridine/ $Al(C2H5)_3$, *Macromolecules*, 32, 4173–4182.

Busico, V., Corradini, P., Martino, L.D., Graziano, F. & Iadicicco, A. 1991. Propene polymerization in the presence of $MgCl_2$-supported Ziegler-Natta catalysts, 4[a)] Effects of Lewis bases on polymer stereochemistry, *Makromol Chem*, 192, 49–57.

Chadwick, J.C. 2001. Advances in propene polymerization using $MgCl_2$-supported catalysts. Fundamental aspects and the role of electron donors, *Macromol Symp*, 173, 21–35.

Chadwick, J.C., Morini, G., Balbonti, G., Camurati, I., Heere, J.J.R., Mingozzi, I. & Testoni, F. 2001. Effects of Internal and External Donors on the Regio- and Stereoselectivity of Active Species in $MgCl_2$-Supported Catalysts for Propene Polymerization, *Macromol Chem Phys*, 202, 1995–2002.
Evangelisti, D. & Colllina, G. 2013. Magnesium dichloride-ethanol adducts and catalyst components obtained therefrom *WO Patent*. 03,082,930.
Fregonese, D., Glisenti, A., Mortara, S., Rizzi, G.A., Tondello, E. & Bresadola, S. 2002. $MgCl_2/TiCl_4/AlEt_3$ catalytic system for olefin polymerisation: a XPS study. *J Mol Catal A: Chem*, 178, 115–123.
Gao, F.T., Xia, X.Z. & Mao, B.Q. 2011. $MgCl_2$-Supported Catalyst Containing Mixed Internal Donors for Propylene Polymerization, *J Appl Polym Sci*, 120,36–42.
Hakim, S., Nekoomanesh, M. & Aghaei, N.M.2008. Investigating the behaviour of a bi-supported SiO2/TiCl4/THF/$MgCl_2$ catalyst in slurry ethylene polymerization: activity and molecular weight, Iran Polym J, 17, 209–216.
Li, S.H., Lan, L.H., Zhang, H.P. & Liu, J.K. 2007. A preparation method of solid catalyst for olefin polymerization. *CN Patent*. 101,054,424.
Li, Y.G. &. Chen, N.G. 1992. "*Liquid phase bulk polymerization of polypropylene production and Application*". Beijing: Sinopec press, 271.
Lou, J.Q., Tu, S.T. & Fan, Z.Q. 2010. Polypropylene Chain Structure Regulation by Alkoxysilane and Ether Type External Donors in $TiCl_{(4)}$/DIBP/$MgCl_{(2)}$-$AlEt_{(3)}$ Ziegler-Natta Catalyst, *Iran Polym J*, 19, 927–936.
Magni, E. & Somorjai, G.A. 1996. Preparation of a model Ziegler-Natta catalyst: electron irradiation induced titanium chloride deposition on magnesium chloride thin films grown on gold, *Surface Sci*, 345, 1–16.
Mao, B.Q., Yang, J.X., Li, Z.L., Yang, A.C., Zheng, Y. & Xia, X.Z. 1994. Spherical catalyst used olefin polymerization. *CN Patent*. 1,091,748.
Mori, H., Hasebe, K. & Terano, M. 1999. Variation in oxidation state of titanium species on $MgCl_2$-supported Ziegler catalyst and its correlation with kinetic behavior for propylene polymerization, *Polymer*, 40, 1389–1394.
Pan, J.L. 1987. "Photoelectron spectroscopy applied in Organic Chemistry". Beijing: Chemical Industry Press, 50.
Talarico, G., Busico, V. & Cipullo, R. 2002. Modeling the active species of $MgC1_2$-supported Ziegler-Natta catalysts: a quantum mechanical(Re-) visitation, *lst Blue sky conference on catalytic* olefin polymerization & 2nd JAIST/JLPO workshop on Ziegler-Natta Catalysts [C], Sorrento (Italy).
Tan, Z., Xu, X.D., Yan, L.A. & Zhou, Q.L. 2011. Advances of compounding internal electron donor used in Ziegler-Natta polypropylene catalyst, *Chem Ind Eng Prog,* 30, 2628–2636.
Tan, Z., Xu, X.D., Yan, L.A., Zhou, Q.L., Song, W.W., Yin, S.S., Yu, J.H., Li, F.K., Wang, Y. & Ren, C.H. 2013. Catalyst carrier for olefin polymerization, solid catalyst component and catalyst. *EP Patent*. 2617739, 2013.
Zhang, H.X., Lee, Y.J., Park, J.R., Lee, D.H. & Yoon, K.B. 2011. Control of molecular weight distribution for polypropylene obtained by a commercial Ziegler-Natta catalyst: Effect of a cocatalyst and hydrogen, *J Appl Polym Sci*, 120, 101–108.
Zhong, C.F., Gao, M.Z. & Mao, B.Q. 2003. Effect of high polymerization temperature on the microstructure of isotactic polypropylene prepared using heterogeneous $TiCl_4/MgCl_2$ catalysts, *J Appl Polym Sci*, 90, 3980–3986.

Modeling and Computation in Engineering III – Zhang & Xie (Eds)
© 2014 Taylor & Francis Group, London, ISBN 978-1-138-02680-3

Non-point source evaluation of Xichong River basin in Xichong County based on GIS

Min Ke
State Key Laboratory of Hydraulics and Mountain River Engineering, College of Water Resource and Hydropower, Sichuan University, Chengdu, China
College of Water Resources and Hydropower and Architecture, Yunnan Agriculture University, Kunming, China

Li Zhou, Xiaodong Li & Tianqi Ao
College of Water Resources and Hydropower, Sichuan University, Chengdu, China
State Key Laboratory of Hydraulics and Mountain River Engineering, Sichuan University, Chengdu, China

ABSTRACT: Extracting aqueous and dividing watersheds for Xichong River (Xichong County) and studying the contribution of the township pollution as well as delineating heavily contaminated areas. This study is based on GIS and integrates standard load method, output coefficient model and source strength coefficient method, etc. Urban sewage, rural sewage and fertilizer pollution of residents in the area are major sources of pollution point source pollution, which is up to 87% the proportion while the most polluted town is Jin town. Using GIS spatial analysis capabilities to directly reflect the spatial variation of non-point source pollution in the township to provide guidance for the subsequent direction of pollution.

1 INTRODUCTION

Point source pollution is also known as non-point source (NPS) pollution, generally refers to contaminants in rainfall, especially in heavy rain, or scouring under the snow, and carrying through natural processes such as runoff and eventually imports the receiving water, including rivers, lakes, reservoirs, the Gulf, etc., and causes eutrophication of water or other forms of pollution. Over years, the rural NPS pollution in some places has become a major source of pollution to local surface and groundwater. Sichuan overall nutritional status of water body was at Intermediate levels at 2009. And the status from excellent to poor nutritional are the Jinsha River, the Yangtze River, Minjiang, Jialing, Tuo, while the Jialing situation is more serious, basically in mild and moderate rich nutrition, respectively 9.2% and 1.4%.

NPS pollution intensity has obvious geographical characteristics including soil, land use and topography etc., which means it is difficult to monitor and evaluate the NPS pollution contribution rate of water pollution. Therefore, the pollution load of the exact area is not only difficult, but also a requirement for contamination assessment. Fast and accurate delineation of agricultural NPS pollution in key areas is the important foundation for the work. GIS combined with the storage, management, analysis, display and application of geographic Information, as well as powerful spatial analysis, has become a powerful tool for studying NPS pollution.

Nengzhong LEI, Dapeng HUANG with GIS have achieved the extraction of data model required and analysis algorithms to obtain multi-scale agricultural NPS pollution degree of risk maps, which provides the targeted area to control agricultural NPS pollution. Shiying

JIANG, Pengli HAN using GIS, pollution load method, the pressure–state–next response (PSR) conceptual model studied in water district evaluate livestock breeding, soil erosion, loss of fertilizer, human waste, rural sewage, aquaculture, pollution garbage as well as analyzing the spatial distribution of the characteristics and key regional pollution. Using GIS spatial analysis, Yanqiong YE has analyzed the spatial and temporal variation of rural NPS pollution in Guangdong Province, which intuitively reflected the spatial and temporal regions of rural NPS pollution dynamic changes.

In this paper, based on spatial analysis of ArcGIS, 2012 as the horizontal, and the application of source strength coefficient and output coefficient model and pollution load, we assessed the NPS pollution of Xichong River (Xichong County territory), which effectively showed the pollution situation in all regions.

2 STUDY AREAS

Xichong, the tributary of right bank of the Jialing River, originates in Xichong County Taiping Bishan southeast, passing through Shunqing District, Xichong County, at 30° 40'–31° 0'36"E, 105° 40'–106° 0'5"N. The river length of approximately is 110 km while upper, middle and lower sections are called Rainbow River, Slippery-beaches River, Constant-sub River. The drainage area is 1000 km^2, flowing together at South Gate dam, Nanchong City, to Jialing River.

Jin, Taiping, Yixing, Fengming, Qingshi, Shuangfeng, Gaoyuan, Renhe, Lianchi, Changlin, Zhanshan, Hongqiao, Jinquan, Zhongling, Fuan, Guanfeng, Shuangluo, Dongdai, Chelong, Xianglong, Shuangjiang, Fenghe, Yongqing, where are the 23 villages Xichong River flows through (Fig. 1). Xichong River water quality monitoring data provided by County Environmental Protection Agency at 2011 showed that ecological function of the river has sig-

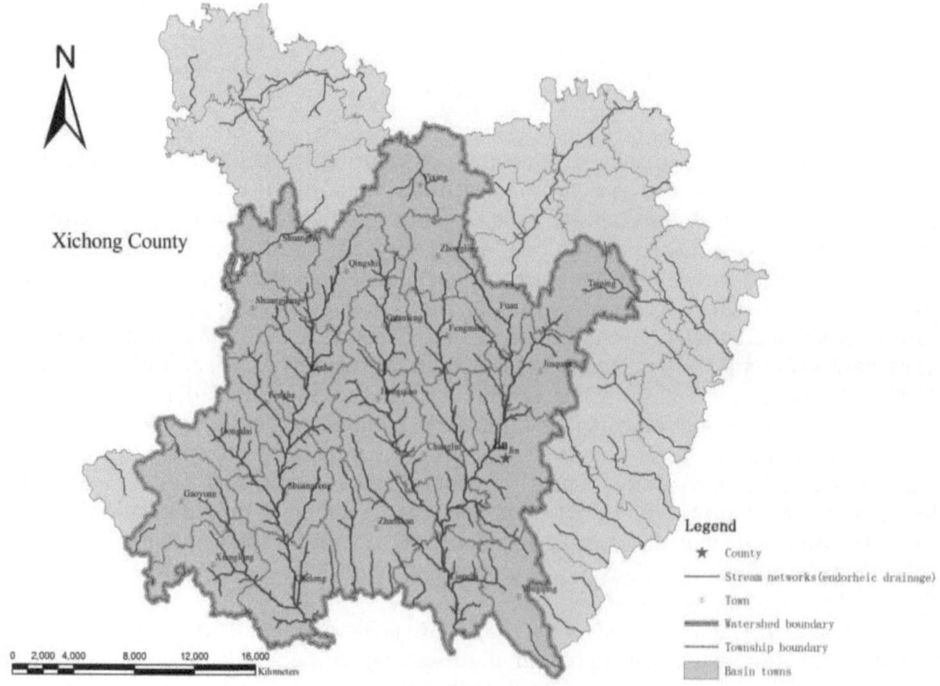

Figure 1. Stream network and division of influence regions.

nificantly weakened, reducing the self-purification capacity, as well as water polluted heavily. Two sections (Yan hometown, West Yang Temple station) monitoring showed that the two sections were worse than the annual total nitrogen water V. The ecological environment along the river basin and watershed residents' lives and production were subject to a great deal of influence.

3 METHODS AND DATASET

3.1 *Methods*

3.1.1 *Pollution load method*

Pollution load method aims at comparing the amount of emission of different sources with the emission criteria, in order to obtain the amount on the same scale, which is simple and comprehensive. The following formula is:

$$p_i = \sum p_{ij} = \frac{G_{ij}}{Si},\ p_i = \frac{p_{ij}}{p_i}$$

Where p_i = I-pollutant pollution load in the entire scope of the evaluation, $10^6 \cdot m^3 \cdot a^{-1}$; p_{ij} = I-pollutant pollution load in the j sources, $10^6 \cdot m^3 \cdot a^{-1}$; G_{ij} = I-pollutant annul discharge into river in the j sources, $t \cdot a^{-1}$; S_i = evaluation criteria for I-pollutant, mg/L; K_{ij} = I-pollutant pollution load ratio in the j sources.

3.1.2 *Method of various types of pollution emissions*

Urban sewage

COD, NH_3-H: Wi $(t \cdot a^{-1}) = 365 \cdot p \cdot a_i \cdot 10^{-6}$ a_i: COD = 40 g/p; NH_3-H = 8 g/p
TN, TP: Wi $(t \cdot a^{-1}) = p \cdot b_i \cdot 10^{-3}$ b_i: TN = 1.87 kg/p; TP = 0.214 kg/p

Emissions of rural residents sewage work as the same formula.

Livestock pollution

COD, NH_3-H: Wi$(t \cdot a^{-1}) = 365 \cdot p \cdot c_i \cdot 10^{-6}$ c_i:COD = 40 g/p; NH_3-H = 8 g/p
TN, TP:

Calculated in accordance with the output coefficient model, specifically the coefficient selected in Table 1.

Urban runoff

COD, NH_3-H: Wi$(t \cdot a^{-1}) = D_i \cdot P \cdot A \cdot R \cdot S \cdot L$ D_i: COD = 50t/a; NH_3-H = 12t/a

Table 1. The export coefficient classification and value.

Type	Dry land	Paddy land	Orchard	Forestry	Waste	Town*	Large animal	Pig	Sheep	Poultry	Population**
TN	0.23	0.15	0.08	0.238	1.49	1.1	73.2	13.9	14	0.6	18.7
TP	0.0092	0.0094	0.005	0.015	0.051	0.024	3.1	1.42	0.45	0.05	2.14

*The unit of all the previous items is $t \cdot km^{-2} \cdot a^{-1}$.
**The unit of all the previous items is $t \cdot (10^4 ca \cdot a)^{-1}$.

Where P = population correction factor, p/150000; A = area correction factor, p/100; R = rainfall correction factor, 1.4; S = rainwater collection pipe network coverage correction factor, 0.6; L = terrain correction factor, 2.5.

TN, TP: Combined with the use of land for construction output coefficient calculation, as shown in Table 1.

Fertilizer

Standard farmland source strength coefficient:

$$\text{COD, NH}_3\text{-H: Wi (t}\cdot\text{a}^{-1}) = E_i \cdot C_j \cdot \text{H} \cdot S \cdot F \cdot R\ E_i: \text{COD} = 0.149\ kg \cdot hm^{-2} \cdot a^{-1};$$
$$\text{NH}_3\text{-H} = 0.03 kg \cdot hm^{-2} \cdot a^{-1}$$

Where C_j = slope correction factor, 1.055~1.090; H = crop correction factor, 1.036; S = soil correction factor, 0.7; F = chemical fertilizers correction factor, 0.8; R = rainfall correction factor, 1.3.

TN, TP: Combined with the use of land for construction output coefficient calculation, as shown in Table 1.

3.2 *Dataset and analysis*

30 m accuracy DEM data of Sichuan Province derives from Computer Network Information Center, Chinese Academy of International Scientific data mirror sites (http://datamirror.csdb.cn)

The township population, including the agricultural and the urban population, land area, the amount of livestock and other data originate from West County Environmental Protection Agency, West County Animal Husbandry Bureau, West County Civil Affairs Bureau and West County Bureau of Statistics respectively.

Using ArcGIS hydrologic analysis tools to extract river basin watershed from DEM, and for the clearly show of the larger flow of the river basin, such as not too dense, selecting 1500 for threshold in grid river generation, we've finally got Xichong River (Xichong County) River Figure (Figure 1).

4 ANALYSIS AND RESULTS

4.1 *Spatial analysis of sources*

According to the calculation method of the respective sources of emissions we can get the results about pollution of five major sources such as urban sewage, rural sewage, livestock, urban runoff and fertilizers, shown in Table 2.

By analyzing the above table, pollution load index for urban sewage is 449.22, and the most serious polluted counties are Jin, ChanL and ShuanF. Pollution load index of rural sewage is 255.14, and the most serious polluted counties are ChanL, ZhanS, CheL. Pollution load index of livestock is 86.09, and the most serious polluted county is CheL. Pollution load index of urban runoff is 49.944, and the most serious polluted counties are Jin and ChanL. Pollution load index of fertilizers is 214.70, and the most serious polluted counties are ChanL, ZhanS, Jin.

The proportion of the various sources is shown in Figure 2 of the total pollution, which the three largest sources are urban sewage, rural sewage and fertilizer pollution.

4.2 *Results*

Using GIS Spatial Analysis function for raster data to distributed simulating NPS pollution in watershed can identify the key polluted areas.

Taking the above data into Xichong River (Xichong County) raster attributes table to obtain the total equivalent standard load diagram, shown in Figure 3.

Table 2. The general contribution of pollution in 2012.

	Pollution load index (t.a^{-1})/(mg.L^{-1})						Ratio%
County name	Urban sewage	Rural sewage	Livestock	Urban runoff	Fertilizers	Total	Cumulative load ratio
Jin	320.47	16.09	4.39	10.692	17.47	369.112	34.98%
TaiP	5.02	3.21	1.07	0.619	2.61	12.529	36.17%
YiX	1.74	1.13	0.38	0.248	0.87	4.368	36.58%
FengM	4.82	13.07	4.93	2.332	10.66	35.812	39.97%
QingS	4.87	18.43	5.63	2.781	12.75	44.461	44.19%
ShuanF	26.36	11.80	4.31	2.146	8.89	53.506	49.26%
GaoY	1.93	5.97	1.86	0.950	4.58	15.29	50.71%
RenH	16.95	19.11	4.72	3.388	14.79	58.958	56.29%
LianC	2.13	14.51	5.18	1.968	12.07	35.858	59.69%
ChanL	37.79	23.07	5.69	5.450	23.79	95.79	68.77%
ZhanS	2.24	20.68	5.46	2.693	18.18	49.253	73.44%
HongQ	2.15	9.18	5.24	1.572	10.77	28.912	76.18%
JinQ	1.64	9.49	3.64	0.148	6.60	21.518	78.22%
ZhonL	2.98	8.30	3.60	1.397	6.50	22.777	80.37%
FuAn	1.87	7.36	4.65	1.459	5.87	21.209	82.38%
GuanF	2.84	11.69	3.83	1.802	8.47	28.632	85.10%
ShuanL	0.13	1.31	0.46	0.214	1.10	3.214	85.40%
DongD	1.87	9.31	4.09	1.388	7.51	24.168	87.69%
CheL	3.76	20.37	5.70	2.717	15.01	47.557	92.20%
XiangL	3.33	12.00	3.41	1.985	9.11	29.835	95.03%
ShuanJ	1.00	2.89	1.02	0.568	2.47	7.948	95.78%
FengH	2.70	11.37	5.13	1.967	8.81	29.977	98.62%
YongQ	0.57	4.84	1.88	1.463	5.80	14.553	100.00%
Total	449.22	255.14	86.09	49.944	214.70	1055.24	

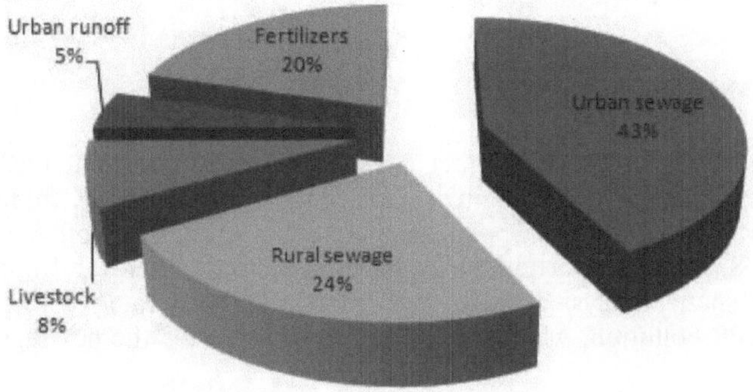

Figure 2. The percentage of each pollution source.

Above all, on integrated pollution the most serious polluted county can be seen is Jin, mainly because of the county location, population, more developed, more intensive river system, so pollution is critical. From the sources analyzed:

1) Affected by urban sewage pollution, the most polluted towns are Jin, ChanL, ShuanF. 2) According to the total standard pollution load of rural sewage, such towns as ChanL, ZhanS and CheL need to pay more attention from local government. 3) As for livestock

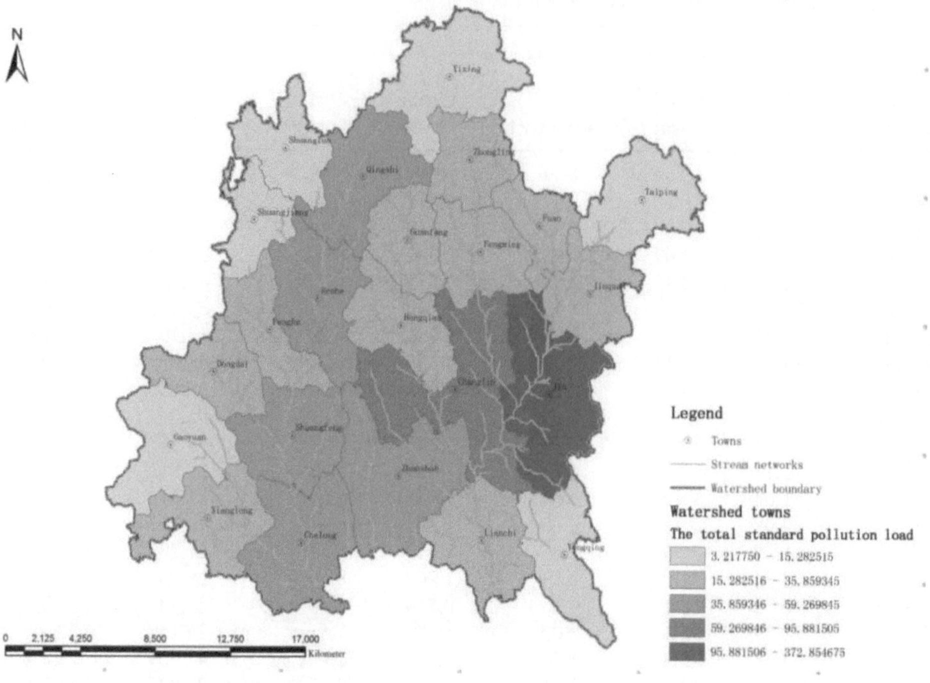

Figure 3. The total standard pollution load.

pollution, the overall amount of pollutants into the river is insignificant in space, the most serious of which is CheL. Excepting YonQ, GaoY, TaiP, ShuanL and YiX in the assessment district, the others should pay more attention from local government. 4) Affected by urban runoff pollution, the most polluted towns are Jin and ChanL. 5) As far as we concerned the fertilizer pollution, the most polluted towns are ChańL, ZhanS, Jin.

5 CONCLUSION

GIS software can quickly and stably extract and classify the river basin, to facilitate research. The most serious sources of pollution in sequence in Xichong River (Xichong County) are urban sewage, rural sewage, fertilizer pollution, livestock and urban runoff, and the first three of these is up to 87% the proportion which should pay more attention from local government. Using GIS spatial analysis function, we can fast and accurately define the key areas of agricultural NPS pollution, which can be some guidance for the next agricultural NPS pollution research.

ACKNOWLEDGEMENT

This study is supported by the National Natural Science Foundation of China (Project No. 50979062); the International S&T Cooperation Projects, the Ministry of Science and Technology of P. R. China (No. 2012DFG21780); the S&T Project of Sichuan Environmental Protection (No.11HBT-01); the National Basic Search Fund Project of Sichuan University (No.2010SCU22005); the Development Project of Sichuan Province's Flash Flood Real-Time Prediction System and the development and application Project of the integrated storm-flood simulation and prediction system of flood disaster mountain basin in Sichuan Province.

REFERENCES

Deng, Q.X. & Zhou, C.Q. (1994). Xichong River fish fauna composition and annual changes. *Sichuan Teachers College*, (1), 71–72.

Du, D.Q. (1992). Nanchong River Pollution Prevention. *Water Resources Protection*, (1), 37–42.

Hao, F.H., Chen, H.G. & Yang, S.T. (2006). *Non-point source pollution model*. Beijing: China Environmental Science Press.

He, Q., Wang, Z.X. & Li, H.P. (2001). Geographic Information System (GIS) applications in non-point source pollution control. *Chongqing Environmental Science*, (1), 24–26, 49.

Huang, J.L., Hong, H.S. & Zhang, L.P. (2006). GIS-based non-point source pollution control and watershed model regionalization. *Environmental Sciences*, (4), 119–124.

Jiang, S.Y., Han, P. & Jia, Z.B., (ed.) (2010). The PSR Evaluation of Agricultural Nonpoint Source Pollution About Water Diversion Project in Danjiangkou Reservoir and GIS-based Spatial Analysis Features. *Journal of Agricultural Sciences*, 29(011), 2153–2162.

Kang, W.Y. (2012). *Estimation and prediction of Poyang Lake area non-point source pollution load*. Nanchang: Nanchang University.

Lei, N.Z. & Huang, D.P. (2007). The risk assessment of agricultural nonpoint source pollution based on GIS. *Chinese Agricultural Science Bulletin*, 23(12), 381–385.

Liu, J., Xu, L. & Zhang, J.Q., (ed.) (2012). Spatial and temporal distribution of Eutrophication Survey in Sichuan Province. *China Environmental Monitoring*, 28(5), 6–8.

Liu, R.M., Shen, Z.Y. & Ding, X.W., (ed.) (2008). Application output coefficient model to estimate the upper reaches of the Yangtze non-point source pollution load. *Journal of Agricultural Sciences*, 27(2), 667–682.

Mo, M.H., Yang, J. & Gu, S., (ed.) (2010). Poyang Lake area of non-point source pollution load. *People Yangtze River*, 41(17), 51–53.

Novotny, V. & Olem, H. (1993). *Water quality: prevention, identification and management of diffuse pollution*. New York: Van Nostrand Reinhold Company.

Wei, X.P., Liu, H.Y. & Li, H., (ed.) (2012). Sources of non-point sources of nitrogen and phosphorus load and control measures in White Deer River in Sichuan Province. *China Rural Water and Hydropower,* (3), 36–38.

Wen, Q.Q. (2012). The Analysis of the Research Progress of Nonpoint Source Pollution in China. *Environmental protection and recycling economy*, 6, 66–68.

Xu, L.K., Wang, Q.J. & Xiang, S.L., (ed.) (2012). Gan River downstream impact studies of various types of non-point source pollution. *East China Transportation University*, 29(1), 48–53.

Yan, Y.Q., Zhang, J.E., Li, Y.X., Li, Y. & Wu, R.S. (2013). The research of spatial-temporal differentiation of agricultural nonpoint source pollution in Guangdong province based on GIS. *Journal of Agricultural Sciences*, (2), 369–377.

Zhang, W.L., Xu, A.G. & Ji, H.J. (ed.) (2004). Problems exist in Chinese agricultural nonpoint source pollution control. *Chinese Agricultural Science*, 37(7), 1026–1033.

Zhong, D.S., & Zhang, H.W., (ed.) (2005). Impact of pollution load method pollution of the water environment. *China Water & Wastewater*, (5), 101–103.

Modeling and Computation in Engineering III – Zhang & Xie (Eds)
© 2014 Taylor & Francis Group, London, ISBN 978-1-138-02680-3

Study on hydrodynamics and salinity environment after a large-scale reclamation project in the Oujiang River Estuary

Ting Xu
Key Laboratory of Engineering Sediment of Ministry of Communications, Tianjin Research Institute for Water Transport Engineering, M.O.T, Tianjin, China
School of Environmental Science and Engineering, Tianjin University, Tianjin, China

ABSTRACT: This paper studies a large-scale reclamation project in the Oujiang River Estuary (ORE), which is a very complex macrotidal estuary with many islands located in Zhejiang Province, China. The large-scale reclamation engineering effects on hydrodynamics and salinity in the ORE are predicted and analyzed by the verified model. The simulated results indicate that the proposed layout of the reclamation is designed in accordance with the laws of flow in the sea area. Such a large-scale reclamation project will have some effects in the ORE but within acceptable limits. The wide range reclamation project describing here is very representative and it has much reference values to other similar projects in other country.

1 INTRODUCTION

In recent years, more and more countries (including Japan, China and Korea) have tried to overcome land-based bottlenecks with the design and construction of marine reclamation projects (Li et al., 2010; T.I. Kim et al., 2006).The large-scale reclamation required is expected to have a considerable impact on the marine environment (Nijkamp and Yim, 2001; Kleija et al., 2003). The tidal circulation and salinity transport processes in estuaries generally are very complex and highly dynamic due to the presence of sinuous coastlines, islands, channels, shoals and some man-made structures (Chen, X., 2004; Yang Z.,2009; Park Y W et al., 2006). Complexities can also arise because of the mixture of salt and fresh water as well as the synchronous impacts of runoff, tides,waves, winds and offshore currents (Chen, S.-N., 2009; Gong, W., Shen, J., 2011). The long-term stability and sustainability of the hydrological salinity environment are necessary for the continued prosperity of aquaculture. Understanding the tidal circulation and salinity transport mechanism in estuaries is of great importance for coastal engineering, disaster prevention, water security and intertidal zone aquiculture.

The 388-km-long Oujiang River, which is located in the Zhejing Province of China, is the second largest river in Zhejiang Province and the fifth largest river in China. The Oujiang River is a mountain river influenced by strong tides with an average range of over 4 m. The Oujiang River Estuary (ORE) is bifurcated into the North and South Branches (Fig. 1). Outside the North and South Branches, there are numerous islands, densely covered shoals, and interlaced shoals and channels with complex topography. Wenzhou Shoal, lying between Lingkun Island and Niyu, is a large-scale, well-developed mouth bar in the ORE. It is significant to reclaim Wenzhou Shoal between the two islands (Fig. 1). First, the reclamation may create abundant land with an area of 88 km^2 to alleviate the pressure of a lack of land in Wenzhou. Second, the reclamation can exploit and utilize the deepwater coastline between Zhuangyuan'ao and Qingshan Island to promote tourism and the fishing industry on Dongtou Island. The reclamation will be a grand territory regulation project incorporating enclosing tideland for cultivation and linking islands to promote harbor industry and tourism.

In this study, hydrodynamic and salt transport model based on the Environmental Fluid Dynamics Code (EFDC) was developed and applied to study tidal circulation and salt

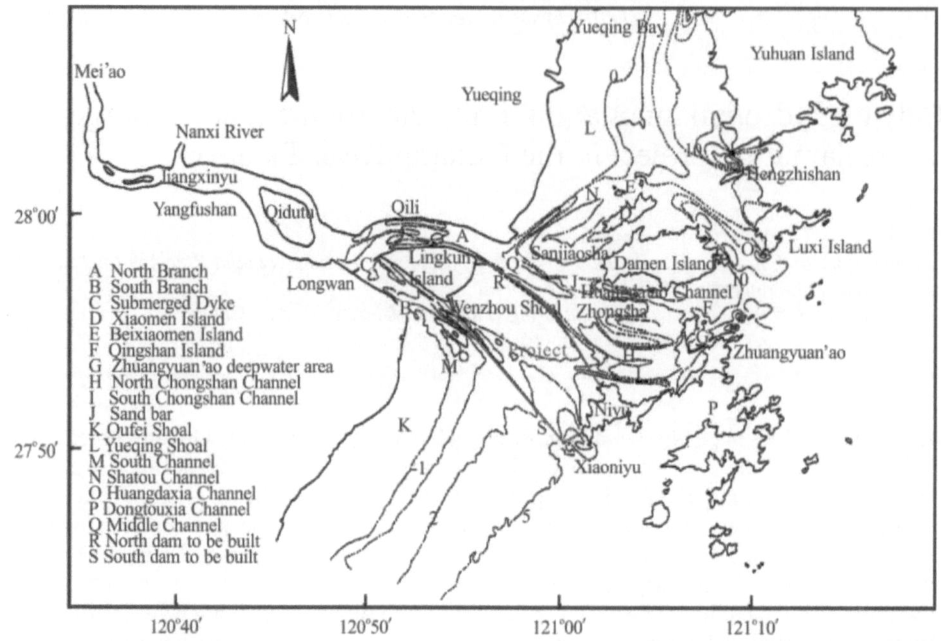

Figure 1. Sketch of the ORE.

transport in the ORE, which contains not only the river reach inside the mouth but also the region outside the mouth. The model was performed to comprehend the influence of the large-scale reclamation on ambient flow pattern and redistribution of tidal prism as well as salinity distribution around the ORE.

2 MATHEMATICAL MODEL SYSTEM

The model used in this study is modified from 'Environmental Fluid Dynamics Code" (EFDC) model, a public-domain modeling package for simulating three-dimensional flow, transport and biogeochemical processes in surface water systems (Hamrick, 1992). The EFDC model can be run in barotropic or baroclinic and in 2D or 3D models. It also has the capability of simulating wetting and drying processes (Ji et al., 2001), and has been tested and verified in a wide range of hydrodynamic and environmental studies (e.g. Kuo et al., 1996; Shen et al., 1999; Jin et al., 2001). Dynamically coupled transport equations for turbulent kinetic energy, turbulent length scale and temperature are also solved. The two turbulence transport equations implement the Mellor-Yamada level 2.5 turbulence closure scheme.

The governing equations for the EFDC hydrodynamic model equations are given below.

The continuity equation is:

$$\frac{\partial(m\zeta)}{\partial t}+\frac{\partial(m_y Hu)}{\partial x}+\frac{\partial(m_x Hv)}{\partial y}+\frac{\partial(mw)}{\partial z}=Q_H \tag{1}$$

The momentum equation is:

$$\frac{\partial(mHu)}{\partial t}+\frac{\partial(m_y Huu)}{\partial x}+\frac{\partial(m_x Hvu)}{\partial y}+\frac{\partial(mwu)}{\partial z}-\left(mf+v\frac{\partial m_y}{\partial x}-u\frac{\partial m_x}{\partial y}\right)Hv$$
$$=-m_y H\frac{\partial(g\zeta+p)}{\partial x}-m_y\left(\frac{\partial h}{\partial x}-z\frac{\partial H}{\partial x}\right)\frac{\partial p}{\partial z}+\frac{\partial}{\partial z}\left(m\frac{1}{H}A_V\frac{\partial u}{\partial z}\right)+Q_u \tag{2}$$

$$\frac{\partial(mHv)}{\partial t}+\frac{\partial(m_yHuv)}{\partial x}+\frac{\partial(m_xHvv)}{\partial y}+\frac{\partial(mwv)}{\partial z}-\left(mf+v\frac{\partial m_y}{\partial x}-u\frac{\partial m_x}{\partial y}\right)Hu$$
$$=-m_xH\frac{\partial(g\zeta+p)}{\partial y}-m_x\left(\frac{\partial h}{\partial y}-z\frac{\partial H}{\partial y}\right)\frac{\partial p}{\partial z}+\frac{\partial}{\partial z}\left(m\frac{1}{H}A_v\frac{\partial v}{\partial z}\right)+Q_v \tag{3}$$

$$\frac{\partial p}{\partial z}=-gH\frac{\rho-\rho_0}{\rho_0}=-gHb \tag{4}$$

The salinity and temperature transport equation is:

$$\partial_t(mHS_a)+\partial_x(m_yHuS_a)+\partial_y(m_xHvS_a)+\partial_z(mwS_a)$$
$$=\partial_z\left(mH^{-1}K_v\partial_zS_a\right)+Q_S \tag{5}$$
$$\partial_t(mHT)+\partial_x(m_yHuT)+\partial_y(m_xHvT)+\partial_z(mwT)$$
$$=\partial_z\left(mH^{-1}K_v\partial_zT\right)+Q_T \tag{6}$$

Where h is the mean water depth; H is the wave height; u, v, w are the velocities in the direction of x, y and z coordinates; ζ is the free-water level; $m = m_xm_y$, m_x *and* m_y are the transformation scale factors in horizontal coordinates. S_a is salinity; T is temperature; Q_S and Q_T is the source and sink of the salinity and temperature.

The Mellor-Yamada Level 2.5 turbulence closure model (Mellor and Yamada 1982) is applied to solve diverse problems in EFDC, and parameters are determined by the following formulas:

$$\begin{cases} A_V=\phi_VA_0ql=(1+36R_q)^{-1}(1+6R_q)^{-1}(1+8R_q)A_0ql \\ A_B=\phi_BK_0ql=(1+36R_q)^{-1}K_0ql \\ R_q=-\dfrac{gH\partial_zb}{q^2}\dfrac{l^2}{H^2} \end{cases} \tag{7}$$

Where q is the turbulence intensity; l is the turbulence mixing-length; R_q is the Richardson number; b is the density gradient; ϕ_A *and* ϕ_R are stability functions that depend on R_q, A_0 and K_0 are constant, usually 0.4 and 0.5 respectively.

3 NUMERICAL SIMULATIONS AND RESULTS

3.1 *Impact on tidal current field*

From the current field at flood and ebb before and after the construction of reclamation project. It was found that after construction, only a small scale of current field near the engineering field is changed and this doesn't change the characteristics of current field of the sea much. Before the reclamation engineering, Wenzhou shoal between Lingkun Island and Niyu Island is a weak flow area, the exchange water flow between Zhongshui and Nanshui channel and the flow across Wenzhou shoal is limited. After the construction of reclamation project, the channel of Zhongshui and Nanshui still keeps the motion of reciprocating flow.

Fig. 2 showed the difference of the flow field engineering. It was found that the impact of the current speed is mainly near the engineering area and maximum change is less than 0.10 m/s. The current change in most areas is weakened and only the current in south channel is increased.

The reclamation engineering is nearby the south and north estuary of Oujiang. After the construction, the in and out tidal capacity from south and north estuary has some changes in

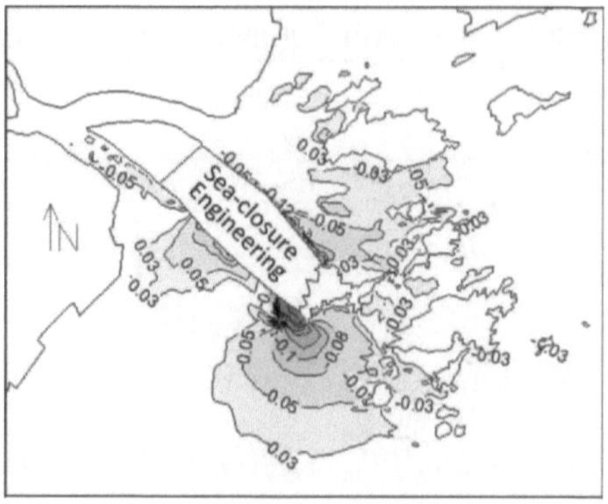

Figure 2. Current change before and after the before and after the construction of reclamation reclamation engineering.

a certain extent. The changes of the in and out tidal capacity in the south and north estuary of Oujiang was shown in Table 1. From Table 1, it was found that the flood tidal capacity increase 1.1% and the ebb tidal capacity decrease 0.6%; The flood amount in the south estuary of Oujiang decreased 2.3% and the ebb tide amount increased 3.9% after the reclamation engineering construction. It was concluded that the volume change of inflow and outflow in south and north estuary of Oujiang is limited after the construction of the reclamation engineering.

If we consider a worst case that the upstream of Oujiang with a discharge runoff of 100 years return period and there will be an astronomical spring tide at the outside of the ORE. After the construction of the reclamation enginering, the maximum tidal level rise on Oujiang costal is only 1–2 cm. It was concluded that the construction of reclamation enginering will not affect the flood discharge of Oujiang.

3.2 *Impact on salinity field*

The distribution of salinity in the ORE changes with the tide fluctuation. Affected by the freshwater runoff flowing into the ORE, the average salinity is the least in the ORE especially in the river section above Qidutu, where the average salinity is between 1 and 8psu. The isoline of 1 psu salinity is up to the west boundary of Jiangxin islet during tidal flood and it is to the east boundary of Qidutu during ebb tide.

In the external sea area of the ORE, the isoline of salinity diffusion is ligules, which is swinging with the flood and ebb tide. Being coincided with the tidal current distribution, there are about four primary salinity ligules spreading to different directions from the ORE when the ebb tide is coming. It includes the spread passing through the south estuary to Niyu, the spread passing through the north estuary to Huangda'ao, the spread passing through Shatou water channel to Yueqing Bay and the spread passing through the inside of Yueqing Bay to the south of Hengzhi mountain. When the flood tide is coming, the four primary salinity ligules are forming in the opposite direction to the ORE from east to west. The isoline of average salinity between 20 and 26 psu spreads to the middle part of Damen island.

The salinity distribution for before and after the reclamation was shown in Fig. 3. It was found the salinity distribution in the ORE sea doesn't change much after the reclamation and the basic form of the salinity distribution is similar to that of pre-engineering construction. The salinity value has only a small change locating in a very small sea area region near the site of reclamation engineering. After the reclamation construction, the salinity isoline has a tendency to distribute along the reclamation embankment. The salinity of Huangda'ao is

Table 1. Tidal capacity change of south and north branch of the ORE before and after the reclamation engineering.

			Unit: 10^8 m^3	
	North		South	
Items	Flood	Ebb	Flood	Ebb
Before	4.18	3.97	1.21	1.53
After	4.23	3.95	1.18	1.59
Tidal capacity change	1.1%	−0.6%	−2.3%	3.9%

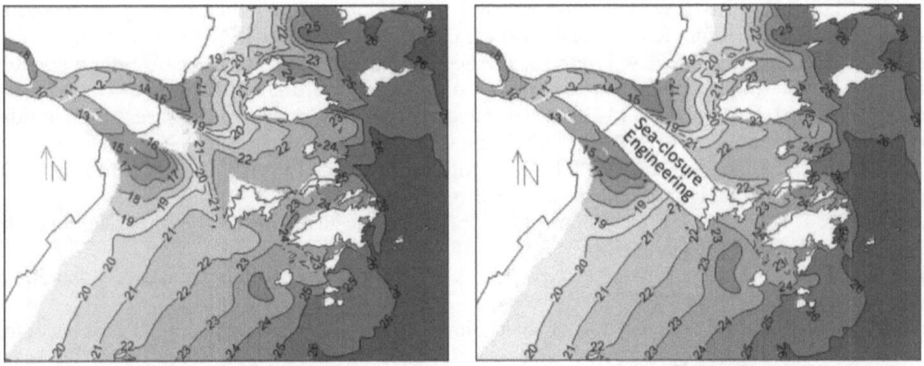

a) before project b) after project

Figure 3. Sanility change before and after the project.

decreased by about 0.2 psu. Besides, there also has some changes of salinity distributions in the south Yueqing bay and north Xiaomen island. The salinity is increased by about 0.4 psu.

4 CONCLUSIONS

The effects of the large scale engineering project of Wenzhou-shoal reclamation project in the ORE on the strong tidal and islands estuary was studied by the EFDC model. The numerical simulation studies focus on the change of velocity field, tidal capacity and salinity field caused by the reclamation engineering. The main conclusions are as the following:

1. The trend of Wenzhou shoal is basically identical with the main trend of the two strands of fluctuation tidal current in the south and north mouth. The reclamation doesn't change the dynamic axis of the ORE tidal current and causes obvious change of tidal structure;
2. The reclamation project of Wenzhou shoal causes a small influence on the peripheral hydrodynamic and salinity field around the project site. It also has few effects on the flood discharge of Oujiang, the mariculture of Yueqing bay, Wenzhou port channel and the deep water resources of Qingshan island and Zhuangyuan'ao;
3. According to the monitoring results of the phase stage during the project performing, the correctness of our numerical results was confirmed. The wide range reclamation project describing here is very representative and it has much reference values to other similar projects in other country.

REFERENCES

Chen, X. 2004. Modeling hydrodynamics and salt transport in the Alafia River estuary, Florida during May 1999–December 2001. Estuarine, Coastal and Shelf Science 61, 477–490.

Chen, S.-N. & Sanford, L.P. 2009. Axial wind effects on salinity structure and longitudinal salt transport in idealized, partially mixed estuaries. Journal of Physical Oceanography 39, 1905–1920.

Gong, W. & Shen, J. 2011. Response of salt intrusion to changing river flow and tidal amplitude during winter season in the Modaomen Estuary, Pearl River Delta area, China. Continental Shelf Research 31, 769–788.

Hamrick, J.M. 1992. A Three-dimensional Environment Fluid Dynamics Computer Code: Theoretical and Computational Aspects. Special Report 317. The College of William and Mary, Virgina Institute of Marine Science, Williamsburg, Virgina, 63 pp.

Ji, Z.-G., Morton, M.R. & Hamrick, J.M. 2001. Wetting and drying simulations of estuarine processes. Estuarine, Coastal, and Shelf Science 53, 683–700.

Kim, T.I., Choi, B.H. & Lee, S.W. 2006. Hydrodynamics and sedimentation induced by large-scale coastal developments in the Keum River Estuary, Korea. Estuarine, Coastal and Shelf Science,515–528.

Kleija, C.S., Hulschera, S.J.M.H. & Loutersb, T. 2003. Comparing uncertain alternatives for a possible airport island location in the North Sea. Ocean Coast. Manag. 46, 1031–1047.

Kuo, A.Y., Shen, J. & Hamrick, J.M. 1996. The effect of acceleration on bottom shear stress in tidal estuaries. Journal of Waterway, Port, Coastal, and Ocean Engineering 122, 75–83.

Li, K.Y., Liu, X.B., Zhao, X.G. & Guo, W.H. 2010. Effects of reclamation projects on marine Ecological environment in Tianjin harbor industrial zone. Procedia Environ. Sci. 2, 792–799.

Mellor G.L. & T. Yamada. 1982. Development of a turbulence closure model for geophysical fluid problems, Reviews of Geophysics and Space Physics, 20, 851–875.

Nijkamp, P. & Yim, H. 2001. Critical success factors for offshore airports a comparative evaluation.J. Air Transp. Manage. 7, 181–188.

Park, Y.W., Cho, Y.K., Sin, Y.S. & Lee, C.H. 2006. Simulation of salt intrusion and mixing influence for Yongsan estuary regarding seawater exchange. In: Proceedings of the 2006 Korea Water Resources Association Conference. Jeju Island, Korea, 18–19 May 2006. 557–561.

Shen, J., Boon, J. & Kuo, A.Y. 1999. A numerical study of a tidal intrusion front and its impact on larval dispersion in the James River estuary. Estuary 22, 681–692.

Yang, Z., Khangaonkar, T. Modeling tidal circulation and stratification in Skagit River estuary using an unstructured grid ocean model. Ocean Modelling. 2009; 28:34–49.

Modeling and Computation in Engineering III – Zhang & Xie (Eds)
© 2014 Taylor & Francis Group, London, ISBN 978-1-138-02680-3

Simulating sediment transport with a coupled hydrodynamic-wave model

Ting Xu
Key Laboratory of Engineering Sediment of Ministry of Communications, Tianjin Research Institute for Water Transport Engineering, M.O.T, Tianjin, China
School of Environmental Science and Engineering, Tianjin University, Tianjin, China

ABSTRACT: This paper realizes the bidirectional coupling of EFDC hydrodynamic model and SWAN wave model, in which the coupled sediment transport model can not only consider the effects of hydrodynamic on waves but also can consider the effects of waves on hydrodynamic. And then taking the simulation of sediment transport in the Oujiang River Estuary for example, predicted results indicated that the distribution of sediment concentration predicted by the coupled model is very reasonable and the calculated values is in good agreement with the measured values. The law of the greater of the water depth, the lower of the sediment concentration is well reflected in the coupled model. So, it is very important to consider hydrodynamic-wave interaction effects on the sediment transport.

1 INTRODUCTION

In estuarine areas, there is a complex interaction between the waves and currents, and the sediment transport is predominantly not only by the currents but also driven by wave motion. For typical estuarine areas interested here, the sediment is mainly suspended by wind waves and transported by current. In shallow water, the wave breaking also plays important roles in the vertical concentration distribution of suspended sediment. A better understanding of wave-current interaction induced sediment transport is crucially important in estuarine area (Zhang et al., 2011; Amoudry L et al., 2010; Dong P et al., 2002). Intensive studies have been conducted in this field using both experimental and numerical approaches. The pioneer wave–current interaction model by Grant and Madsen(1979) used this assumption applying a time-invariant eddy viscosity to account for the turbulence. Since then a range of similar models for the wave–current interaction within the sea bed boundary layer have been published; see e.g. Holmedal et al. (2003) for a review.

We are using a community approach to develop the model as a tool for both research and practical applications. The need and value for this approach was elucidated in a community sediment-transport modeling workshop (Sherwood et al., 2002). A community effort enables us to include a broad range of processes and scales, more than would be feasible for individuals or small groups. We have started with a model that is being used and developed actively by a large research community. We are incorporating proven methodologies from other models such as ECOMSed, EFDC, POM, ADCIRC, COHERENS, and Delft3D. Scientists and engineers may contribute to the model according to their expertize, and users (including scientists from other disciplines, students, resource managers, engineers, and operational personnel) may draw from well-tested, state-of-the-art algorithms. Incorporation of alternative parameterizations for similar processes allows us to compare them in identical frameworks. Collaborative work on a community model helps identify key research and modeling issues, and efficiently focus research efforts, minimizing duplication and preventing critical components from being overlooked. Wide use and broad participation in model development, along with extensive testing and peer review, will produce a robust model that can serve the scientific community.

Therefore, the objective of the present paper is to develop a three dimensional numerical model of sediment transport under the combined action of wind, waves and current on the basis of the existing model. A 3D numerical model for sediment transport based on the coupling of EFDC model and SWAN model is developed and applied in the present paper to investigate sediment transport process.

2 HYDRODYNAMIC-WAVE MODEL

For circulation modeling purposes, the US Environmental Protection Agency's "Environmental Fluid Dynamics Code" (EFDC) model is utilized, which solves the 3D, vertically hydrostatic, free surface, turbulent-averaged equations of motion for a variable density fluid (Hamrick, 1992). The EFDC model can be run in barotropic or baroclinic and in 2D or 3D modes. It also has the capability of simulating wetting and drying processes (Ji et al., 2001), and has been tested and verified in a wide range of hydrodynamic and environmental studies (e.g. Kuo et al., 1996; Shen et al., 1999; Jin et al., 2001). It is thus a comprehensive circulation model, which enables us to operate it with different levels of physics to ascertain the appropriate level of effort needed for eventual wave–current modeling and forecasting.

SWAN is a third generation wave model based on the wave action balance equation that is formulated for use in coastal applications (Booij et al., 1999; SWAN Team, 2009a). SWAN models the effects of wind wave generation, refraction, shoaling, dissipation by bottom friction, white capping, nonlinear wave–wave interactions, and ambient currents on the wave properties (SWAN Team, 2009b). In this study, the SWAN model calculates the two-dimensional and time-dependent waves for simulating wave induced sediment resuspension.

Coupling between the wave and circulation models can be performed in two ways—(1) off-line coupling and (2) on-line coupling (commonly used terminology when coupling earth system models). In general, off-line coupling entails running the two models separately (in no particular order) for the entire duration of simulation (e.g. 2 days). The information can then either be transferred from one model to the other model (one-way approach), or it could be exchanged among both the models (two-way approach). In the one-way approach, the second model is executed using the output of the first model, whereas the two-way approach requires both models to be executed multiple times, while exchanging information, until the solution con-verges. On-line coupling, in contrast, invokes both models simultaneously. After a suitable time interval (e.g. 1 h), the two models exchange information and continue the simulation until the next communication time step. This technique avoids the use of multiple iterations as there is dynamic exchange of information among the two models during the simulation process. However, in between two information exchange time-steps, conditions are assumed to be steady. In contrast, in the off-line approach, the required information at all time-steps from the other model is precalculated. Thus, there are advantages and disadvantages associated with both the methods and the effects of these on the model accuracy and efficiency must be investigated. Gaurav et al., 2013 points out that not only is off-line coupling more efficient than on-line coupling, the on-line coupling approach may not be necessary because the effect of the waves on the currents was found to be marginal at the spatial scales (~1.5 km) investigated.

In this study, we have used the off-line coupling. Firstly, the simulation is started with EFDC which provides, by way of output, water surface elevations (WSEs) and flow velocities. SWAN is then initiated with the EFDC output along with the other forcing terms (winds, the wave open boundary conditions (WOBCs), etc.). At last, EFDC restarts with the SWAN output along with the forcing terms (wave height, wave direction, wave period etc.). The coupling process of EFDC and SWAN can be seen in Fig. 1. So, the sediment transport not only considers the effect of tides and currents on waves but also consider the effect of waves on currents.

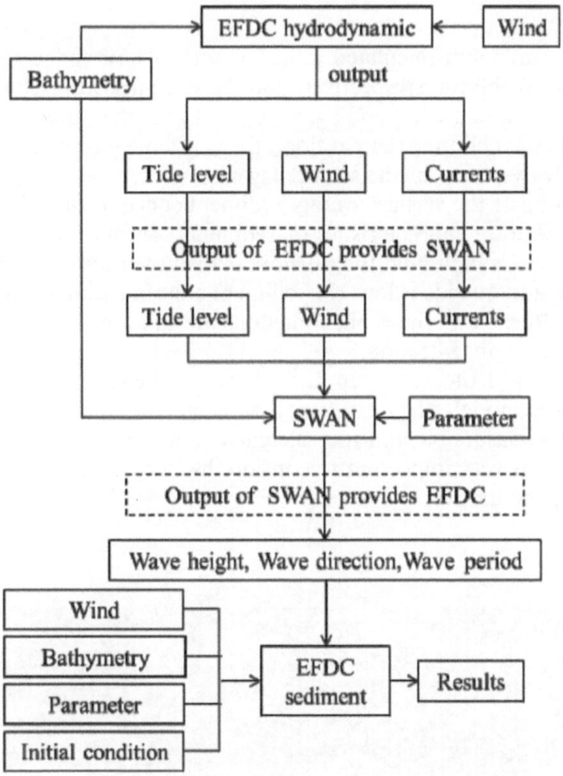

Figure 1. Flow diagram of the EFDC hydrodynamic and SWAN wave models.

3 NUMERICAL SIMULATIONS OF SEDIMENT TRANSPORT

3.1 *Model setup*

The Oujiang River Estuary is bounded by a complex shoreline. It is further complicated by its many islands and complex terrain. Based on the features of the study area, a curvilinear orthogonal grid was used to represent the complex geometry. The model grid consists of 12161 grid cells in the horizontal direction, with grid size ranging from 10 to 1000 m. In the vertical direction, 6 layers are divided, according to the relative water depth, i.e., surface, 0.2, 0.4, 0.6, 0.8 and the bottom layers. The time step for the circulation module is 5 min with 1 min as sub time step for the transport module, and coupling time step between wave and current is set as 1 h for computational efficiency. In the whole computational domain to account for the effect of Coriolis force, and Manning's coefficients take values of 0.01–0.025 according to the distribution of medium diameters of bed materials measured in recent years. Smagorinsky constant for the horizontal eddy viscosity coefficient takes a typical value 0.12. The settling velocity was set to 1.0×10^{-4} ms^{-1}; the critical shear stress for resuspension was set to 0.4 Nm^{-2}; the critical shear stress for deposition was set to 0.2 N m^{-2}; the bed sediment porosity was set to 0.5; and the bed sediment specific weight was set to 2.5.

In this paper, SWAN is used to compute the wave fields under the forcing of wind. The depth induced wave breaking was enabled with default options and parameters. The bottom friction was computed by the Madsen scheme with default equivalent bottom roughness length scale. The model had 18 uniformly distributed directions, and the frequency resolution was determined by fi + 1 = 1.9 fi with fmax = 1.0 Hz and fmin = 0.04 Hz. The model domain had a uniform spatial grid of 1 km. Incoming waves at the southern open boundary was assumed to be zero.

3.2 *Model results*

Fig. 3 shows the measured and calculated values of vertical suspended sediment concentration at flood time and at ebb time respectively and the station see Fig. 2. From Fig. 3 we can see that the calculated values are in good agreement with the measured values. The sediment concentration is higher at ebb time than at flood time, and the sediment concentration in the bottom layer is nearly two times of the surface layer.

From the distribution of the vertical average sediment concentration(Fig. 4), it can be seen that, the spatial distribution characteristics of sediment concentration is that the sediment concentration of coastal is high, and the offshore is low; the sediment concentration inside of the estuary is high and outside is low; the sediment concentration in shallow water is high and in deep water is low; all of these phenomenons are in good agreement with the actual distribution of sediment concentration.

The law of the greater of the water depth, the lower of the sediment concentration is well reflected in the coupled model. The distribution of sediment concentration predicted by the coupled model is very reasonable and the calculated values is in good agreement with the measured values. So, it is very important to consider hydrodynamic-wave interaction effects on the sediment incipient motion.

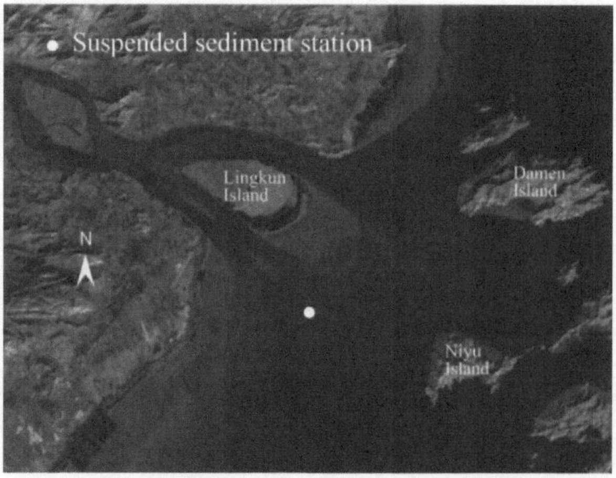

Figure 2. Sketch of Oujiang River Estuary.

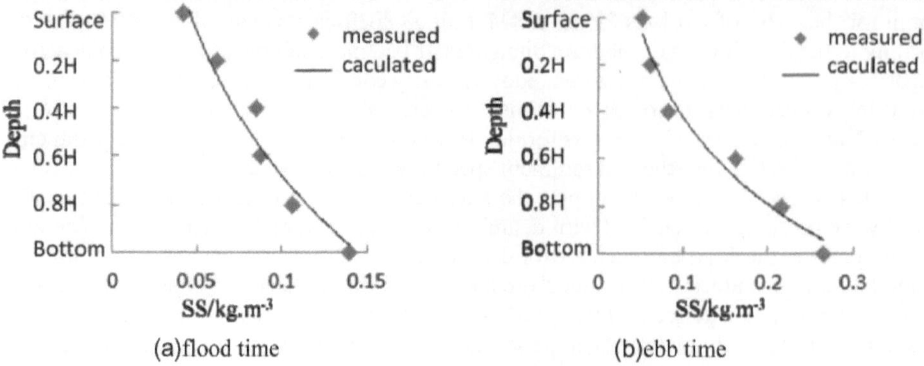

Figure 3. Vertical distribution of sediment concentration at the different time.

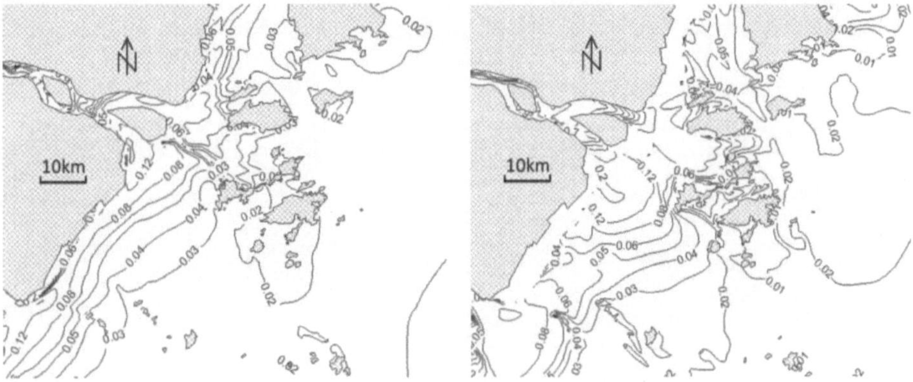

(a)flood time (b)ebb time

Figure 4. The distribution of the sediment concentration field.

4 CONCLUSIONS

This paper realizes the bidirectional coupling of EFDC hydrodynamic model and SWAN wave model, which means that in the coupled sediment transport model can not only consider the effects of hydrodynamic on waves but also can consider the effects of waves on hydrodynamic. And then taking the simulation of sediment transport in the Oujiang River Estuary for example, predicted results indicated that the distribution of sediment concentration predicted by the coupled model is very reasonable and the calculated values is in good agreement with the measured values. The law of the greater of the water depth, the lower of the sediment concentration is well reflected in the coupled model. So, it is very important to consider hydrodynamic-wave interaction effects on the sediment transport.

REFERENCES

Amoudry, L. & Liu P. L.-F. Parameterization of near-bed processes under collinear wave and current flows from a two-phase sheet flow model. Continental Shelf Research, 2010, Vol. 30, No. 13, pp. 1403–1416.

Booij, N., Ris, R.C. & Holthuijsen, L.H. 1999. A third-generation wave model for coastal regions 1. Model description and validation. Journal of Geophysical Research 104, 7649–7666.

Dong, P. & Zhang, K. 2002, Intense near bed sediment motions in waves and currents. Coastal Engineering, Vol. 45, No. 2, pp. 75–87.

Grant, W.D. & Madsen, O.S. 1979. Combined wave and current interaction with a rough bottom. Journal of Geophysical Research 84, 1797–1808.

Holmedal, L.E., Myrhaug, D. & Rue, H. 2003. The sea bed boundary layer under random waves plus current. Continental Shelf Research 23 (7), 717–750, erratum 1035.

Ji, Z.-G., Morton, M.R. & Hamrick, J.M. 2001. Wetting and drying simulations of estuarine processes. Estuarine, Coastal, and Shelf Science 53, 683–700.

Jin, K.-R., Hamrick, J.H. & Tisdale, T. 2001. Application of a three-dimensional hydrodynamic model for Lake Okeechobee. Journal of Hydraulic Engineering 126, 758–771.

Kuo, A.Y., Shen, J. & Hamrick, J.M. 1996. The effect of acceleration on bottom shear stress in tidal estuaries. Journal of Waterway, Port, Coastal, and Ocean Engineering 122, 75–83.

Shen, J., Boon, J. & Kuo, A.Y. 1999. A numerical study of a tidal intrusion front and its impact on larval dispersion in the James River estuary. Estuary 22, 681–692.

Sherwood, C.R., Harris, C.K., Geyer, W.R. & Butman, B. 2002. Toward a community coastal sediment-transport modeling system: Report of the Second Workshop. EOS, Transactions, American Geophysical Union 83 (51).

SWAN Team, 2009a. SWAN User Manual Version 40.72. Delft University of Technology.

SWAN Team, 2009b. SWAN Scientific and Technical Documentation 40.72. Delft University of Technology.

Zhang, C., Zheng, J., Wang, Y. & Demirbilek, Z. 2011. Modeling wave–current bottom boundary layers beneath shoaling and breaking waves. Geo-Marine Letters 31, 189–201.

Modeling and Computation in Engineering III – Zhang & Xie (Eds)
© 2014 Taylor & Francis Group, London, ISBN 978-1-138-02680-3

The mechanical analysis of a long-span multiple bar steel tower on temperature loads

Jing Cai & Siying Wang
Airport School of Civil Aviation University of China, Tianjin, China

ABSTRACT: The influence of temperature to the long-span multiple bar steel structural transmission tower of through various aspects is joint action. The influence of different connection, span number and temperature is different. In order to study the mechanics and deformation characteristics of the large statically indeterminate tower structure on temperature load, ANSYS software is used to establish the large span steel structure model of the transmission tower system and the deformation and internal force of the most adverse conditions are obtained under different temperature load and gravity. Further the safety factor and the relative deformation are defined here to analyze the strength security and stability of the tower components.

1 INTRODUCTION

For large span steel transmission tower, temperature is one of the prominent problems. Under temperature field, for bars cannot expand and contract freely when hearted and cooled, which produces additional stress and deformation in the bar. According to references [Cao Weidi 2011], many transmission tower accidents were due to temperature which causes compressive deformation in the main and inclined bars the tower, even due to disaster collapse. At present, research [Fan Zhong 2007, Fan Zhong 2007, Lin Cuocuo 2010] on the effects of temperature on long-span steel structure as the national stadium and Guangzhou new railway station shows that the impact of temperature on long-span steel structure is prominent, and enough attention should be paid on it in construction and design. Therefore, a transmission steel tower which is 55.9 m high, 49 m span length is researched on the stress and deformation characteristics on the different temperature loads. One-span, three-span and five-span towers are analyzed respectively over different temperature loads. On the other hand, in order to study the influence of connection form the connection forms are divided into rigid connection and hinged connection. At last, the most unfavorable stress distribution and deformation characteristics of the tower bars are obtained here.

2 ANSYS ANALYSIS MODEL

2.1 *Model parameters*

In this paper, the transmission tower is 750 KV ultra-high voltage substations; Substation is 55.9 meters tall, and each cross is 49 meters long. Tower pillar, beam and brace are made up of 23 different sections, as shown in table 1. In order to compare the influence of different connection forms of the structure, models of beam and column rigid connection and hinged connection are built. At the same time, in order to study the influence of the span, three spans tower models are respectively established as shown in figure 1. Rigid model: For bars in the structure BEAM188 space beam element is adopt. The bottom of the tower and beam-column joints are also rigid connection. Hinged model: hinged form transmission

Table 1. Mar sectional dimension in the tower.

Number	Diameter/ mm	Thickness/ mm	Number	Diameter/ mm	Thickness/ mm	Number	Diameter/ mm	Thickness/ mm
1	φ351	12	9	φ146	5	17	L56	5
2	φ351	10	10	φ121	5	18	φ219	10
3	φ325	8	11	φ68	5	19	φ219	8
4	φ194	6	12	φ140	5	20	φ102	5
5	φ203	6	13	L180	12	21	φ95	5
6	φ203	7	14	L140	12	22	L80	8
7	L160	12	15	L100	8	23	φ152	6
8	φ83	5	16	L75	8			

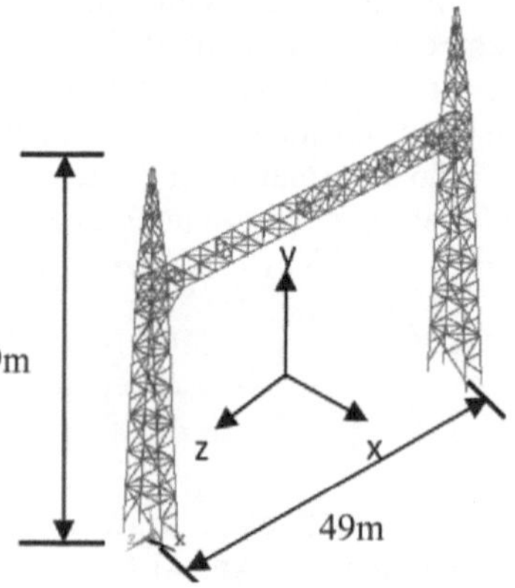

Figure 1. One-span tower.

tower consists of BEAM188 and LINK8 elements. LINK8 element is used to simulate the beam bar. The bottom of the tower and beam-column joints are hinged connection. Here the direct model method is used [Xiao Qi 2011, Wang Yu 2011], and a single tower consists of 102 key points, 267 member bars, one-span tower consists of 296 points, 803 member bars, three-span consists of 684 points, 1875 member bars, five-span consists of 1072 key points, 2947 member bars. The parameters are elastic modulus 2.06×10^{10} Pa, Poisson's ratio 0.25, and the linear expansion coefficient of steel 12×10^{-6}/°C, the density of steel 7850 kg /m^3.

2.2 *Project site daily temperature variations*

The tower is in the field and area temperature difference between day and night is larger. According to the research of Barber and Yanzuoren [F.S. Barber 1957, Yan Zuoren 1982], the maximum temperature difference in the project site is 50°C. the ladder temperature[Sun Ruirui 2008, Teng Runchun 2010] 10°C, 30°C and 50°C are used to analyze the deformation and stress performance of steel tower under gravity and uniform temperature together.

3 THE ANALYSIS OF STRESS AND DEFORMATION UNDER TEMPERATURE LOAD

Using the above ANASYS model, the most unfavorable stress and deformation of the bar under gravity and uniform temperature 10°C, 30°C and 50°C respectively are obtained. (Table 2 to Table 4). Figure 2 shows the direction of the internal stress and displacement of the bar.

X, Y, Z is displacement directions. N is force at X axial (worth to pull and negative pressure), My is bar bending stress at Y direction, Mz is bar bending stress at along the Z direction.

Table 2, 3 and 4 show that the displacement at X direction is affected by the span number within 30%, and displacement at Y direction is not affected by span number basically. But with the increase of temperature the displacements at X, Y directions have a tendency to increase. The displacement at Z direction is affected by span number and temperature. Axial force is affected by the span number and temperature at the same time, as the span number and the temperature increases, the pressures of some bars at the bottom are the most value. The bending stress at Y direction is affected by span number slightly, but it is greatly influenced by temperature, and increases with temperature rise. Connection forms have an effect on the bending stress along Y direction, and hinged case part of the bar bending stress along the Y direction reduces under the action of temperature tower bar safety analysis.

Table 2. The most unfavorable displacement and internal force of the bar under gravity and uniform temperature 10°C.

Connection form	X (mm)	Y (mm)	Z (mm)	N (kN)	My (MPa)	Mz (MPa)
one-span rigid connection	0.664	−10.061	4.672	−77.683	14.6	13.2
one-span hinged connection	0.645	−10.568	5.064	−74.415	14.3	14.4
three-span rigid connection	0.711	−7.424	8.897	−94.545	14.7	16.1
three-span hinged connection	0.709	−7.735	9.112	−90.729	14.6	17.7
five-span rigid connection	0.762	−7.192	13.502	−112.671	14.8	20.1
five-span hinged connection	0.778	−7.477	13.711	−108.523	14.5	22.2

Table 3. The most unfavorable displacement and internal force of the bar under gravity and uniform temperature 30°C.

Connection form	X (mm)	Y (mm)	Z (mm)	N (kN)	My (MPa)	Mz (MPa)
one-span rigid connection	1.533	19.206	9.653	−100.06	18.6	16.9
one-span hinged connection	1.474	19.202	9.878	−93.655	15.2	18.9
three-span rigid connection	1.72	19.203	29.851	−158.976	19.0	29.2
three-span hinged connection	1.705	19.186	28.751	−150.872	15.4	32.6
five-span rigid connection	1.907	19.231	51.885	−215.811	19.4	41.6
five-span hinged connection	1.933	19.213	50.52	−206.615	15.7	46.7

Table 4. The most unfavorable displacement and internal force of the bar under gravity and uniform temperature 50°C.

Connection form	X (mm)	Y (mm)	Z (mm)	N (kN)	My (MPa)	Mz (MPa)
one-span rigid connection	2.502	32.647	15.275	−122.437	29.6	20.7
one-span hinged connection	2.354	32.638	15.392	−112.895	16.1	23.3
three-span rigid connection	2.813	32.71	54.621	−223.406	30.3	42.2
three-span hinged connection	2.746	32.685	53.03	−211.014	16.6	47.6
five-span rigid connection	3.119	32.756	91.633	−318.958	30.9	63.1
five-span hinged connection	3.127	32.729	89.587	−304.706	17.1	71.3

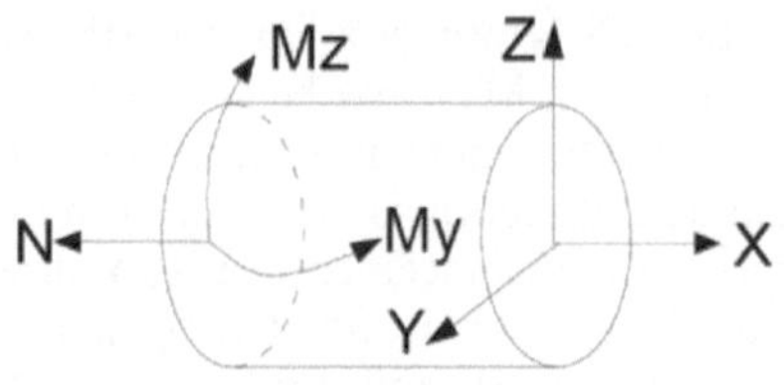

Figure 2. Member internal force and displacement direction.

Table 5. The biggest relative extensions of the bars.

Connection form	10°C	30°C	50°C
one-span rigid connection	0.131‰	0.130‰	0.130‰
one-span hinged connection	0.137‰	0.137‰	0.137‰
three-span rigid connection	0.134‰	0.145‰	–0.186‰
three-span hinged connection	0.141‰	0.151‰	–0.186‰
five-span rigid connection	0.138‰	–0.169‰	–0.238‰
five-span hinged connection	0.144‰	–0.183‰	–0.278‰

Note: the value in the table is the most value in various conditions, namely the largest permillage expansion amount.

4 BAR RELATIVE EXTENSIONS ANALYSIS

According to data in table 2 to table 4, the ratio of the variation of bar length after heating and its original length is defined as relative extension. The biggest relative extensions of the bars are presented in table 5 (plus shows elongation, minus shows compression).

The table 5 shows that the biggest relative extensions of the bars are within 0.3 ‰. So the expansion amount can be ignored.

5 SAFETY FACTOR OF STRENGTH ANALYSIS

Using the data in table 2 to table 4 and selecting various working condition to analyze, the ratio of the bar work stress and the design stress is defined as stress level β, design stress is 215 MPa. Then F = 1-β is defined as the safety factor of strength. This paper presents three direction stress safety factor, axial stress safety factor, Y direction bending stress safety factor, Z direction bending stress safety factor. Tables 6–8 show the worst location bar safety factors of three spans towers under different temperature conditions. Figures 3–5 show the axial stress and bending stress safety factors curve changing with temperatures.

Tables 6–8 and Figures 3–5 show that for axial stress safety factors, the single span has a small change under different temperatures. Safety factors of three-span and five-sp1crease obviously when the temperature increases. When the temperature is 50°C, five-span bar safety factor is 70%. Both connection and temperature affect the safety factors, and as the temperature increases, the safety factors of the single span or multiple are falling. The safety factors of bending stress at Y direction change especially. So the safety factors of hinged connection are higher than that of rigid connection. Safety factors of bending stress at Z direction decline as the temperature increases, and the changes between each span is becoming more and more big. Under the condition of temperature 50°C, safety factors of axial stress and bending stress at Z direction of five-span demote to 70%.

Table 6. Gravity +warming 10°C three spans towers safety factors.

Connection form	F_N	F_{My}	F_{Mz}
one-span rigid connection	87.47%	93.21%	93.86%
one-span hinged connection	86.91%	93.35%	93.30%
three-span rigid connection	87.13%	93.16%	92.51%
three-span hinged connection	86.46%	93.21%	91.77%
five-span rigid connection	86.82%	93.12%	90.65%
five-span hinged connection	86.19%	93.26%	89.67%

Table 7. Gravity +warming 30°C three spans towers safety factors.

Connection form	F_N	F_{My}	F_{Mz}
one-span rigid connection	87.53%	91.35%	92.14%
one-span hinged connection	86.91%	92.93%	91.21%
three-span rigid connection	86.15%	91.16%	86.42%
three-span hinged connection	85.57%	92.84%	84.84%
five-span rigid connection	83.84%	90.98%	80.65%
five-span hinged connection	82.49%	92.70%	78.28%

Table 8. Gravity +warming 50°C three spans towers safety factors.

Connection form	FN	FMy	FMz
one-span rigid connection	87.58%	86.23%	90.37%
one-span hinged connection	86.91%	92.51%	89.21%
three-span rigid connection	82.21%	85.91%	80.37%
three-span hinged connection	82.22%	92.28%	77.86%
five-span rigid connection	77.22%	85.63%	70.66%
five-span hinged connection	73.35%	92.05%	66.84%

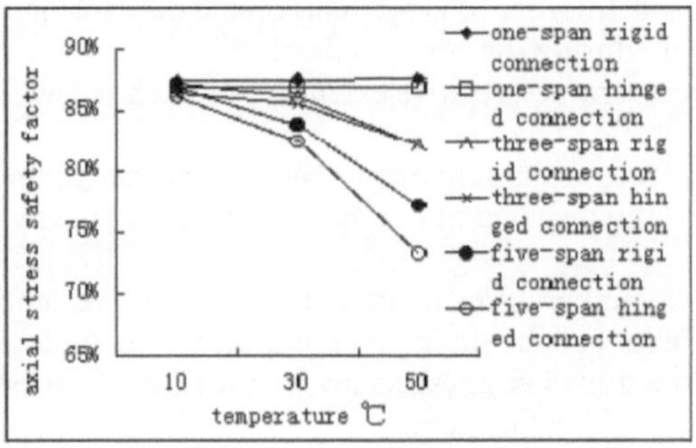

Figure 3. Axial stress safety factor.

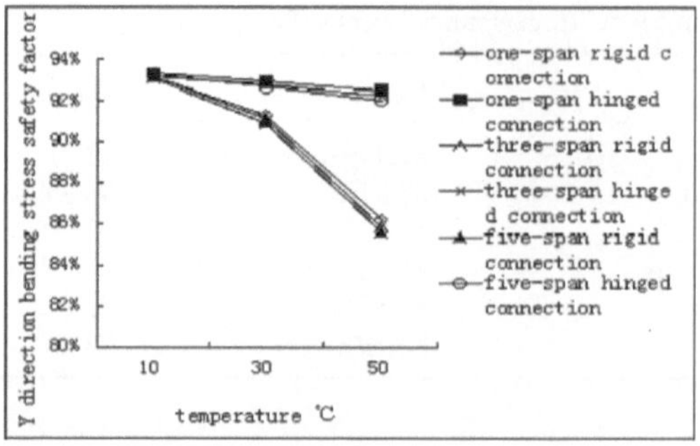

Figure 4. Y direction of the bending stress safety factor.

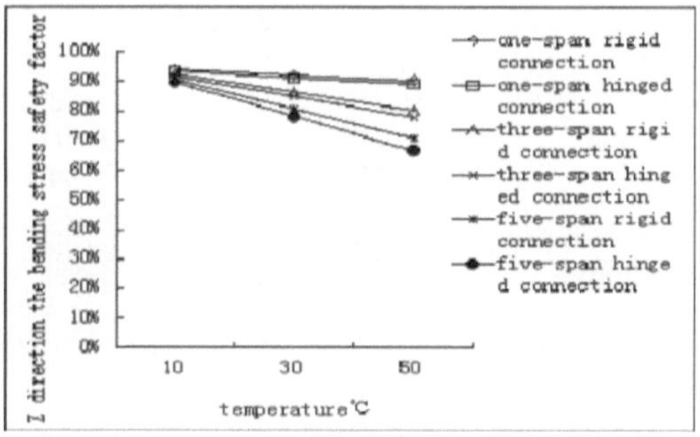

Figure 5. Z direction of the bending stress safety factor.

6 CONCLUSION

The deformation and stress distribution of a large span multiple bar steel transmission tower is analyzed under the temperature loads, and the influence of different temperature loads and connections are compared in this paper. Such conclusions are obtained here.

1. The impact of temperature is larger on the long-span steel structure, which should be paid enough attention on in construction and design of the structure.
2. The greater the structural span is, the greater the displacement along the span direction under temperature loads. In order to avoid remarkable displacement at beam-end, expansion joints can be set up during the construction.
3. Under the effect of temperature, relative extension of each bar within 0.3‰, so it can be ignored.
4. For axial stress safety factors, the single span has a small change under different temperatures. Safety factors of three-span and five-span decrease obviously when the temperature increases.
5. Both connection and temperature affect the safety factors, and as the temperature increases, the safety factors of the single span or multiple are falling. The safety factors of bending stress at Y direction change especially. So the safety factors of hinged connection are higher than that of rigid connection.
6. Safety factors of bending stress at Z direction decline as the temperature increases, and the changes between each span is becoming more and more big. Under the condition of temperature 50°C, safety factors of axial stress and bending stress at Z direction of five-span demote to 70%.

REFERENCES

Cao Weidi, Liang Qingxiang, Bai Ke. Research on the impact of ice to transmission tower [J]. Journal of Taiyuan science and technology university, 2011, 32 (2): 138–141.

Fan Zhong, Liu Xianming, Fan Xuewei etc. Study and design of national stadium large-span steel structure [J]. Journal of building structures, 2007, 28 (2): 1–16.

Fan Zhong, Wang Zhe, Tang Jie etc. The national stadium large-span steel structure research on temperature field analysis and closed temperature [J]. Journal of building structures, 2007, 28 (2): 32–40.

F.S. Barber. Calculation of Maximum Pavement Temperature from Weather Report [J]. H.R.B. Bull, 168, 1957.

Lin Cuocuo, Wang Yuanqing, Shi Yongjiu, Shi Gang. Effect of temperature on the influence of the steel structure design and construction analysis [J]. The 10th national conference on modern structural engineering, 2010: 948–956.

Sun Ruirui. Large span steel arch structure temperature effect and the welding effect analysis [D]. Tianjin: tianjin university construction college. 2008.

Teng Runchun. Discuss the influence of temperature on the structure [J]. Journal of ShanXi building, 2010, 4 (5): 72–73.

Wang Yu, Liu Jing, Liu Kai. Based on the ANSYS analysis methods to set up the finite element model of aerial ropeway tower [J]. Design and research, 2011, 5 (38): 40 and 42.

Xiao Qi, Zhou Lingfeng, Cai Jingsu, Deng Yi. Compact narrow base ANSYS modeling and dynamic analysis of transmission tower [J]. Water and electricity energy science, 2011, 29 (2): 167–168.

Yan Zuoren. The layered pavement temperature field analysis [D]. Tongji university graduate thesis, 1982.

Modeling and Computation in Engineering III – Zhang & Xie (Eds)
© 2014 Taylor & Francis Group, London, ISBN 978-1-138-02680-3

The application of study on Chaos Ant Colony Optimization in cascade reservoir power generation optimized operation

Wenlin Yuan & Chengjun Wu
School of Water Conservancy and Environment Engineering, Zhengzhou University, Zhengzhou, China

ABSTRACT: The traditional ant colony optimization has the defect of the stagnation during last runs and trapping in local optimal solution. Combining of the characteristics of the optimal generation of cascade reservoir, the chaos ant colony optimization has been proposed basing on increasing the field search mechanisms of the algorithm and using of advantages of ergodicity in chaos algorithm. The ant colony algorithm parameters have been verified in the paper. A practical example calculation shows that the results of the algorithm is rational and reliable when it is used to solve the problem of optimal operation of cascade reservoirs, so a new way for searching the solution to the optimal operation of cascade reservoirs with high dimension and complexity is provided.

1 INTRODUCTION

Along with the construction and operation of basin and inter-basin large-scale giant reservoirs in our country, it has very important scientific research and practical application significance in carrying out the study on high dimension, complex cascade reservoirs power generation optimal operation for full utilizing water resources and ensuring the safety and stable operation of the power system. The system has the characteristics of big size, complex model solution space and is difficult to get the solution quickly, and the traditional optimization algorithm or the simple intelligent algorithm has the disadvantages of being difficult to handle many constraints, long computer run hour, which is easy to fall into the local optimal solution (João Luiz B. & Brandão, 2010, Raheleh A et al. 2008). In order to solve this problem, this paper, by using the advantages of feedback and parallelism of ant colony algorithm, combining with the characteristics of optimal operation for power generation of cascade reservoirs, has put forward the application of Chaos Ant Colony Optimization (CACO) algorithm to the solution of optimal operation model for power generation of cascade reservoirs (Xiaohui Yuan et al. 2008).

2 THE MODEL OF JOINT OPTIMIZATION SCHEDULING OF CASCADE RESERVOIRS

Taking joint operation As a whole into consideration, cascade reservoirs not only have the compensation benefits of the reservoir capacity and compensation benefits of hydrology, but have the compensation benefits of electric power. When the process of runoff is known in the scheduling period, the objective function of long-term optimization scheduling model is the largest generate electricity capacity of the cascade reservoirs in the scheduling period in this paper.

2.1 *Objective function*

The objective function is the largest annual output of electricity of the cascade reservoirs in this paper, which is described as follows:

$$E = Max\sum_{t=1}^{T}\sum_{i=1}^{I} N(i,t)\times \Delta t \tag{1}$$

where E is the total power generation of the cascade reservoirs in the scheduling period. t and T respectively represent the time numbers and the total number of hours in the scheduling period. i and I respectively represent the reservoir numbers and the total number of reservoirs. $N(i,t)$ represents the average output of the i-th reservoir in the t-th time interval. Δt represents the time step.

2.2 *Constraint conditions*

1. Water balance constraints

$$V(i,t+1) = V(i,t) + (Q_{In}(i,t) - Q_{Out}(i,t))\times \Delta t \tag{2}$$

where $V(i,t)$ and $V(i,t+1)$ respectively represent the capacity of the i-th reservoir at the early stage and last stage in the t-th time interval. $Q_{In}(i,t)$ and $Q_{Out}(i,t)$ respectively represent the inflow and outflow of the i-th reservoir in time t.

2. Flow balance constraints

$$Q_{In}(i+1,t) = Q_{Out}(i,t) + q(i,t) \tag{3}$$

where $q(i,t)$ represents the local inflow between the i-th and $(i+1)$-th reservoirs in time t.

3. Storage level constraints

$$Z_{\min}(i,t) \le Z(i,t) \le Z_{\max}(i,t) \tag{4}$$

where $Z(i,t)$ represents the mean water level of the i-th reservoir in time t. $Z_{\max}(i,t)$ and $Z_{\min}(i,t)$ respectively represent the top and bottom limitation of the allowed storage level of the i-th reservoir in time t. $Z_{\min}(i,t)$ is generally determined by dead water level and $Z_{\max}(i,t)$ is obtained by comprehensive consideration of the flood control requirements.

4. Discharged volume constraints

$$Q_{\min}(i,t) \le Q(i,t) \le Q_{\max}(i,t) \tag{5}$$

where $Q(i,t)$ represents the average outflow of the i-th reservoir in time t. $Q_{\max}(i,t)$ and $Q_{\min}(i,t)$ respectively represent the top and bottom limitation of the discharged volume of the i-th reservoir in time t. $Q_{\min}(i,t)$ is determined by comprehensive downstream water utilizing, shipping requirements, etc. $Q_{\max}(i,t)$ is determined by the requirements of downstream flood control, etc.

5. Power output constraints

$$N_{\min}(i,t) \le N(i,t) \le N_{max}(i,t) \tag{6}$$

where $N(i,t)$ represents the average output of the i-th power station in time t. $N_{min}(i,t)$ and $N_{\max}(i,t)$ respectively represent the top and bottom limitation of the allowable output of the i-th power station in time t. They are determined by the rated output, blocked capacity and expected output, etc.

6. Variable non-negative constraints.

3 DESIGN OF ANT COLONY OPTIMIZATION ALGORITHM

3.1 *The basic principle of Ant Colony Optimization algorithm*

Ant Colony Optimization algorithm (ACO), proposed by Italian scholar Dorigin M. who was enlightened by the motivation of the biological evolution, is a new bionic simulated evolutionary algorithm to solve the complex problems of optimization by simulating the behavior of finding food for ants in nature.

M ants are initialized randomly and S is set as a set of initial paths, if the ant $m(m=1,2,\ldots,M)$ is always in the path s at any time, then the path s is considered as the current path of ant m. Assuming that the amount of pheromone on each path is equal at the initial time, that is $\tau_i(0)=\tau_0(i=1,2,\ldots,M)$ (τ_0 is a positive constant). Ant m transfers its direction according to the amount of pheromone on each path during activity and the state transition rules used by ant colony are called random ratio rules, that is randomly selecting a number q in a given interval [0, 1], which will compare to q_0 (q_0 is a threshold parameter which decides whether to search or use it). If $q \leq q_0$, then update the current path, and put the updated path into the set of paths S. If $q > q_0$, then select the path j in the set of paths S on probability P_{ij} according to the principle that one has higher pheromone concentration and higher probability. P_{ij} can be calculated as:

$$P_{ij}=\begin{cases}\dfrac{\tau_i(t)^{\alpha}\times\eta_{ij}(t)^{\beta}}{\sum\limits_{s\in S}\tau_s(t)^{\alpha}\times\eta_{is}(t)^{\beta}}, & i,j,s\in S\\ 0 & s\notin S\end{cases} \tag{7}$$

where $\tau_i(t)$ and $\tau_s(t)$ respectively represent the pheromone concentration of path i and s during the t-th time interval, after each iteration the pheromone concentration is updated according to Eq. (8). $\eta_{ij}(t)$, which changes when the fitness changes during the research of ants, represents the heuristic function which comes into being while the path i transfers to path j during the t-th time interval. Its specific expression is described as Eq. (9) and Eq. (10). α and β respectively represent the relative importance of accumulated information and heuristic information to choose the path during the activity of ants.

$$\tau_i(t+1)=(1-\rho)\times\tau_i(t)+\rho\times\Delta\tau_i \tag{8}$$

$$\Delta\tau_i=\sum_{m=1}^{M}\Delta\tau_i^m \tag{9}$$

$$\eta_{ij}(t)=E_i-E_j \qquad i,j\in S \tag{10}$$

where $\Delta\tau_i^m$ represents the pheromone amount on the i-th path the m-th ant leaves, its value depends on the degree of ants' performance, and the higher of the fitness, the more the pheromone releases. Its specific expression is that if i is the optimal path, then $\Delta\tau_i^m=E_i/C$, and if i is not the optimal path, then $\Delta\tau_i^m=0$. E_i is the objective function value corresponded to path i. C is a positive constant number, which is generally 100~500 times to the estimated optimal objective function value. $\Delta\tau_i$ represents the incremental pheromone amount in this cycle from path i. ρ is the pheromone residue coefficient and $(1-\rho)$ is the pheromone attenuation coefficient.

3.2 *Parameters setting of ACO*

1. Pheromone residue coefficient ρ

Just like other bionic evolutionary algorithms, ACO also has the problem of the rate of convergence and getting into local optimal solution. The size of pheromone residual coefficient ρ is directly related to the global searching ability and the rate of convergence in ACO, especially when the scale of issues to deal with is large. Because of the existence of pheromone attenuation coefficient $(1-\rho)$, the pheromone will reduce to be close to 0 for those who has never been searched. So the global searching ability of this algorithm is reduced. Conversely, if the pheromone residue coefficient ρ increases and the pheromone residual coefficient $(1-\rho)$ decreases, the random performance and the global searching ability of the algorithm can be improved, but the algorithm convergence speed will be reduced. So, the selection of pheromone residue coefficient ρ must be comprehensively considered with two performance index—the global searching ability and the rate of convergence, and we should make reasonable or the compromise choice to a certain problem (Gao Shang & Yang Jingyu, 2006).

2. Population scale M
Population scale M of ACO is a constant parameter that does not change with the process of evolution and it affects the stability and the rate of convergence of the algorithm (Jalali MR et al. 2006). If the population scale is too big, the global searching ability and the stability of ant colony algorithm can be improved, but the information feedback effects are weakened, which makes the rate of convergence slow down. Conversely, if the population scale is too small, although it accelerates the rate of convergence of the algorithm, the stability of the algorithm is reduced, and it is likely to get into premature stagnation.

3. Heuristic factors
Heuristic factors include information heuristic factor α and expectations heuristic factor β. Information heuristic factor α reflects the strength of random factors in the process of searching optimization for ant colony, the bigger of the value, the weaker of the random performance, which will make the algorithm get into local optimum early. Expectations heuristic factor β reflects the strength of apriority and certainty factors in the process of searching optimization for ant colony, the bigger of the value, the higher possibility that the ants choose the local shortest path in a certain local point. Although the rate of convergence can be improved, it is likely to get a local best answer because of its weaker random performance.

ACO is still at the start of the study, though the parameters in algorithm is clear in physical meaning, the exact value which is used in solving the practical problem can only be obtained through the simulation experiment.

3.3 *Mechanism of neighborhood search*

When the ant colony are executing the rule of state transition, if $q \le q_0$ (where q is random number and q_0 is a threshold parameter), then we update the current path according to the mechanism of neighborhood search, and compare the objective function values between current path and the original path, and add the better path to the set of paths S to increase the diversity of the set of paths. Mechanism of neighborhood search is described as follow.

Suppose that the current path i is $Z_i = (Z_{11}, Z_{12}, \ldots, Z_{1T}, Z_{21}, Z_{22}, \ldots Z_{2T}, \ldots, Z_{I1}, Z_{I2}, \ldots Z_{IT})$. Its iteration number is l. The neighborhood search radius r is calculated as follows in this paper:

$$r_{it} = \gamma \times \min\{(Z_{it,\max} - Z_{it}), (Z_{it} - Z_{it,\min})\} \tag{11}$$

where, Z_{it}, $Z_{it,\max}$ and $Z_{it,\min}$ respectively represent the current water level, upper limit water level and lower limit water level of the i-th reservoir in the t-th time interval. γ represents the coefficient of water scaling according to the reservoir' condition, $\gamma = 0.25$ in this paper.

r_{it} represents the step used to change the upward and downward level of the current water. That is:

$$\begin{cases} Z'_{it} = Z_{it} + r_{it} \\ Z''_{it} = Z_{it} - r_{it} \end{cases} \tag{12}$$

$2^{I\times T}$ new paths are generated based on the current path through the neighborhood search, at the same time the generated path needs to be examined by the constrain. If the reservoir which the search spot belongs to has next grade cascade, it is needed to recalculate the discharged volume of the following influence reservoir until to the last grade, and at the same time we should amend the realm of the searching space of water level of the follow-up influence time. To those that meet the demand of constrain conditions and above requirements, the optimized ones are selected to join the set of paths S after recalculation. S includes the paths searched before.

Although PSO uses the principle of the positive feedback, to some extent it accelerates the optimization process, there are some defects, such as the stagnation behavior and being easy to fall into the local optimal solution in the search process. Many scholars have proposed some improvement measures, such as the fusion of genetic algorithm and the ant colony algorithm proposed by Ding Jianli (Ding Jianli et al. 2003). But the way of selection, hybridization and variation of genetic algorithm and the selection of parameters are all determined by experience, besides, it is easy to fall into premature convergence. According to the problems above, the chaos ant colony optimization is proposed in this paper based on chaos optimization algorithm which has the characteristics of "random", "ergodicity" and "regular". In the search process of ACO, we put the chaos perturbation into local optimal solution and adjust the information to make the solution escape from local extremum, and ultimately obtain the approximate optimal solution.

4 DESIGN OF CHAOS OPTIMIZATION ALGORITHM

Chaos Optimization Algorithm (COA), as a new optimization algorithm, is proposed with the development of chaos discipline in recent years. The basic idea is to make the chaotic variable map to the space of optimization variable, and then create a sequence of chaotic variable, and make full use of the ergodicity, randomicity and regularity of chaotic variables in chaotic motion to look for the global optimal solution. Logistic mapping is a typical chaotic system, the iterative formula is shown as Eq. (13) below.

$$x(k+1)=\mu\times(1-x(k)),\quad \mu\in(2,4] \tag{13}$$

where μ is a controlled variable, if $\mu = 4$, $0\leq x\leq 1$, logistic is in chaotic state absolutely. Because chaotic motion has the characteristic of random, ergodicity and sensitivity to the initial conditions and so on. This paper will use the chaotic property of $\mu = 4$ (Qiu Lin et al. 2005), and make Logistic mapping in Eq. (13) as chaotic signal generator, and create chaos perturbation to do the local search of current optimum solution in ACO in avoiding getting into the local optimal solution. Optimization step is shown as follows:

Step 1: Suppose that $i=1$, $t=1$, $k=1$, i, t and k respectively represent the reservoir numbers, the number of the time interval, maximum iteration number of chaos algorithm. Then we take the current optimum solution of ACO as initial solution of chaos algorithm, of which the current optimal solution can be expressed as $Z_{bext}^{L}=(Z_{11}, Z_{11}, \ldots, Z_{1T}, Z_{21}, Z_{22}, \ldots, Z_{2T}, \ldots, Z_{I1}, Z_{I2}, \ldots, Z_{IT})$. We apply the water level Z_{it} of the i-th reservoir in the t-th time interval to calculate the initial variables in Eq. (13). The transformation equation is shown as below.

$$x_{it}^{0}=\frac{Z_{it}-Z_{it,\min}}{Z_{it,\max}-Z_{it,\min}} \tag{14}$$

where, $Z_{it,\max}$ and $Z_{it,\min}$ respectively represent upper limit and lower limit of Z_{it}.

Step 2: Apply Eq. (13) to calculate x_{it}^{k+1}. Interval [0, 1] is converted to interval [−1, 1] according to the following equation. Then we can get variable y_{it}^{k+1}.

$$y_{it}^{k+1}=2\times x_{it}^{k+1}-1 \tag{15}$$

Step 3: Calculate the new reservoir water level variables using y_{it}^{k+1} according to the equation below.

$$Z_{it}^{k+1} = Z_{it}^{k} + \lambda \times x_{it}^{k+1} \times \Delta Z_{it} \tag{16}$$

where, λ is a control coefficient of search step. It usually takes the value between 0.001~0.1 (Jia Renfu et al. 2008), and it needs to take a test with constraint. ΔZ_{it} can be calculated as follows:

$$\Delta Z_{it} = Z_{it,\max} - Z_{it,\min} \tag{17}$$

Step 4: Calculate the objective function value F, then compare it with the current optimal objective function value F_{best}, if $F > F_{best}$, then $F_{best} = F$, $Z_{it} = Z_{it}^{k+1}$.

Step 5: Judge whether it meets the demand of iteration stop conditions $k = K$, if not, return to *Step* 2. Otherwise, go to next step.

Step 6: Judge whether it meets the demand of interval number $t = T$, if not, return to *Step* 2. Otherwise, go to next step.

Step 7: Judge whether it meets the demand of reservoir number $i = I$, if not, return to *Step* 2. Otherwise, output optimal results.

5 DESIGN OF APPLYING CACO ALGORITHM TO CASCADE RESERVOIRS' POWER GENERATION OPTIMIZATION SCHEDULING

According to the basic principles of ant colony algorithm and chaos optimization algorithm, taking into account the characteristics of cascade reservoirs' power generation optimization scheduling, the design procedure of applying CACO algorithm to cascade reservoirs' power generation optimization scheduling is as follows:

Step 1: Set CACO optimization parameters. The main parameters include population scale M, maximum iteration number L, pheromones volatile parameter ρ, information heuristic factor α, expectation heuristic factor β, pheromones trace initial value τ_0, constant q_0 and C.

Step 2: Initialize ant colony. Use cascade reservoirs upstream water level as decision variables, randomly generate M groups water level change sequences $(Z_1^0, Z_2^0, \ldots, Z_M^0)$ which are treated as artificial ant path allowed by the water level of permitted scope for every reservoir in any time, and at the same time put the path in the set of paths S, Where $Z_m^0 = (Z_{m,11}^0, Z_{m,12}^0, \ldots, Z_{m,1T}^0, Z_{m,21}^0, Z_{m,22}^0, \ldots Z_{m,2T}^0, \ldots, Z_{m,I1}^0, Z_{m,I2}^0, \ldots Z_{m,IT}^0)$. l represents the iteration number in $Z_{m,it}^l$. m is the ant colony number. t is the time interval number. At the same time initialize the cumulative pheromone of the path $\tau_i = \tau_0 (i = 1, 2, \ldots, M)$.

Step 3: Update the set of paths. For every ant in the current, generate random numbers $q \in [0,1]$, and compare it with constant q_0, if $q \le q_0$, then conduct local search to the path of the ant in the area of neighborhood radius r, and put the local optimal solution in the set of paths S. If $q > q_0$, then calculate ants' transition probability by Eq. (7), select the path S_j according to the principle of high probability in the set of paths S.

Step 4: Update the pheromone. Update the current pheromone according to the Eq. (8), and make Eq. (1) as fitness function, evaluate the population fitness value, and name the current optimal path as Z_{bext}^l.

Step 5: Optimize current optimal solution by chaos. Use the current optimal path Z_{bext}^l as initial value, apply Eq. (12) to chaos optimization, then replace Z_{bext}^l with the optimization result.

Step 6: Judge whether it meets the demand of maximum iteration number, if not, turn to *Step* 3, until it meets the termination iteration conditions. Otherwise, output the current optimal path Z_{bext}^L as approximate optimal solution.

The framework of CACO is shown as Figure 1.

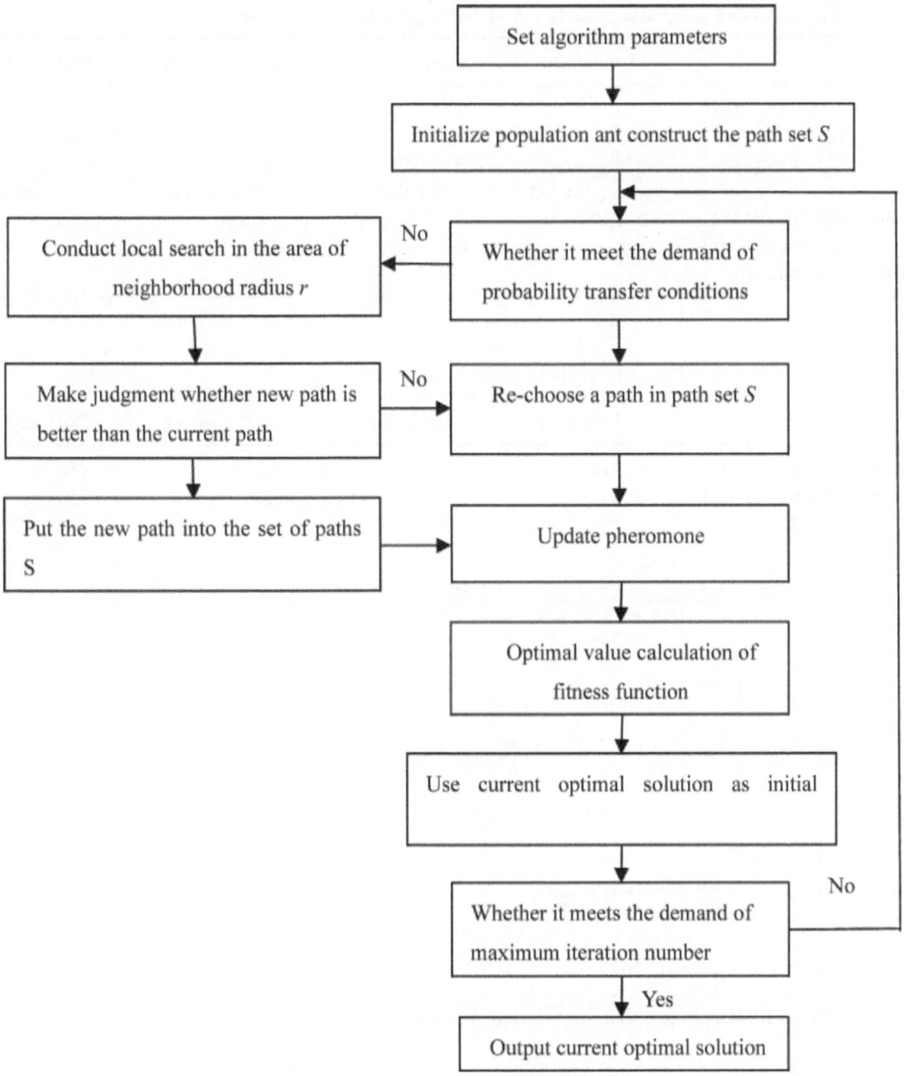

Figure 1. The framework of CACO.

6 CALCULATION EXAMPLE

6.1 *Basic information*

Cascade reservoirs along upstream Wujiang river include carry-over storage reservoirs and many annual flow regulations. They are one of the most complex reservoirs in our country at present and one of the twelve great hydropower bases in our country, which plays an important role in GuiZhou power grid. At present, HongJiadu, DongFeng, SuoFengying and WuJiangdu have already been completed. The main characteristic parameters of Wujiang cascade reservoir is shown in Table 1.

In this paper, we conduct the simulation calculation adopting actual measurement annual runoff series of a year with month as the time interval. Eq. (1) is used as objective function, incoming runoff data is shown in Figure 3. Priming level of each reservoir is shown in Table 2.

Table 1. The main characteristic parameters of Wujiang cascade reservoir.

	HongJiadu	DongFeng	SuoFengying	WuJiangdu
Normal storage level (m)	1140	970	837	760
Level of dead water (m)	1076	936	822	720
Regulation performance	Many-year regulation	Annual regulation	Daily regulation	Annual regulation
Installed capacity (MW)	3×200	$3 \times 190 + 125$	3×200	5×250
Power station integrated efficiency coefficient	8.5	8.35	8.5	8.17

Table 2. Begin and end water level during dispatching period.

	HongJiadu	DongFeng	SuoFengying	WuJiangdu
Starting water level (m)	1100	960	829.5	750
Terminate water level (m)	1115	960	829.5	750

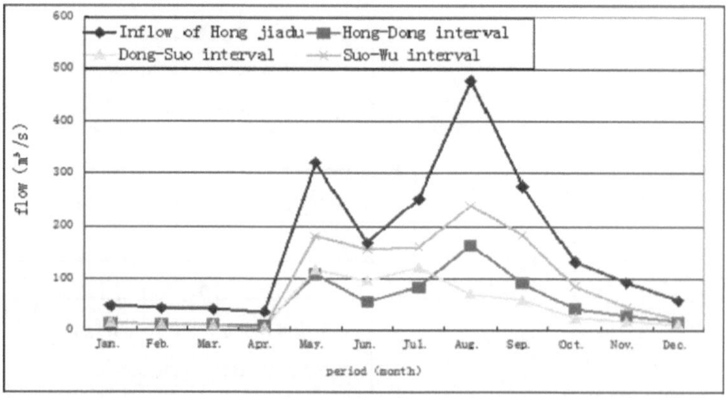

Figure 2. The inflow hydrograph of the cascade reservoirs.

6.2 *Analysis and determination of algorithm parameters*

Conduct simulation calculation by Visual Basic 6.0 according to above basic information. When applying intelligent optimization algorithm to solve practical problem, the parameters in the algorithm have great influence on the calculation result and the parameter setting should be confirmed based on the actual problems. Therefore, this paper analyzes the main parameters related in the algorithm on the basis of simulation calculation.

1. Population scale and iteration number

When we conduct the simulation calculation, the population scale are respectively taken 40, 80, 120, and iteration number are respectively taken 50, 100, 200, 500, 800, 1000, 2000 and other parameters are fixed. Calculate ten times and record the optimal value, average and the corresponding computer hours of optimal value to every case, which are shown in Table 3.

It can be seen from Table 3 that along with the increase of population scale and iteration number, the objective function value much approach to approximate optimal solution, but the increase range of computer hours is much bigger than the increase range of objective function value. Therefore, considering the calculation precision and computational efficiency with compromise, this paper uses 80 as the population scale and uses 800 as iteration number.

Table 3. The impact analysis of the algorithm performance of the population size and the number of iterations.

	Population scale								
	40			80			120		
Iteration number	Optimal value (Hundred million KW·h)	Average (Hundred million KW·h)	Computer time (seconds)	Optimal value (Hundred million KW·h)	Average (Hundred million KW·h)	Computer time (seconds)	Optimal value (Hundred million KW·h)	Average (Hundred million KW·h)	Time consuming (seconds)
50	81.95	81.88	69	85.77	84.67	89	88.56	87.53	98
100	83.36	83.27	98	87.23	86.89	123	89.97	89.07	168
200	87.58	87.47	108	92.78	92.21	189	93.68	93.25	256
500	89.9	89.76	199	94.66	94.25	245	95.09	89.98	450
800	91.37	91.21	257	95.88	95.86	317	95.86	95.83	647
1000	93.22	93.09	355	95.91	95.89	420	95.92	95.9	836
2000	95.88	95.79	770	95.93	95.91	980	95.93	95.91	1544

2. Other main parameters
When we use 80 as the population scale and 800 as iteration number, the information heuristic factor α, expectations heuristic factor β, pheromones residue coefficient ρ and positive normal number C are calculated and analyzed. The results are shown in Table 4.

From that Table 4, heuristic factor α and β have little influence on the objective function value. According to the simulation results, they respectively take the value 1 and 3. Along with the increase of pheromone residue coefficient ρ, the objective function values fist increases and then decreases, and it is found that value 0.4 is close to the approximate optimal solution. It is can be seen from the table that the value of C has great influence on the objective function value by analyzing C as different times coefficient to the total average annual energy generation. C takes the value 300 in this paper.

Through the above analysis of main parameters in CACO, and in order to ensure the stability and convergence of the algorithm, the parameters selected of calculated example in this paper are shown as follow: the population scale M is 80, the iteration times L is 800, the heuristic factors α is 1 and the heuristic factors β is 3, the pheromones residue coefficient ρ is 0.4, the normal number C is $300\overline{E}$ ($\overline{E}$ is cascade reservoirs' annual average generating capacity), the threshold value q_0 is 0.6, the initial value of pheromones trace τ_0 is 0.0001. Through 10 times simulation calculation, we take the optimal as approximate optimal solution. The relation between objective function values and the iteration times is shown in Figure 3. The changing process of the reservoir water level is shown in Figures 4–6, electric energy production capacity and ACO are shown in Table 5.

From Table 5 and Figures 4 to 6 we can obtain the following conclusions:

1. By combining ACO with chaos algorithm, the calculation precision and stability of the solution of the algorithm are improved and the generating capacity of each reservoir is increased. Cascade reservoirs' total generation is increased by 18%.

Table 4. The impact analysis of the algorithm performance of the main parameters.

Heuristic factor α	Optimal value (Hundred million KW·h)	Heuristic factor β	Optimal value (Hundred million KW·h)	Times coefficient C	Optimal value (Hundred million KW·h)	Pheromones residue coefficient ρ	Optimal value (Hundred million KW·h)
0.1	92.34	0.1	93.23	1	45.89	0.2	92.12
0.5	93.19	0.5	93.1	100	56.27	0.3	94.89
1	95.79	1	94.44	200	85.12	0.4	95.91
2	95.22	2	95.03	300	95.59	0.5	95.08
3	94.89	3	95.89	400	93.14	0.6	94.16
4	94.56	4	95.34	500	89.77	0.7	93.22
5	93.22	5	94.21			0.8	89.34
6	92.78	6	92.36			0.9	87.11

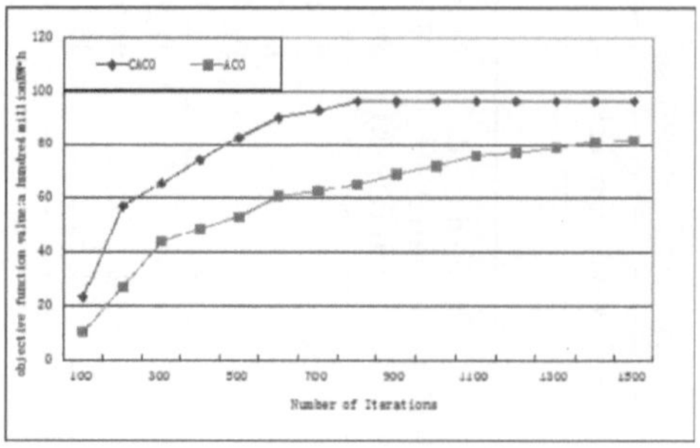

Figure 3. The objective function value with changes of iterations.

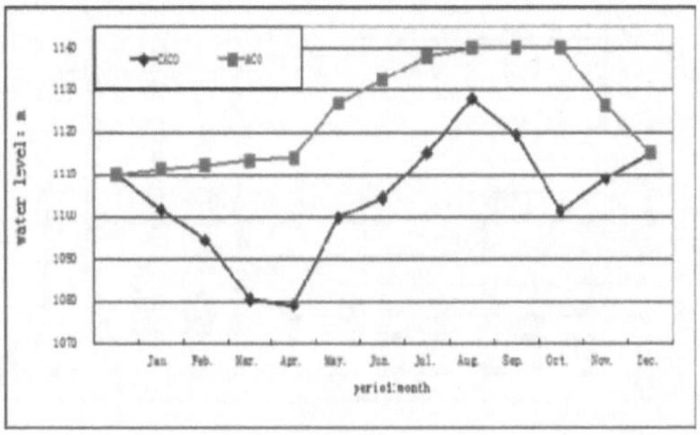

Figure 4. The process of water level of Hongjiadu reservoir.

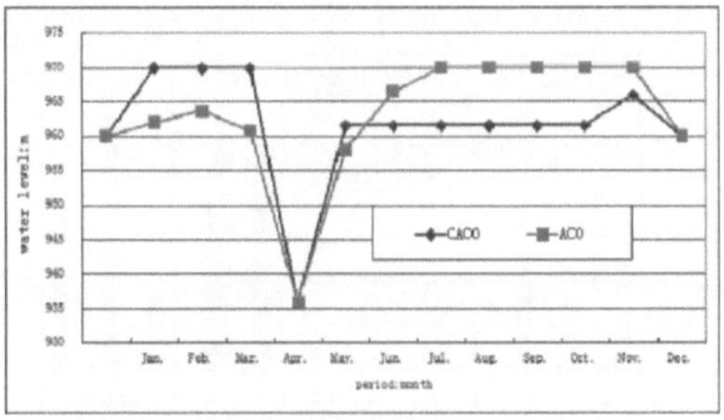

Figure 5. The process of water level of Dongfeng reservoir.

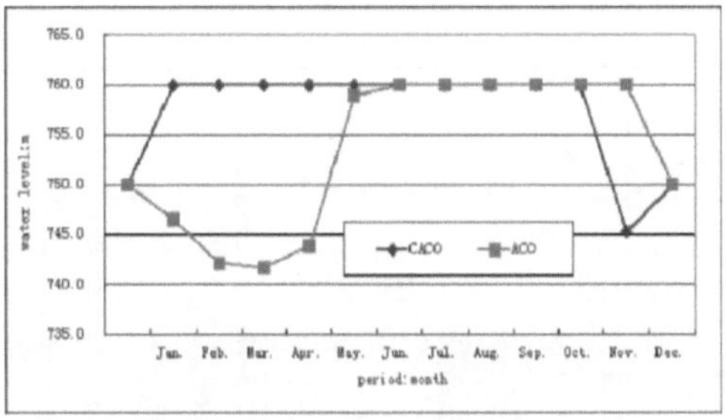

Figure 6. The process of water level of Wujiangdu reservoir.

Table 5. The computation result of the model of maximum of generation benefits of Wujiang cascade reservoir.

Algorithm	HongJiadu (Hundred million KW·h)	DongFeng (Hundred million KW·h)	SuoFengying (Hundred million KW·h)	WuJiangdu (Hundred million KW·h)	Sum total (Hundred million KW·h)	Compute time consuming (second)
ACO	17.26	19.01	12.63	32.12	81.03	274
CACO	18.48	22.68	16.64	38.07	95.87	318

2. In the aspect of calculation efficiency, the compute hours of CACO is 318 seconds compared to ACO 274 seconds under the same condition. The calculation efficiency declines to some extent. This is because CACO adopted chaos algorithm to conduct ergodicity search, which solves the problem of stagnant in the later of ACO, improves the search efficiency in the later of algorithm, and avoids the disadvantage that it is easy to fall into the local optimal solution.

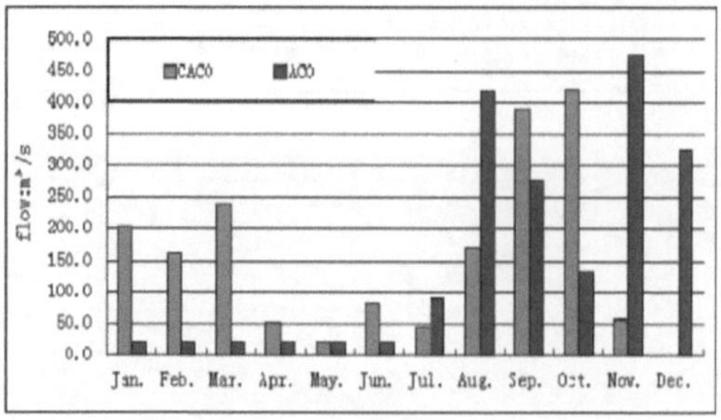

Figure 7. The process of outflow of Hongjiadu reservoir.

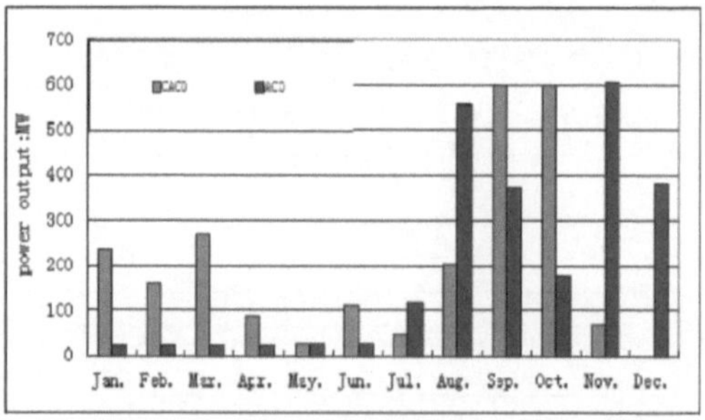

Figure 8. The process of output of Hongjiadu reservoir.

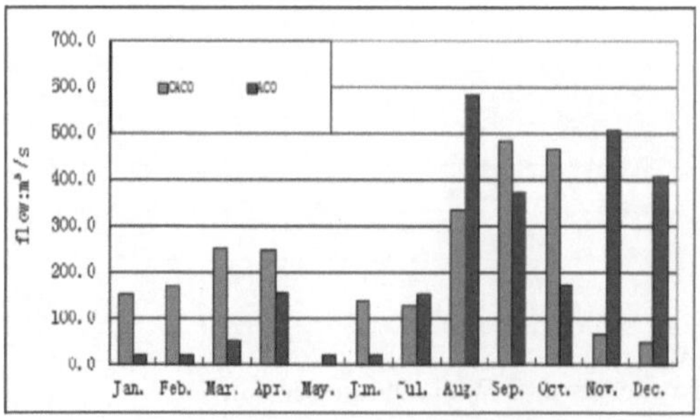

Figure 9. The process of outflow of Dongfeng.

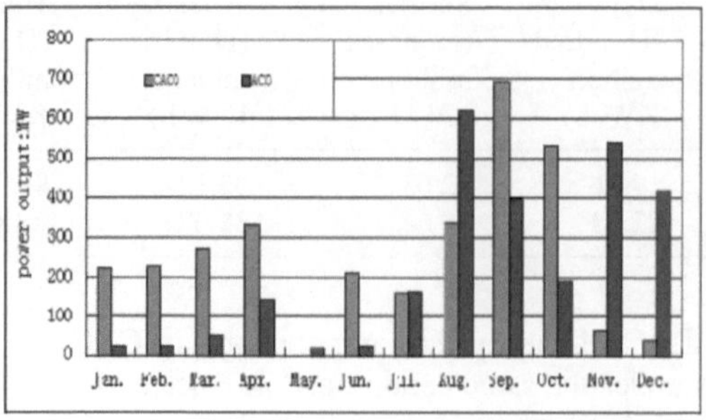

Figure 10. The process of output of Dongfeng.

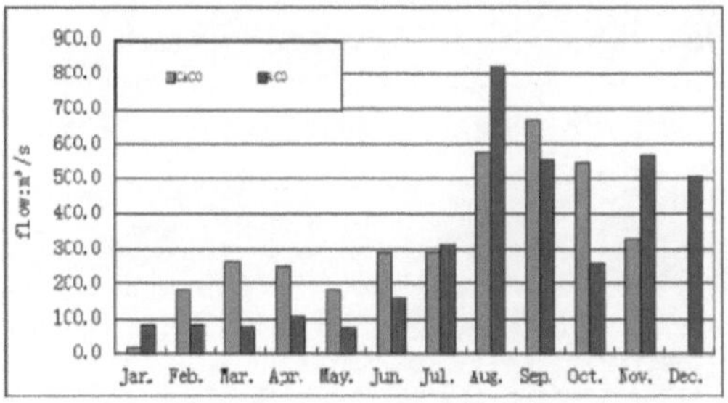

Figure 11. The process of outflow of Wujiangdu reservoir.

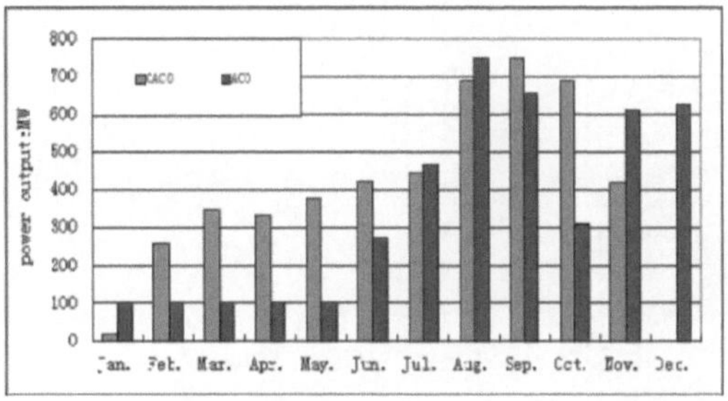

Figure 12. The process of output of Wujiangdu reservoir.

7 SUMMARY

For ACO has the disadvantages of stagnant phenomenon and being easy to fall into local optimal solution, CACO has been proposed in cascade reservoir power generation optimized operation in this paper. The WuJiang river cascade reservoirs shows that by using simulation methods to analyze and calibrate the main parameters involved in algorithm, the satisfactory approximate optimal solution is obtained, besides, the calculation precision and the stability of the solution of the algorithm is improved. So, CACO provides a new way for searching the solution to the optimal operation of cascade reservoirs with high dimension and complexity.

REFERENCES

Ding Jianli, Chen Zengqiang, Yuan Zhuzhi. On the combination of genetic algorithm and ant algorithm[J]. Journal of Computer Research and Development, 2003, 40(9):1351–1356.

Gao Shang, Yang Jingyu. Swarm Intelligence Algorithms and Application [M]. Beijing: China Water-Power Press, 2006:63–68.

Jalali M R, Afshar A, Marino M A. Improved ant colony optimization algorithm for reservoir operation [J].Scientia Iranica, 2006, 13(3):295–302.

Jia Renfu, Chen Shoulun, Liang Wei. Long term optimal dispatch of large scaled hydropower station groups based on chaos optimal algorithm [J]. Journal of Hydraulic Engineering, 2008, 39(9):1131–1135.

João Luiz B. Brandão. Performance of the equivalent reservoir modelling technique for multi-reservoir hydropower systems [J]. Water Resour Manage, 2010, 24:3101–3114.

Raheleh A, Seyed J M, Abbas G. Reliability-based simulation optimization model for multi-reservoir hydropower systems operations: khersan experience [J]. Journal of Water Resources Planning and Management, 2008, 134(1):24–33.

Qiu Lin, Tian Jinghuan, Duan Chunqing ect. The application of chaos-optimization algorithm in reservoir optimal operation [J]. China Rural Water and Hydropower, 2005, (7):17–20.

Xiaohui Yuan, Yongchuan Zhang, Yanbin Yuan. Improved self-adaptive chaotic genetic algorithm for hydrogeneration Scheduling [J]. Water Resour. Plann. Manage. 2008, 134(4):314–319.

Modeling and Computation in Engineering III – Zhang & Xie (Eds)
© 2014 Taylor & Francis Group, London, ISBN 978-1-138-02680-3

Study of small grinding wheel with multi-grains diamond

L. Wang, X.L. Tian, W.L. Wang, Z.Y. Wu & X.J. Tang
National Key Laboratory for Equipment Remanufacturing, Academy of Armored Forces Engineering, Beijing, China

ABSTRACT: In this paper, the grit trajectory model of grinding engineering ceramics in axial deep creep-feed grinding with small grinding wheel is established and the relationship between cutting parameters and the law of diamond grits' motion is revealed. The relation between feeding speed and the crushing of edge, grinding force and the diamond grit wear was analyzed according to the actual machining of ceramics by changing processing parameters, and the experimental results are consistent with the simulation results. Experiments is based on appropriate experimental schemes and force measuring system, the changing of the diamond grain's curve radius was real-time detected by edge detection and curve fitting of profile. This paper also provides references on how to use the feed speed that is reasonable to enhance the processing efficiency by removing ceramic chunks via the splitting and to reduce the damage on the workpiece and the grinding wheel.

1 INTRODUCTION

According to the theory of indentation fracture mechanics, there is Hertz stress zone between abrasive and workpiece, which leading the initiation and propagation of crack on the ceramics workpiece surface. It mainly produces two kinds of crack system: median/radial crack system and the lateral crack system (Huang, 2003). When engineering ceramics is machinned in axial feed grinding with a small multi-grains diamond grinding wheel, direction of normal grinding force and axial grinding force is pointed to waiting process, which main causes median/radial crack, and the tangential grinding force is caused lateral cracks. What is more, the propagation of crack is along the axial feed direction, and the material removal caused by a a complex combination of two kinds of crack system. Therefore, the axial deep creep-feed grindindg with a small grinding wheel can be an efficient method for rough machining (Guo, 2010).

This paper provides references on how to use the feed speed that is reasonable to enhance the processing efficiency by removing ceramic chunks via the splitting and to reduce the damage on the workpiece and the grinding wheel. In this paper, the relation between feeding speed and the crushing of edge, grinding force and the diamond grit wear was analyzed according to the simulation experiments and the actual machining of ceramics by changing processing parameters.

2 THEORETICAL MODEL AND SIMULATION OF THE AXIAL FEED MOVEMENT

2.1 *Structure design of a small multi-grains diamond grinding wheel*

As shown in Fig. 1, two rows of coarser diamond grit are welded on a small grinding wheel end surface. The structure characteristics of a grinding wheel is that head is inverted cone type, gap of between diamond grits on the end surface edges is less than 1 mm, and the matrix materials is 45 steel. diamond grit tip angle 2θ and cutting edge radius ρg is used to characterize the diamond grit' cutting performance. The cone angle of single diamond grit is prepared to 60°, 120° and 150° welded in symmetrical parts, initial tip radius is 4 mm.

2.2 *The basic kinetical vector model of the axial feed grinding*

The axial feed grinding motion model shown in Fig. 2, R represents the circle radius of machined part, and r represents the radius of diamond grit location. Workpiece axis is selected as Oxyz origin for the absolute coordinate system, and small grinding wheel axis is selected as origin for three relative coordinate origin like O'uvw, O'x'y'z', O "u'v'w'.

Grinding wheel axially vector C specifies the origin position of the O'uvw relative coordinates in the absolute Oxyz coordinate system. as shown in Fig. 2. The revolutive angle of grinding wheel relative to the workpiece is θ_w, then small grinding wheel axially vector C can be expressed as follows, in which R stands for circle radius of machined part and r stands for radius of diamond grit location

$$C(\theta_w) = (R+r)\begin{bmatrix} \cos\theta_w \\ \sin\theta_w \\ 0 \end{bmatrix} \tag{1.1}$$

Vector T describes the position vector of single grain located on grinding wheel circumference surface in the relative coordinate system O'uvw. If the rotative direction of workpiece is clockwise, the down-grinding rotative direction of grinding wheel is counter-clockwise, and the up-grinding rotative direction of grinding wheel is clockwise. The expression of Vector T can be expressed as follows, in which θ_t stand for the rotative angle of grinding wheel, "+" represents down-grinding, "–" represents up-grinding

$$T(\theta_t) = r\begin{bmatrix} \cos\theta_t \\ \pm\sin\theta_t \\ 0 \end{bmatrix} \tag{1.2}$$

Figure 1. Structure diagram of small grinding wheel.

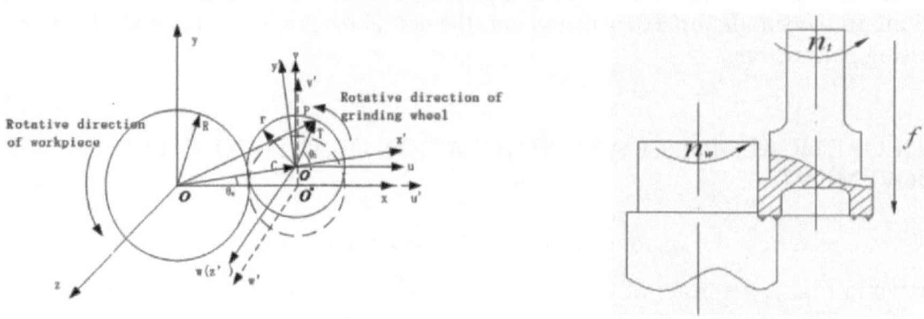

a) Circumferential movement view **b) axial movement view**

Figure 2. Grinding wheel axial—feed movement view.

The relative vector in the O'x'y'z' coordinate system is transformed into a relative vector in O'uvw, which need a transformation matrix:

$$M(\theta_t)=\begin{bmatrix}\cos\theta_w & \mp\sin\theta_w & 0\\ \pm\sin\theta_w & \cos\theta_w & 0\\ 0 & 0 & 1\end{bmatrix} \tag{1.3}$$

In the above expression, the above symbolic represents the transformation matrix for down-grinding, the following symbolic represents the transformation matrix for up-grinding.

2.3 *Vector model of diamond grit motion*

According to the vector sum between the grinding wheel axially vector C and particle vector T, the diamond grit position vector P expression in the absolute coordinate system can be got:

$$P(\theta_w,\theta_t)=C(\theta_w)+M(\theta_t)\cdot T(\theta_t) \tag{1.4}$$

Diamond grits are distributed on the circumference surface, and the adjacent grit track just have a phase angle differ. The expression can be extended to this, in which k represent Sequence of diamond grit and m represent total number of diamond grit

$$T_k=r\begin{bmatrix}\cos(\theta_t+(2\times\pi(k-1))/m)\\ \pm\sin(\theta_t+(2\times\pi(k-1))/m)\\ 0\end{bmatrix} \tag{1.5}$$

Expression:

In the actual processing, grinding wheel has feed movement along the axis, so the vector expression of the same circumference of each grain motion:

$$P(\theta_w,\theta_t)=C(\theta_w)+M(\theta_t)\cdot T_k(\theta_t)+F \tag{1.6}$$

$$F=[0\ \ 0\ \ -f\cdot t]^T \tag{1.7}$$

2.4 *Simulation of diamond grit trajectories*

The trajectory of the axial feed grinding of engineering ceramic materials was simulated with MATLAB, grinding wheel diameter is 20 mm, the diameter of the workpiece is 30 mm, grinding wheel speed is 5500 r/min, workpiece speed is 150 r/min, the axial feed speed V = 50 mm/min. diamond trajectory simulation when cylindrical surface is down grinding machining shown in Fig. 3.

Numerical dot plot shown in Fig. 4 is the time from 0 to 0.4, every 0.004 seconds to take the position of diamond grits and a straight line connecting the simulation trajectories. The unit of workpiece speed nw and wheel speed nt is r/min, the feed speed f is mm/min.

Trajectory of diamond grit will affect the grinding force, processing quality, wheel wear and so on. Down grinding shown in Fig. 4 show the speed change of ceramic grinding wheel and the workpiece impact dense of lines, which leads to the change of the number of contact points between the two, the more contact points indicated that diamond grit participate more times. From the simulation results can be seen that the rotational speed of grinding wheel has an effect on the dense degree of contact point, it is contact less when the rotational speed of grinding wheel is 5500 r/min than the 3000 r/min,it means more small debris, and leading to smaller grinding force. what is more, it improve the surface quality of machined surface.

The axial trajectory of diamond grit is a spiral line shown in Fig. 4. The change of axial feed speed and workpiece speed will directly affect helix pitch. When the axial feed speed is constant, rotational speed of workpiece has a great influence on the axial trajectory move-

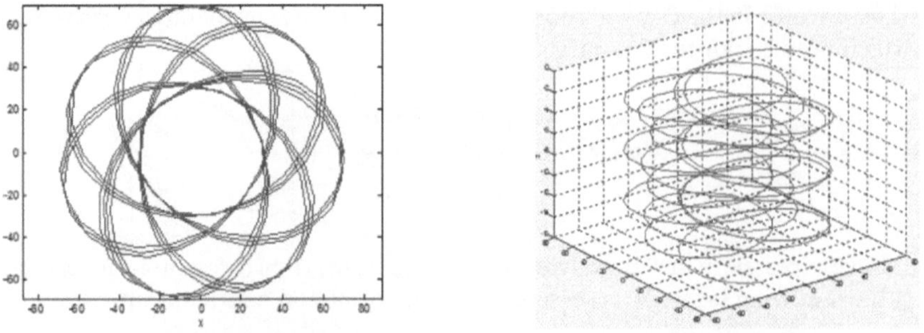

a) **Circumferential trajectory** b) **axial trajectory**

Figure 3. Simulation of diamond grit trajectory.

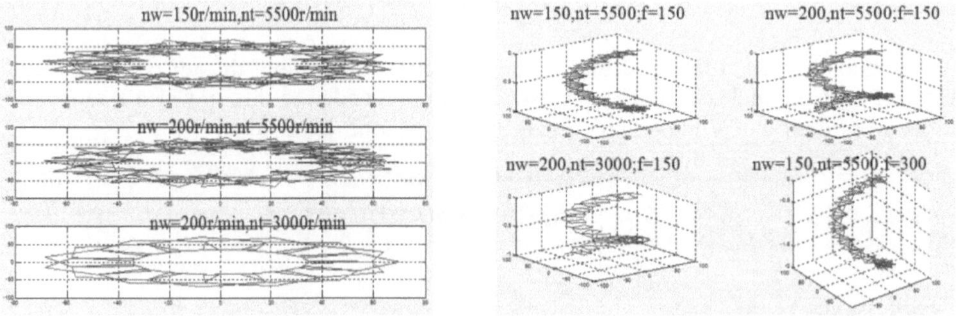

a) **Circumferential trajectory** b) **axial trajectory**

Figure 4. Numerical trajectory of diamond grit at each time.

ment of diamond grit, the faster speed cause more smaller helix pitch. when the rotational speed of workpiece is constant, the faster feed speed of grinding wheel causes the larger helix pitch and the more sparse contact point. In this case, it means abrasive dust becomes large, which results in the grinding force becomes larger, and diamond abrasive is easier to wear, and the workpiece the broken pieces of surface edge will become bigger, and the surface roughness increases. The next article will focus on experiments about the relationship of the axial feed speed connected with grinding force, processing quality, wheel wear and tear and others.

3 SCHEME FOR EXPERIMENTAL STUDY ON THE AXIAL FEED SPEED

The experimental system. The experiment uses the platform to carry out experiments shown in Fig. 5, which mainly include BV75 vertical machining centers, low-speed motor, dynamometer, etc. Small grinding wheel is clamped on the spindle machining center. Ceramic workpiece is clamped in three jaw which chuck on the low speed motor. Force measuring instrument type SDC-C4F, which developed from the Beijing University of Aeronautics and Astronautics.

Material parameters. Material dimensions of Si3 N4 ceramics is 26 mm in diameter, height 60 mm. The main performance parameters of materials: density $\rho = 2.73$ g/cm^3, hardness HRA = 85, the elastic modulus E = 160 Gpa, breaking strength KIC = 2.85 MPa.m$^{1/2}$.

Experimental parameters. Three sets of process parameters of the experiment shown in Table 1.

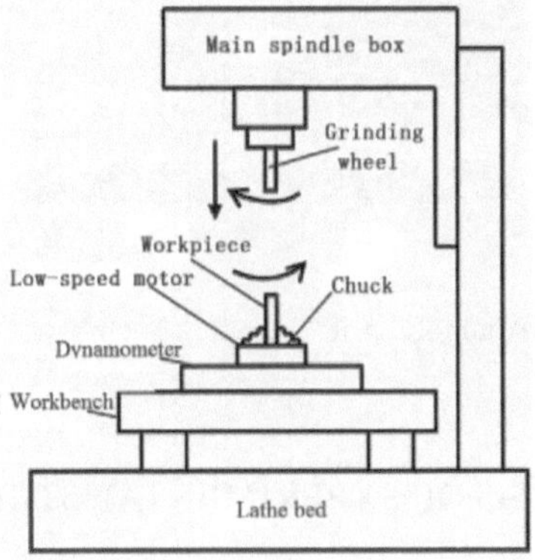

Figure 5. The experimental platform.

Table 1. The experimental parameters.

No.	nt(r/min)	nw(r/min)	a_p/mm
1	5000	200	2
2	7000	200	2
3	7000	200	1

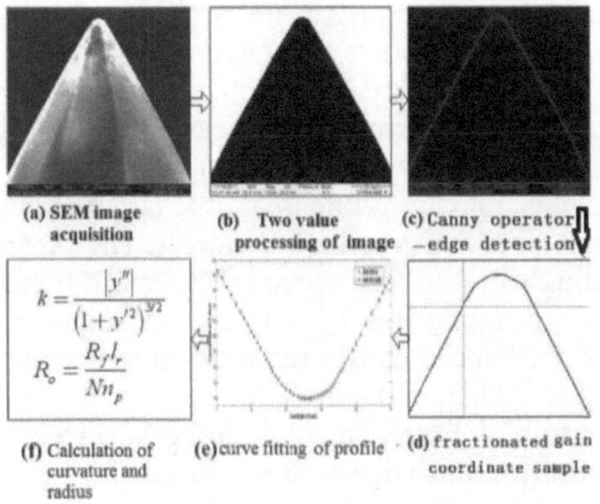

Figure 6. Detecting the radius of diamond tip curvature.

Measurement of diamond grit wear. The diamond grit will endure friction and compression of the ceramic workpiece in the actual grinding process, it occurs in the form of broken off from the matrix form when grinding force exceeds the strength of materials. If so, diamond grit tip is worn, broken or caved, and the tip radius will gradually increase according to a specific law. Meanwhile, the passivation diamond make grinding force significantly increase, surface quality will become bad. So, the wear properties of diamond grit can be measured by real-time detecting the radius of diamond tip curvature (Tian, 2013), shown in Fig. 6. Through the real-time tracing observation of labelled diamond grains by SEM, the wear process of single diamond grit is shown in Fig. 7.

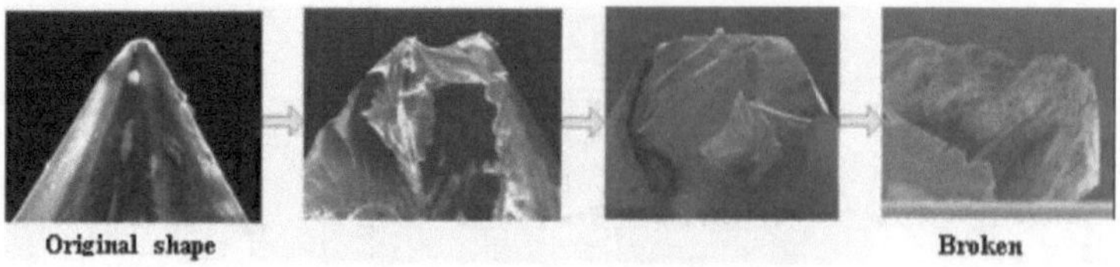

Figure 7. Wear process of single diamond grit.

4 RESULTS AND ANALYSIS OF EXPERIMENTAL STUDY ON THE AXIAL FEED SPEED

The relation between feed speed and surface quality. As shown in Fig. 8, when the axial feed rate is large, the edge will produce flake off, namely there is crushing of edge. This is due to the rapid crack propagation, the feed speed is the critical of crack control. Crushing of edge occurs mainly in the inlet and outlet, this is because the larger impact force exist at the entrance, on the other hand, the main reason is the lack of material support at the exit. In addition, it can be seen the depth of broken edge is shallow than the depth of grinding, median crack can less reduce material strength, but mainly cause material removal. Meanwhile, with the feed rate increases, the trajectry of grain becomes sparse, and the processing debris particles formed becomes large, and machined surface becomes rough.

The relation between feed speed and grinding force. As shown in Fig. 9, the normal grinding force F_n, the tangential grinding force F_τ and the axial grinding force F_a have to change with the variation of the feed speed. Under three different groups of processing parameters, the trend that feed speed impacts on the grinding force is consistent. In any group of processing parameters, axial grinding force is the largest, followed by normal grinding force, and the tangential grinding force minimum. With the increase of feed speed, three grinding force are more larger. This is due to feed speed increases which causes the feed per revolution increases, corresponding to the maximum undeformed chip thickness of single grain must thicken, thus grinding force acting on the workpiece and diamond grit is increased. The growth of grinding force, especially in the median crack, resulting in a more larger material removal.

The relation between feed speed and grinding wheel wear. The experimental results is shown in Fig. 10 when the grinding wheel to feed speed of V were 100 mm/min, 200 mm/min. As feed speed becomes higher, grinding travel is shorter, and diamond grit wear easier. This is because the increase of feed speed causes the increasing normal and tangential friction between grain and workpiece. Diamond grit is subjected to the action of a high reaction force, so that high temperature of grain contact with the workpiece causes the diamond grit local softening, therefore the grinding edge wear increased.

5 SUMMARY

The paper establishs the motion vector model of diamond grit. simulation experiments research the diamond grit movement law by changing processing parameters. Appropriate experimental program and force measurement systems are used in this experiment. In addition, the changing of the diamond grain's curve radius was real-time detected by edge detection and curve fitting of profile can realize the qualitative analysis of diamond grit wear. Actual machining experimental results are basically consistent with the simulation analysis. It proves that the axial helical pitch is elongated when axial feed speed becomes fast, which will make the dense degree of contact point thinner and make the surface roughness worse. Moreover, abrasive dust and grinding force become larger, therefore it accelerates diamond grit wear further.

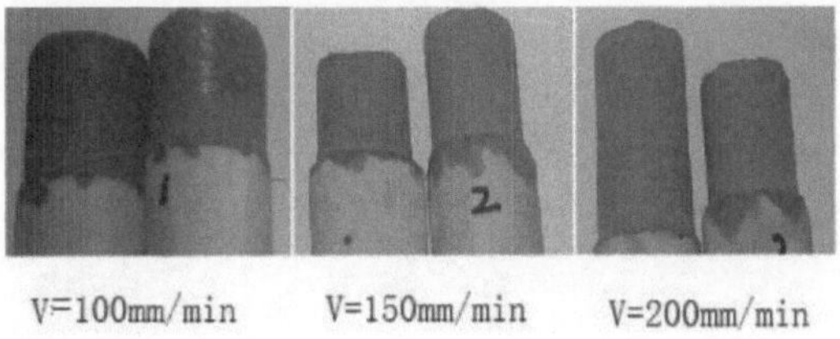

Figure 8. Processed workpieces.

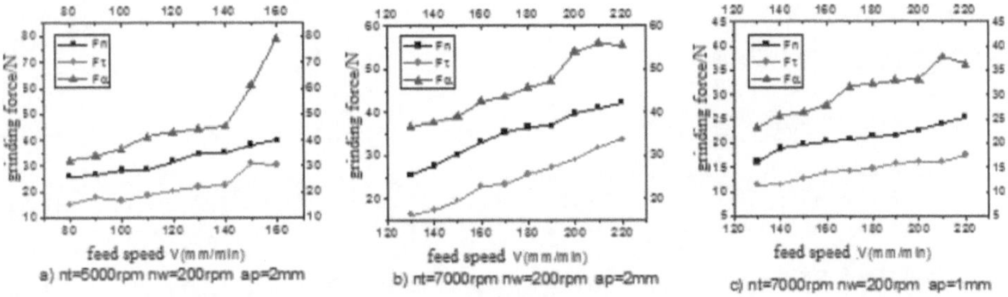

Figure 9. The relation between feed speed and grinding force.

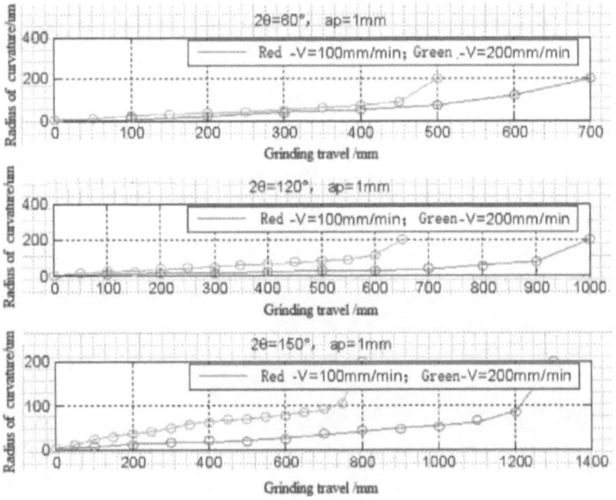

Figure 10. The relation between feed speed and grinding wheel wear.

REFERENCES

Guo F., Tian X.L., Mao Y.T., Wang J.Q., Xu Sen and SH. Zhang. Investigation on axial turning-grinding of engineering ceramics[J]. Advanced Materials Research. 2010, Vol. 154–155: 1027–1032.

Huang H., Liu Y.C. Experimental investigations of machining characteristics and removal mechanisms of advanced ceramics in high speed deep grinding [J]. International Journal of Machine Tools & Manufacture, 2003, 43: 811–823.

Tian X.L., Wang J.Q., Zhang B.G., et al. Research of radii of curvature measurement for micro-accessory in precision and ultra-precision machining[J].6th International Symposium on Advanced Optical Manufacturing and Testing Technologies, 2012, 8418(17): 1–6.

Modeling and Computation in Engineering III – Zhang & Xie (Eds)
© 2014 Taylor & Francis Group, London, ISBN 978-1-138-02680-3

GPS elevation fitting based on improved least squares collocation

Shuibing Li, Shichao Ge & Yongxiang Xu
NARI Group Corporation (State Grid Electric Power Research Institute), Nanjing, China

ABSTRACT: The signals' covariation is the key point in GPS elevation fitting based on least squares collocation. The covariation is difficult to confirm because the data is limited. The semivariable function solves the problem of the covariation, establishes the model of GPS elevation abnorma based on improved least squares collocation, and applies in a local area. Through comparison among the polynomial surface, the polyhedral function, kriging interpolation function, proves effectiveness of the model of improved least squares collocation for GPS elevation fitting.

1 INTRODUCTION

The measurement accuracy of plane position of GPS can meet the requirements of practice, but the abnormal height between GPS geodetic height and normal height, need to transform. The correct transformation method, to obtain of accuracy GPS leveling height for meeting the practical requirement, can be widely applied to engineering, deformation monitoring and other aspects. It has the important theoretical and practical significance to research on the GPS height fitting for promoting the application of GPS technology.

In 1967 Moritz proposed the least squares collocation based on covariance estimation, which is a basic method in physical geodesy. The covariance estimation method to calculate the gravity field elements in physical geodesy is called configuration. The principle assumes a random process with zero mean between the signal to be measured and the measured signal. By the signal to be measured, the measured signal, variance and covariance between each other, calculate the linear estimation of signal to be measured. The least squares collocation, except for the estimation of the gravity field elements, has been widely used in the coordinate transformation, adjustment of control network and so on.

2 GPS ELEVATION FITTING BY LEAST SQUARES COLLOCATION

The least square collocation method bases on statistical linear optimal unbiased estimation, and considers non random variables and random variables in the course of estimation, which makes estimation of height anomaly has a higher accuracy.

The function of least squares collocation:

$$L = S + BX + \Delta \tag{1}$$

In the function, L = observation vector, S = observed signal, X = non random parameters to be determined, Δ = observation noise, $\Delta \sim (0, D_\Delta)$, S' = non observed point observation signal.

The observation equation bases on generalized least squares:

$$L = CZ + BX + \Delta \tag{2}$$

In the function, $C=[E,0]$, $Z=[S,S']^T$
$\tilde{X}, \tilde{Z}, V$ = the most probable value of X, Z, Δ

$$L = B\tilde{X} + C\tilde{Z} + V \quad (3)$$

$$D_{LL} = D_{\Delta\Delta} + CD_{ZZ}C^T \quad (4)$$

According to $D_{ZZ} = \begin{bmatrix} D_{SS} & D_{SS'} \\ D_{S'S} & D_{S'S'} \end{bmatrix}$, $CD_{ZZ}C^T = D_{SS'}$.
According to the principle of least square collocation method, $D_{S\Delta} = 0, D_{\Delta S} = 0$, and the covariance matrix is known.

$$V^T P_{\Delta\Delta} V + \tilde{Z}^T P_{ZZ} \tilde{Z} = \min \quad (5)$$

The extremal function:

$$\phi = V^T P_{\Delta\Delta} V + \tilde{Z}^T P_{ZZ} \tilde{Z} + 2K^T (B\tilde{X} + C\tilde{Z} + V - L) \quad (6)$$

Partial derivatives of $\tilde{X}$, $\tilde{Z}$, V are equal to zero:

$$B^T K = 0 \quad (7)$$

$$P_{ZZ}\tilde{Z} = C^T K \quad (8)$$

$$K = P_{\Delta\Delta} V \quad (9)$$

Assume unit weight error equal to one, $\sigma_0 = 1$:

$$V = D_{\Delta\Delta} K \quad (10)$$

$$\tilde{Z} = D_{ZZ} C^T K \quad (11)$$

Insert the formula 10, 11 into the formula 3:

$$B\tilde{X} + CD_{ZZ}C^T K + D_{\Delta\Delta} K - L = 0 \quad (12)$$

Insert the formula 4 into the formula 12:

$$B\tilde{X} + D_{LL} K - L = 0 \quad (13)$$

Multiply the both sides of the formula 13 by $B^T D_{LL}^{-1}$:

$$B^T D_{LL}^{-1} B\tilde{X} + B^T K - BD_{LL}^{-1} L = 0 \quad (14)$$

Insert the formula 7 into the formula 14:

$$\tilde{X} = (B^T D_{LL}^{-1} B)^{-1} B^T D_{LL}^{-1} L \quad (15)$$

Insert the formula 15 into the formula 13:

$$K = D_{LL}^{-1}(L - B\tilde{X}) \quad (16)$$

Insert the formula 16 into the formula 11:

$$\tilde{Z} = D_{ZZ} C^T D_{LL}^{-1}(L - B\tilde{X}) \quad (17)$$

Break the formula 17:

$$\tilde{S} = D_{SS} D_{LL}^{-1} (L - B\tilde{X}) \tag{18}$$

$$\tilde{S}' = D_{S'S} D_{LL}^{-1} (L - B\tilde{X}) \tag{19}$$

Adjusted value of unknown points:

$$\tilde{L}' = B'\tilde{X} + \tilde{S}' \tag{20}$$

In the formula 20, $B' = \begin{bmatrix} 1 & x_1' - x_0 & y_1' - y_0 \\ 1 & x_2' - x_0 & y_2' - y_0 \\ \vdots & \vdots & \vdots \\ 1 & x_m' - x_0 & y_m' - y_0 \end{bmatrix}$, m is the number of unknown points, (x_0, y_0) is the mean value of all coordinates, (x_i', y_i') is the coordinate of unknown point.

From the formula 17, 18, 19, defining D_{ZZ} is the key point to solving $\tilde{L}'$, $\tilde{S}$, $\tilde{S}'$. Strictly speaking, the elements of covariance matrix should be compute through a lot of survey data. Due to the actual conditions, it is difficult to get the covariance matrix. This paper adopts variation function to determine, because the covariance function and variation function usually choose some functions of distance to determine. When the signal meets the two order stationary or intrinsic hypothesis, the covariance function can be obtained by variation function, so the variation function can be regarded as another form of covariance function. And this paper regards the height anomaly as no error observation for processing.

3 IMPROVED LEAST SQUARES COLLOCATION

The covariance function $D(h)$ is a degree of correlation reflection of regional variables $Z(x)$ and $Z(x+h)$, which usually is a decreasing function of distance h. Variation function $\gamma(h)$ is a degree of variation reflection of regional variables $Z(x)$ and $Z(x+h)$.

The relationship between the variation function and the covariance function:

$$\gamma(h) = D(0) - D(h) \tag{21}$$

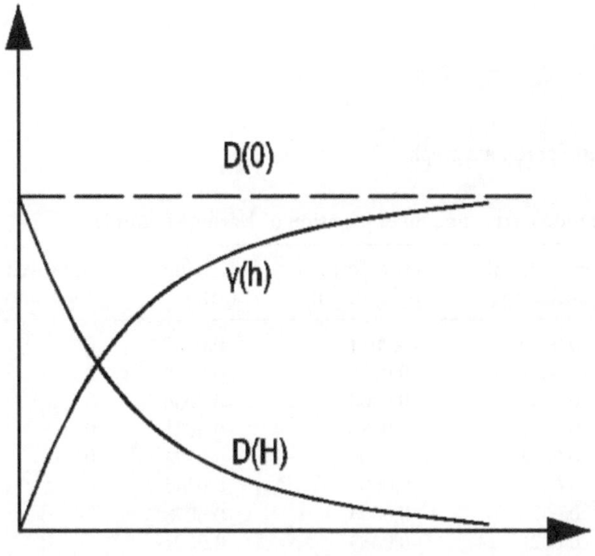

Figure 1. The relationship between variation functions and the covariance function.

According to the formula 21:

$$D(h)= D(0)- \gamma(h) \tag{22}$$

In the formula 22, select the spherical function as theoretical model, the spherical function is $\gamma(h)=C_0+C\left(\frac{3h}{2a}-\frac{1}{2}\cdot\frac{h^3}{a^3}\right)$, $D(0)=C_0-C$, C_0, C, a are unknown parameters. The experimental semivariogram is $\gamma^*(h)=1/2N(h)\sum_k(\zeta(x_i,y_i)-\zeta(x_j,y_j))^2$. Using least squares fits parameters C_0, C, a by the experimental semivariogram data. The obtained parameters substituted into formula 21 can get the covariance function expressions. Inserting the covariance function into the formula 13, 17, 18 can get parameter estimations. Inserting estimations into the formula 19 can get height anomaly of the unknown points.

4 ANALYSIS ON PROJECT

In this paper, select 24 GPS points with leveling measurement in flat area as the experimental sample, 8 GPS points as the elevation the fitting learning sample, 16 GPS points as the elevation fitting prediction sample.

Compare the accuracy between polynomial surface, the polyhedral function, kriging interpolation function and improved least squares collocation, $\sigma=\pm\sqrt{[VV]/(n-1)}$ is the formula of the accuracy, n is the number of the sample, the results as follows:

From Tables 1 and 2, polynomial surface, polyhedral function, Kriging interpolation method can achieve high precision in GPS elevation fitting in the flat area. The precision

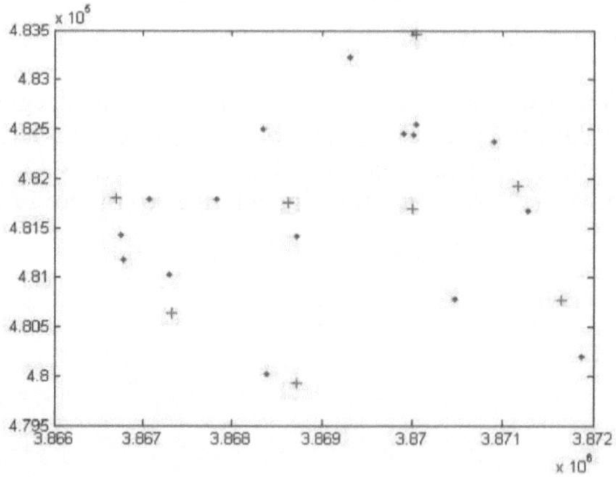

Figure 2. GPS points coordinate graph.

Table 1. The residual error and inner precision of learning example.

Learning sample	Polynomial surface (m)	Polyhedral function (m)	Kriging (m)	Improved least squares collocation (m)
1	0.0009	–0.0011	0.0185	0
2	–0.0011	0.0016	0.0049	0
3	0.0001	0.0008	–0.0006	0
4	–0.0031	0.0086	–0.0025	0
5	–0.0004	0.0009	–0.0323	0
6	–0.0005	0.0004	0.0145	0
7	0.002	–0.0044	–0.0078	0
8	0.0021	–0.0069	0.0021	0
σ	0.0017	0.0046	0.0155	0

Table 2. The residual error and outer precision of forecasting example.

Prediction sample	Polynomial surface (m)	Polyhedral function (m)	Kriging (m)	Improved least squares collocation (m)
1	0.0221	0.0525	0.023	0.0297
2	0.0182	0.04	0.0291	0.0099
3	0.0218	–0.0157	–0.0254	0.008
4	0.0027	–0.005	0.0145	0.0021
5	–0.0017	0.001	0.0182	–0.0018
6	–0.0107	–0.0027	–0.0055	–0.0126
7	0.0027	–0.0097	0.0012	0.0051
8	0.0036	0.0079	0.0298	–0.001
9	0.0044	0.0054	–0.0116	0.0102
10	0.0035	0.0157	–0.0006	0.0112
11	0.0074	0.0019	0.0067	0.0194
12	0.0243	0.0471	0.0251	0.0303
13	0	0.0053	–0.0046	–0.0013
14	0.0121	0.0152	–0.0078	0.0218
15	0	–0.0062	0.0139	0
16	–0.0045	–0.0123	0.0069	–0.0057
σ	0.0100	0.0202	0.0154	0.0120

of GPS elevation fitting by improved least square collocation can reaches centimeter level, therefore, improved least square collocation *t* is feasible in GPS height fitting.

5 CONCLUSION

This paper regards the height anomaly as no error observation value, adopts variation function to confirm the covariance between each signal, by variation function and the covariance all having a functional relationship with distance.

Adopting least squares method obtains parameters of the covariance function by the experimental semivariogram data. Inserting the known covariance function into the formula of improved least squares collocation can get height anomaly of the unknown points.

Through comparison among the polynomial surface, the polyhedral function, kriging interpolation function, proves effectiveness of the model of improved least squares collocation for GPS elevation fitting.

REFERENCES

Li Junhai. 2011. Combination of kriging and least squares collocation and its application in GPS height fitting. Science of Surveying and Mapping 36(1):99–100.

Sha Yuejin. 2000. Using the least square collocation method in GPS height fitting, Engineering of Surveying and Mapping 3:5–6.

Xu Shaoquan. 2008. GPS measurement principle and Application. Wuhan. Wuhan University press: 58–168.

Yao Daorong. 2008. The clock wave, compared with ordinary Kriging method and least squares collocation, Geodesy and Geodynamics 28(3):28.

Zou Xiancai. 2004. The study of local geoid determination, the least square collocation method. Journal of Wuhan University 29(3):219.

Zhang Shubi. 2008. Measurement adjustment, Xuzhou: China University of Mining and Technology press: 9–34.

Modeling and Computation in Engineering III – Zhang & Xie (Eds)
© 2014 Taylor & Francis Group, London, ISBN 978-1-138-02680-3

Research on dynamic full level of progressive cavity pumps

D. Jiang
College of Petroleum Engineering, China University of Petroleum, Qingdao, China
Oil Production Technology Research Institute, SINOPEC Shengli Oilfield Company, Dongying, China

M.Z. Li & G.D. Zhang
College of Petroleum Engineering, China University of Petroleum, Qingdao, China

H.C. Li
MOE Key Laboratory of Petroleum Engineering in China University of Petroleum, Beijing, China

R. Han
College of Petroleum Engineering, China University of Petroleum, Qingdao, China

ABSTRACT: In order to determine the mechanism of PCP full level, three sections were chosen, and Bernoulli equations were established between the adjacent sections. The height of the liquid in the opened cavity was confirmed, the numerical models of the opened cavity volume and the pumping liquid volume were established, and then the dynamic full level of PCP was determined. The results indicate that: the full level curve is separated into two sections due to the influence of the reciprocating motion of rotor. At the first half of the rotor motion, as the cavity open, the full level increases to a stable value. At the second half of the rotor motion, as the cavity close, the pumping liquid volume is less than the opened cavity volume, the full level decrease. Compared to viscosity and water content, the influence of dynamic liquid level, revolving speed and production gas-oil ratio upon full level is great.

1 INTRODUCTION

Progressive cavity pumps are widely used in oil production both at home and abroad for its well adaptability and convenient management (Wu B.C. and Li X., 2010). Compared to the conventional sucker rod pumps, the PCP has higher pump efficiency which mainly depends on the high full level of the pump. It is known that the PCP cavity can be filled with liquid fully, however, according to the analysis of mass field production data, it is find that the pump is not always full, especial in the heavy oil reservoirs. Due to this poor cognition, the PCP and the production parameters chosen by petroleum engineers may not coordinate with each other well. Therefore, it is important to determine the mechanism of PCP, especial the full level, the ratio of the pumping liquid volume and the opened cavity volume, which represents the imbibition ability of the pump. A great number of scholars have studied the full extent of conventional sucker rod pumps, and researched the variation law of full level under different conditions (Xia H.F. and Li F.J., 1995; Wang W.Y., et al, 2012; Zheng J.D., et al, 1999; Di M.Y., et al, 2011). However the theoretical study (Zhou D.S. and Yuan H., 2008; Jorge Robles, et al, 2011) of PCP is incomplete due to its complex meshing relationship between stator and rotor, and the geometric nonlinearity as well as its material nonlinearity. In order to explore the work mechanisms of PCP further, a dynamic full level model is established, and different influence factor is analyzed. The paper can provide guides for petroleum engineers.

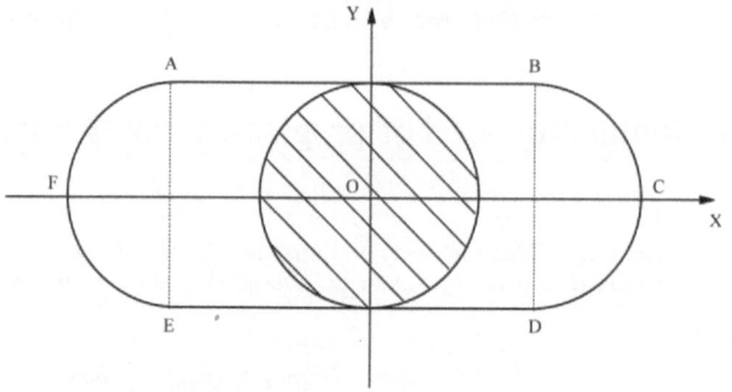

Figure 1. Schematic diagram of opened cavity.

2 PUMPING LIQUID VOLUME

The rotor reciprocates in the cavity of the PCP when it is rotating. The pump inlet opens and closes regularly, so the fluid can be raised to the ground. As shown in Figure 1.

The reciprocating velocity of the rotor in the stator lining is described as follows:

$$v = 2e\omega \sin\left(\omega t + \frac{\alpha}{2}\right) \tag{1}$$

The distance for the movement of the rotor is defined as:

$$l = 2e\left[1 - \cos\left(\omega t + \frac{\alpha}{2}\right)\right] \tag{2}$$

To be simple, an angle $\theta\,(0 \leq \theta \leq 2\pi)$ was used, and the opened cavity volume as follows:

$$V = \frac{TeD}{\pi}(\theta - \sin\theta) \tag{3}$$

Assume the liquid level is L, when the rotor turns to the angle θ, the volume of the opened cavity can be calculated by the equation (4).

$$V' = \frac{TeD}{\pi}\left[\theta - \frac{2\pi L}{T} - \sin\left(\theta - \frac{2\pi L}{T}\right)\right] \tag{4}$$

Then, pumping liquid volume can be described as equation (5):

$$V_L = V - V' = 2eDL + \frac{TeD}{\pi}\left[\sin\left(\theta - \frac{2\pi L}{T}\right) - \sin\theta\right] \tag{5}$$

3 DYNAMIC FULL LEVEL MODEL

Because the cavity structure is complex and the element that affects the full level is diverse, some assumption were made: (1) The fluid in the pump is in ideal conditions. (2) There is no loss between the two neighbouring cavities. (3) The temperature in the pump is constant. (4) The fluid in the pump conforms to the separated fluid model.

Three sections were chosen, a section (1–1) of the tube under the pump, the section (2–2) of pump inlet and the interface (3–3) between oil and gas in the cavity, see Figure 2.

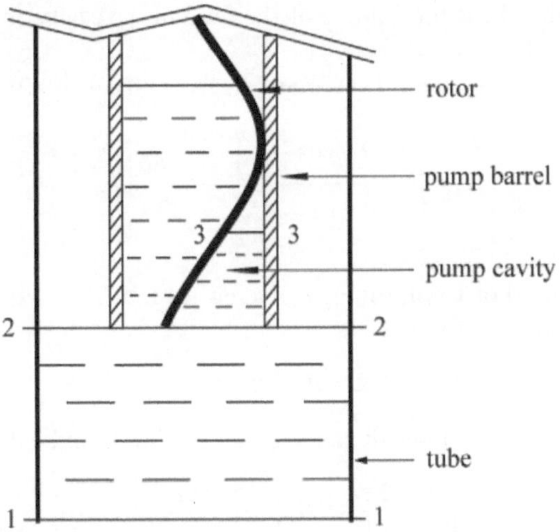

Figure 2. Schematic diagram of inlet of progressive cavity pumps.

The energy conservation equation between the section (1–1) and (2–2) described as equation (6).

$$z_1 + \frac{P_1}{\rho g} + \frac{U_1^2}{2g} = z_2 + \frac{P_2}{\rho g} + \frac{U_2^2}{2g} + \Delta H_1 + \Delta H_2 \tag{6}$$

$$\Delta H_1 = \zeta \frac{U_2^2}{2g} \qquad \Delta H_2 = \lambda \frac{l}{d}\frac{U_1^2}{2g} \tag{7}$$

where z_i is the height of the section *i-i*, m. P_i is the pressure of the section *i-i*, Pa. U_i is the velocity of fluid in the section *i-i*, m/s. ΔH_1 is the throttle loss of section *i-i*, J. ζ is flow coefficient. A_i is the flow area of section *i-i*, m^2.

The energy conservation equation between the section (2–2) and (3–3) described as equation (8).

$$z_2 + \frac{P_2}{\rho g} + \frac{U_2^2}{2g} = z_3 + \frac{P_3}{\rho g} + \frac{U_3^2}{2g} + \Delta H_2' \tag{8}$$

$$\Delta H'_2 = \int_0^{X-L} \frac{\mu U_3\left[1-\cos\frac{2\pi(x-L)}{T}\right]\left[\pi D + 4e\left(1-\cos\frac{2\pi x}{T}\right)\right]^2}{2\rho g e^2 D^2\left(1-\cos\frac{2\pi x}{T}\right)^3}dx \tag{9}$$

where $\Delta H_2'$ is the frictional loss in the cavity, J. l is the length of the runner, m. d is the equivalent diameter, m. U_L is the velocity of the fluid at one section, m/s. λ is the flow coefficient.

The pressure in the cavity can be determined by gas state equation.

$$P_3 = Z\frac{TP_0V_L(R_P - R_S)(1-f_w)}{T_0B_0(V-V_L)} \tag{10}$$

where Z is the gas compressive coefficient, T is the temperature inside the pump, °C. P_0 is the pressure in standard conditions, MPa. T_0 is the temperature under standard condition,°C; R_S is solution gas-oil ratio, m^3/m^3; R_P is produced gas oil ratio, m^3/m^3; f_w is water-cut; B_0 is gas volume factor, m³/m³. V_L is the volume of the fluid in the cavity, m³; V-V_L is the volume of the gas in the cavity, m³.

The height of the fluid in the pump cavity can be described as follows:

$$L = \int_0^x U_3 x \frac{1-\cos\frac{2\pi(x-L)}{T}}{1-\cos\frac{2\pi x}{T}} \frac{60}{NT} dx \tag{11}$$

where N is the rotate speed of rotor, r/min. T is screw lead, m.

The starting condition is as follows:

$$L = 0 \qquad x = 0 \tag{12}$$

From the equation (4) and (5), the dynamic liquid full level can be described as follows:

$$\beta = \frac{V_L}{V} = \frac{\frac{2\pi L}{T} + \sin\left(\theta - \frac{2\pi L}{T}\right) - \sin\theta}{\theta - \sin\theta} \tag{13}$$

4 CALCULATED RESULTS

The dynamic full level of PCP under different viscosity is shown in Figure 3. It is obvious that the curves can be separated into two sections. As the cavity open ($0° \le \theta < 180°$), the throttle loss of the pump inlet decreases, the full level increases, and a balance was obtained when $\theta = 180°$. As the cavity close ($180° \le \theta < 360°$), the throttle loss of the pump inlet increases, the frictional in the cavity increases, which causes the full level of PCP decreases. The increase of viscosity reduces the flow ability of the fluid, the resistance resisting the fluid flow into the pump, and the full level decreases, however this variation is not obvious, which indicate the well adaptability of PCP in lifting heavy oil.

The dynamic liquid level represents the energy of the fluid at the pump inlet. The effect of dynamic liquid level on full level is shown in Figure 4. The effect of dynamic liquid level on full level is great. The full level decreases with the lower of dynamic liquid level, and when the dynamic liquid level is relatively low, the full level come to a balance in a short time, the phenomenon reflects the great effect of the energy of the fluid at the pump inlet upon its flow ability into the pump.

The effect of revolving speed upon full level is great. As can be seen in Figure 5, in a period of the opening and closing of the pump cavity, the higher of revolving speed, the shorter of

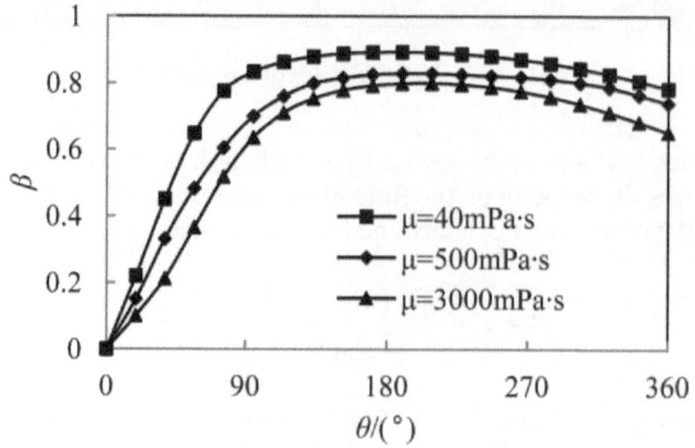

Figure 3. Dynamic full level of different viscosity.

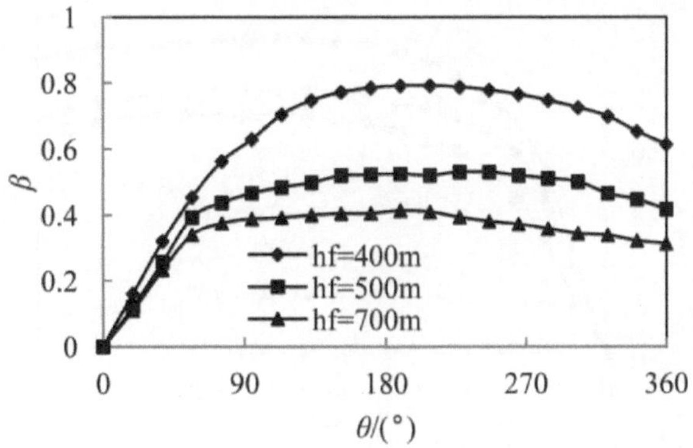

Figure 4. Dynamic full level of different pump depth.

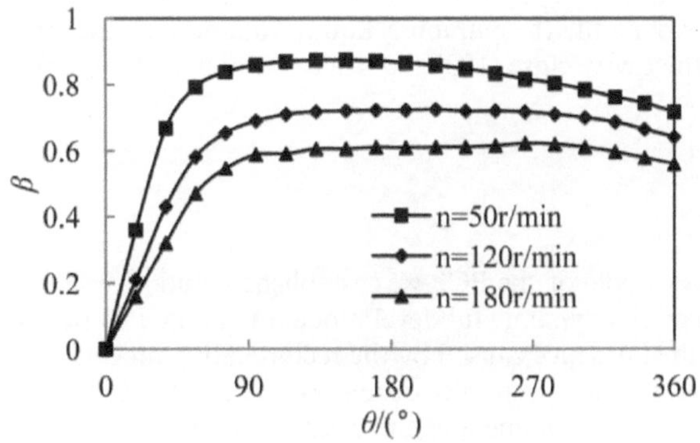

Figure 5. Dynamic full level of different speed.

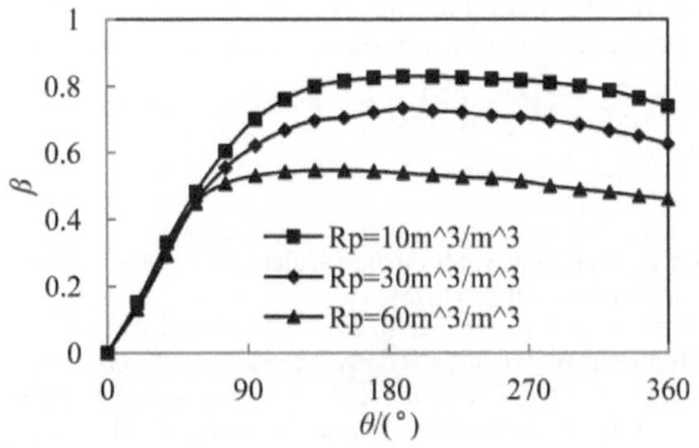

Figure 6. Dynamic full level of different producing gas-oil ratio.

the pumping time, which causes the full level decreases, and the full level comes to balance quickly.

With the production gas-oil ratio rising, the free gas at the pump inlet increase, so the full level of the progressive cavity pumps decrease, as shown in Figure 6. As can be seen in the figure, when the production gas-oil ratio is high, the full level comes to the balance quickly, which also reflects the great influence of the production gas-oil ratio upon the full level.

The effect of water content upon full level is similar (Fig. 7) to that of viscosity, which causes the viscosity decrease with the increase of water content. Due to the perfect flow

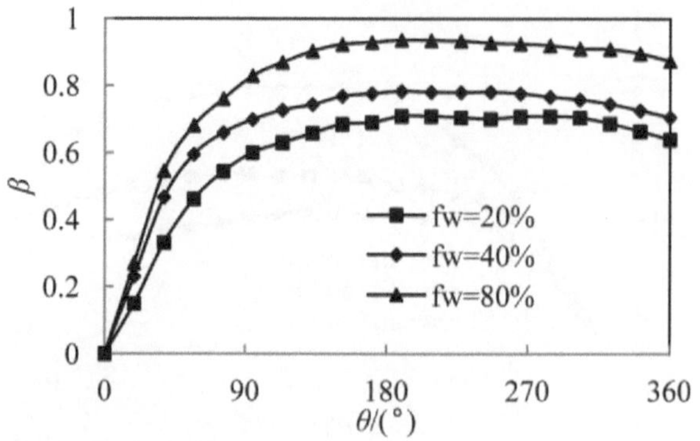

Figure 7. Dynamic full level of different water content.

ability at high water content, the pumping liquid volume and the opened cavity volume come to balance as the cavity close (180° ≤ θ ≤ 360°), which cause the full level curve almost horizontal.

5 CONCLUSION

The dynamic full level model of the PCP was established, and several influence factors were analyzed. The curve of the dynamic full level model of the PCP is obviously separated into two sections due to the influence caused by the reciprocating motion of rotor. As the cavity open, the throttle loss of the pump inlet decreases, the full level increases, and the balance between the pumping liquid volume and the opened cavity volume is obtained shortly. As the cavity close, the throttle loss of the pump inlet and the friction drag in the pump cavity increase, which cause the pumping liquid volume is smaller than the opened cavity volume, thus the full level decrease. The main factors upon the PCP full level are the height of dynamic liquid level, revolving speed and the produced gas oil ratio. Relatively the influence of the viscosity and the water content is not obvious, which indicate the well adaptability of PCP in lifting heavy oil.

REFERENCES

Di M.Y., Yang H.B., Li H.Z., et al. 2011. Study on the calculation of fulling degree of sucker rod pump [J]. *Complex Hydrocarbon Reservoirs*, 4(4): 71–75.

Jorge Robles, Matias Perez, John Bettenson, et al. 2011. Design and Application of Charge PCP Systems in High GVF Heavy Oil Wells [C]//*SPE Progressing Cavity Pumps Conference.*

Wang W.Y., Wan G.Q., Lv X.B. 2012. The dynamic simulation-based new method to calculate the full level of sucker rod pump [J]. *China Petroleum Machinery*, 40(3): 67–71.

Wu B.C., Li X. 2010. The Special Successful PCP Applications in Heavy Oilfield [C]//*SPE Progressing Cavity Pumps Conference.*

Xia H.F., Li F.J. 1995. The calculation for the dynamic full level of the sucker-rod pump [J]. *ACTA*, 16(4): 167–173.

Zheng J.D., Yan X.Z., Cao G.S., et al. 1999. The calculation and analysis of discharge coefficient for suck rod pump nuder polymer flooding [J]. *ACTA*, 20(1): 88–92.

Zhou D.S., Yuan H. 2008. Design of Progressive Cavity Pump Wells [C]//*SPE Progressing Cavity Pumps Conference.*

Modeling and Computation in Engineering III – Zhang & Xie (Eds)
© 2014 Taylor & Francis Group, London, ISBN 978-1-138-02680-3

Method of 3D complex geological modeling

H.H. Wei, X.H. Li & Y. Bao
Beijing University of Technology, China

ABSTRACT: It is very difficult to build a 3D complex geological model using conventional modeling software. Based on geometric theories, the paper puts forward a new method to compute point cloud of a complex 3D geological model for a hillside landscape design. First according to the model pattern, the model is decomposed into several parts, and the point set of each part is calculated with the model paramaters. Then all the point sets are gathered to form the point cloud of the designed model, and a surface is generated with the point cloud using 3D modeling software. Finally the surface is used to incised the hillbody to obtain 3D landscape effect. By changing the pattern of the design and calculating of excavated and filled volume, the landscape engineering can be optimized.

1 INTRODUCTION

Modern design philosophy of project gives full play to the designer's imagination and creativity to meet different aspects and levels of people's demand for appreciation, this needs to innovate engineering design method. The goal of landscape engineering design is not only to meet the functional requirements, security, but also to pursuit beautiful and chic scene, and harmonious collocation with surrounding environment. Therefore complex 3D modeling is involved.

At present, 3D modeling technology has been widely used in water conservancy and hydropower, mining, geological survey, geophysical prospecting, mathematical geology, etc (Zhou, 2005; Zhong & Li & Yang, 2005; Wang & Bai, 2006; Hao & Ma & Zhou, 2008; Chen & Wang, 2009; Wang, 2011). Some applications have also been found in landscape engineering design (Song, 2003), but the applications are mainly involved in large range of overall planning, rather than in specific local design. Many landscape engineerings are very complicated in pattern. But complex 3D model cannot be achieved with simple application of conventional geological modeling software. So some specific solution must be proposed for specific requirements to achieve the final design.

Based on the specific circumstances of an actual landscape engineering, the paper puts forward a practical method to construct a complex 3D model. Firstly mathematic model is set up for calculating the coordinates of all the points on the surface according to the morphology and the designed tilt angle of the surface, then the discrete points are developed into 3D curved surface by conventional three-dimensional modeling software. Finally the geological body is incised with the 3D surface to obtain the designed 3D landscape effect map. Furthermore the earthwork of the engineering can be optimized with the 3D effect map.

2 THE ENGINEERING SITUATION

The engineering is conducted in a upward valley. And a stepping walkway shown in Figure 1 is designed along the valley. The walkway has many steps and each step has its dip angle different from others. And at both side of the walkway some byroads are planned. It shows that the walkway is very complex in structure. It is key that whether the pattern as shown in

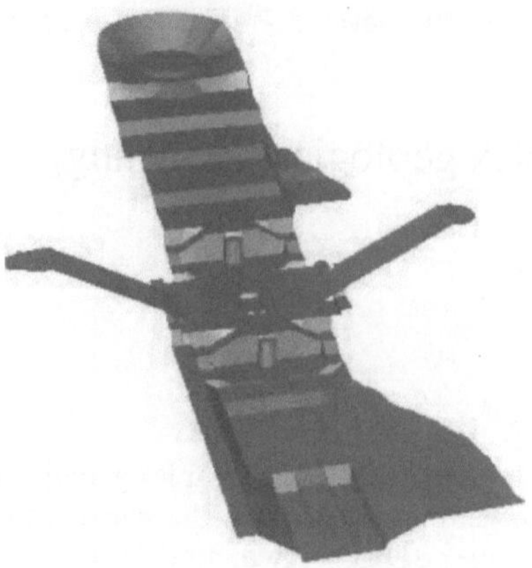

Figure 1. Designed walkway.

Figure 1 is harmonious with arrounding terrain? How can it be the best in shape and for the earthwork? For these questions digital 3D modeling should be used. There are many commercial 3D modeling softwares available, but it is very difficult to construct a complex model by direct using of conventional modeling softwares.

In allusion to the problem, first the walkway is divided into several parts according to its structure, and the point coordinates of each part is calculated separately. Then all points are gathered to form the whole 3D surface by using traditional 3D modeling software.

3 COORDINATE COMPUTATION

Assume that the tilt angle of one step of the walkway is δ shown as in Figure 2, and the step are divided into m points along horizontal direction. The slope of bottom line at point P on the horizontal plane is k

$$k = atn\frac{\Delta Y}{\Delta X} \tag{1}$$

Then the slope of normal line at point P is

$$s = k \pm \pi/2 \tag{2}$$

P′ is the upper point corresponding to point P, then the horizontal distance from P to P′ is

$$\Delta D = \Delta h * ctn(\delta) \tag{3}$$

Δh is the level interval between two adjacent points along height. The increment of horizontal coordinates from P to P′ is

$$\begin{aligned} \Delta x &= \Delta D * \cos(s) \\ \Delta y &= \Delta D * \sin(s) \end{aligned} \tag{4}$$

Based on δ and Δh, the horizontal displacement ΔD of line PP′ can be calculated by equation (3), and ΔD and s are put into equation (4) to get Δx and Δy. And finally coordinates $(x_{p'}, y_{p'})$ can be calculated as follow

$$x_{p'} = x_p + \Delta x \qquad y_{p'} = y_p + \Delta y \qquad h_{p'} = h_p + \Delta h$$

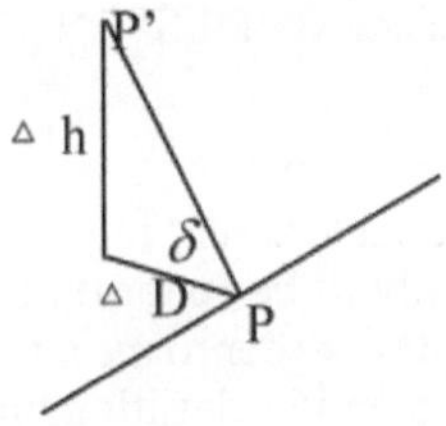

Figure 2. Relation between adjacent points.

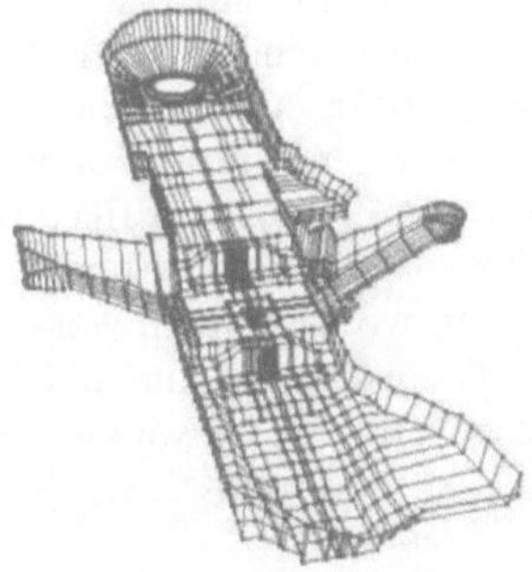

Figure 3. Designed 3D surface gridnet.

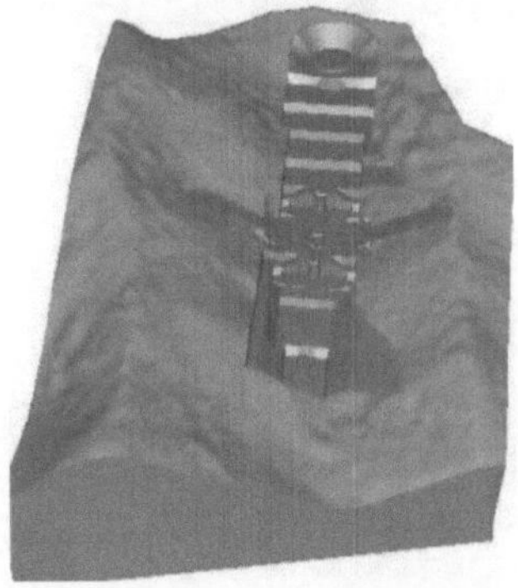

Figure 4. 3D landscape effect.

According to the principle of the processing method above, all the data of the lines at each elevation are processed. then all the point sets are compounded to form the space point set of the designed surface (shown in Fig. 3).

4 GENERATION OF WALKWAY EXCAVATION EFFECT

After the dataset of the walkway excavation surface is established, GEMCOM, a conventional 3D geological modeling software system is used to create a 3D surface model (shown in Fig. 1) with the dataset.

The walkway surface and the valley which was generated in advace are displayed together in window. And then the valley is cut with the surface under the use of GEMCOM again. After the part above the surface is removed, the whole 3D effect of the designed walkway is obtained shown in Figure 4, which is very useful for optimizing hillside landscape design.

5 OPTIMIZATION OF ENGINEERING DESIGN

5.1 *Optimization of the position*

Whether an engineering design is resonable is very important for a landscape design. A landscape design needs the pattern not only to be beautiful in shape but also to be harmonious with surrounding natural environment. In order to meet these needs, the pattern of the design should be changed both in orientation and in elevation until it is satisfied.

5.2 *Optimization of earthwork*

Earthwork is very important for construction engineering. When the excavated volume is equal to filled volume, earthwork amount is tend to be the minimum. Based on the 3D effect map of the walkway, earth volume can be calculated. The body above the designed walkway surface (shown in Fig. 5) is excavated volume and the body below the walkway surface but above the topographic surface (shown in Fig. 6) is the filled volume. So the the earth volume can been obtained by calculating the volume of those bodies.

In order to reduce the construction work of earth work, it is needed that the excavated volume is equal to filled volume. The excavated and filled volumes can be calculated by constant upraising or lowering the elevation of the walkway surface. when the both volumes

Figure 5. Excavated volume.

Figure 6. Filled volume.

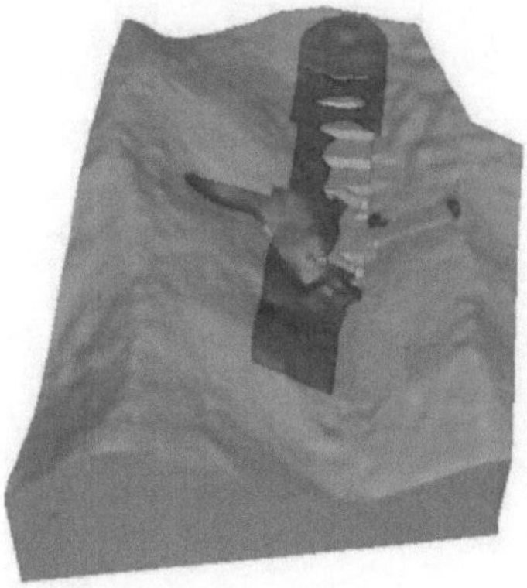

Figure 7. Excavated and filled 3D effect.

are approximately equal, the earth volume of the project is the least. The compound of excavated and filled volumes is shown in Figure 7.

6 CONCLUSION

Digital 3D geological engineering design method takes full use of the advantages of 3D geological modeling and plays an important role in complex landscape engineering design. Based on strict mathematical method, it not only carries out quickly and accurately design idea during the process of the design, but also reproduces the whole scene of design effect. It is very important for optimizing landscape engineering design.

In addition, digital 3D geological engineering design method is extremely convenient for earthwork calculation. It can quickly calculate the excavated and filled volumes by making full use of its volume calculation function, which is superior to any other conventional method for optimizing construction work and reducing the construction cost as far as possible.

REFERENCES

Bingling Chen & Xiaopeng Wang, 2009. Real Three-dimensional geological body modelingtechnology and its application to coal geological prospecting. Coal geology of china, 21(Z2).

Bo Wang, 2011. Research on dynamic reserve calculation of ore deposit based on 3D geological model. Master's thesis (Nanjing normal university), 2011.

Denghua Zhong & Mingchao Li & Jianmin Yang, 2005. Three-dimension visual construction forcomplex engineering rock mass structure and its application. Chinese Journal of Rock Mechanics and Engineering, 24(4):575–580.

Minghua Wang & Yun Bai, 2006. Current research situation and glowing trend of three-dimension geological modeling. Soil Engineering and Foundation, 20(4):68–70.

Xiaodong Song, 2003. Authenticity and Generalizability of Computer Landscape Elumation technology. City Planning, (8):25–27 ext.32.

Xiao Zhou, 2005. GIS Application in Planning and Development of Landscape. Science & Technology Information, (27):19–21.

Xiuqiang Hao & Bin Ma & Yaodong Zhou, 2008. Three-dimension visual modeling technology and its application to underground mining. China Mining Magazine, 17(5):68–70.

Modeling and Computation in Engineering III – Zhang & Xie (Eds)
© 2014 Taylor & Francis Group, London, ISBN 978-1-138-02680-3

The prediction of wall factor of particle settling in rectangular ducts using Support Vector Machine

M.Z. Li, J.B. Li, G.D. Zhang & C.T. Liu
College of Petroleum Engineering, China University of Petroleum, Qingdao, China

ABSTRACT: In order to characterize wall effect of rectangular ducts on settling particle, a prediction method of wall factor was presented using Support Vector Machine. Extracted ρ, ρ_s, d_p, d/a, a/b, K and n as feature parameters, which affect particle settling velocity greatly. Culled 70 data of Machač to train and test the model, the results show that: Support Vector Machine sufficiently accurate in engineering applications, the error is less 5% for 90% calculated results and can be used in parallel plate and rectangular ducts case, the application range of Support Vector Machine are broader, which can meet complicated engineering applications.

1 INTRODUCTION

Particles will settle in liquid and this process is very common in the industry, especially the petroleum producing. It is necessary to get the settlement laws of particles. We were using the traditional stork's law and it can not meet the complicated industrial conditions. One additional drag force is found in the boundary-limited fluid medium, too.

There is few study of particles' setting in rectangular ducts. In recent years, the unconventional reservoirs were developed and hydraulic fracturing technology has been widely used. In this process, the fracture surface will prevent particles' settling and transport, thus affecting the final fracturing result. Therefore, it is important to accurately determine the settling velocity of propane particles.

Wall factor is defined as the ratio of particle setting velocity in infinite and finite boundary fluid. It is widely used to describe the effect of walls to particles setting velocity. Chhabra and Uhlherr, Lali, et al presented wall effect on settling velocity of spheres in non-Newtonian solutions at high Reynolds number ($1.0 \le Re \le 10^3$ and low Reynolds number (2×10^{-3} $Re \le 150$) separately, and corresponding correlations were established (Chhabra R.P., Uhlherr P.H.T.,1980). Miyamura experimental studied the setting possess of spherical particles in the triangular pipe, square pipe and the parallel plate. geting the parallel plate model which is generally recognized nowdays (Miyamura A., Iwasaki S., Ishii T., 1981). Ivan Machač and Zdeněk Lecjaks, Malhotra and Sharma established wall factor correlations for spheres settling through power law fluid and surfactant-based shear thinning viscoelastic fluids in rectangular ducts and parallel plates (Machač I., Lecjaks Z.,1995). Kawase and Ulbrecht, Missirlis, et al, Daoyun Song et al theoretically studied and numerically simulated the settling velocity of a sphere in bound non-Newtonian fluid (Kawase, Y., Ulbrecht, J.J., 1983). It is now widely accepted that elasticity and shear tinning behavior of non-Newtonian fluid reduce the retardation effect of the confining walls.

Due to the limitations of the experiment condition, the result of above study are only applicable a particular channel shape, fluid type and flow area. There is no single model can characterize how the wall face affect the setting velocity in any fluid and any area. This article based on SVM and a new wall factor perdition model is established.

2 DATA BASE

2.1 *Characteristic parameters*

Characteristic parameters are the input of SVM and they will affect the accuracy of the SVM model. There are 3 rules to get the characteristic parameters: (1) easy to get; (2) have great impact on output; (3) little correlation between the parameters. In the Stokes law, for Newtonian fluid, settling velocity of particle depends on the fluid viscosity, fluid density, particle density and particle diameter while related with the flow index and consistency index for the None-Newtonian fluid.

The settling velocity of particles will slow down in a finite fluid medium because of a additional resistance from the container walls. This resistance is related with the container shape and size. In a rectangular channel, the ratio of diameter and width (d/a) and the width to length ratio (a/b) are selected as the characteristic parameters to characterize the resistance of walls. Above all, ρ, ρ_s, d_p, d/a, a/b, K and n are choose to be the input variable for the case of rectangle channel.

2.2 *Basic dates*

There are few experimental studies about the settling process in the rectangle or parallel plate channel. The SVM model need 7 variable while some study do not provide them.

This article based on the experimental dates of Ivan Machač. A SVM model of wall factor. The basic dates comes from some figures. Point-recognition software is used to get them, thus some deviation are unavailable. There 70 dates can be used, as Figure 1.

The date is divided in to 46 training sets and 24 test sets. Network model is established based on training sets, then tested by test sets.

2.3 *Data pretreatment*

The rang of characteristic parameters are widely, Normalization process is required to reduce the negative impact of features' widely distribution. The parameters are normalized to [−1, 1] using the following equation.

$$y = 2\frac{y - y_{\min}}{y_{\max} - y_{\min}} - 1 \tag{1}$$

where y is normalized parameters. $y_{\min}$, is the maximum date; $y_{\max}$ is the minimum date.

3 MODELING

Vapnik put forward the SVM model based on VC-dimensional statistical theory and Structural risk minimization principle (V. Vapnik,1983). This model has unique advantages, especially for small-sample size problems, which has been widely used.

Consider a set of training points, $\{(x_1, z_1)\ldots(x_N, z_N)\}$, where $x_i \in R^n$ is a feature vector and $z_i \in R^n$ is the target output. Under given parameters $C > 0$ and $\varepsilon > 0$, the standard form of support vector regression is:

$$\min_{w,b,\xi,\xi^*} \quad \frac{1}{2}w^T w + C\sum_{i=1}^{l}\xi_i + C\sum_{i=1}^{l}\xi_i^* \tag{2}$$

Subject to $w^T\phi(x_i) + b - z_i \le \varepsilon + \xi_i$,
$z_i - w^T\phi(x_i) - b_i \le \varepsilon + \xi_i^*$,
$\xi_i, \xi_i^* \ge 0,\ i = 1, \ldots, l.$

Table 1. Measured wall factors.

ρ (kg/m^3)	ρs (kg/m^3)	dp (mm)	d/a (–)	a/b (–)	K (Pa·s^n)	n (–)	Fw (–)
Traning sets (46)							
1003	2506	1.465	0.122	1	1.12	0.741	0.745
1003	2506	1.465	0.122	0.421	1.12	0.741	0.808
1003	2506	1.465	0.122	0.309	1.12	0.741	0.835
1003	2506	1.465	0.122	0.231	1.12	0.741	0.849
1003	2506	1.465	0.122	0.192	1.12	0.741	0.868
1003	2506	1.465	0.122	0.15	1.12	0.741	0.880
1003	2836	2.67	0.223	1	1.12	0.741	0.569
1003	2836	2.67	0.223	0.421	1.12	0.741	0.735
1003	2836	2.67	0.223	0.309	1.12	0.741	0.764
1003	2836	2.67	0.223	0.231	1.12	0.741	0.789
1003	2836	2.67	0.223	0.192	1.12	0.741	0.799
1003	2836	2.67	0.223	0.15	1.12	0.741	0.809
1003	2857	3.22	0.268	1	1.12	0.741	0.473
1003	2857	3.22	0.268	0.421	1.12	0.741	0.702
1003	2857	3.22	0.268	0.309	1.12	0.741	0.731
1003	2857	3.22	0.268	0.231	1.12	0.741	0.745
1003	2857	3.22	0.268	0.192	1.12	0.741	0.760
1003	2857	3.22	0.268	0.15	1.12	0.741	0.778
1003	7516	1.001	0.083	1	1.12	0.741	0.788
1003	7516	1.001	0.83	0.421	1.12	0.741	0.872
1003	7516	1.001	0.083	0.309	1.12	0.741	0.911
1003	7516	1.001	0.083	0.231	1.12	0.741	0.925
1003	7516	1.001	0.083	0.192	1.12	0.741	0.934
1003	7516	1.001	0.083	0.15	1.12	0.741	0.946
1248	2506	1.465	0.122	1	0.548	1	0.776
1248	2506	1.999	0.167	0.309	0.548	1	0.809
1248	2506	1.999	0.167	0.231	0.548	1	0.829
1248	2506	1.999	0.167	0.192	0.548	1	0.827
1248	2506	1.999	0.167	0.15	0.548	1	0.842
1248	2506	1.465	0.122	0.421	0.548	1	0.844
1248	2506	1.465	0.122	0.231	0.548	1	0.858
1248	2506	1.465	0.122	0.192	0.548	1	0.854
1248	2506	1.465	0.122	0.15	0.548	1	0.865
1248	2596	4.117	0.343	0.309	0.548	1	0.611
1248	2596	4.117	0.343	0.231	0.548	1	0.619
1248	2596	4.117	0.343	0.192	0.548	1	0.625
1248	2596	4.117	0.343	0.15	0.548	1	0.630
1248	2836	2.67	0.223	1	0.548	1	0.580
1248	2836	2.67	0.223	0.421	0.548	1	0.697
1248	2857	3.22	0.268	1	0.548	1	0.502
1248	2857	3.22	0.268	0.421	0.548	1	0.640
1248	2857	3.22	0.268	0.231	0.548	1	0.670
1248	2857	3.22	0.268	0.192	0.548	1	0.677
1248	2857	3.22	0.268	0.15	0.548	1	0.690
1248	7516	1.001	0.083	1	0.548	1	0.804
Test sets (24)							
1003	2506	1.999	0.167	1	1.12	0.741	0.677
1003	2506	1.999	0.167	0.421	1.12	0.741	0.765
1003	2506	1.999	0.167	0.309	1.12	0.741	0.791
1003	2506	1.999	0.167	0.231	1.12	0.741	0.802
1003	2506	1.999	0.167	0.192	1.12	0.741	0.806
1003	2596	4.117	0.343	1	1.12	0.741	0.324
1003	2596	4.117	0.343	0.421	1.12	0.741	0.634

(Continued)

Table 1. (Continued).

ρ (kg/m^3)	ρs (kg/m^3)	dp (mm)	d/a (–)	a/b (–)	K (Pa·s^n)	n (–)	Fw (–)
1003	2596	4.117	0.343	0.309	1.12	0.741	0.667
1003	2596	4.117	0.343	0.231	1.12	0.741	0.683
1003	2596	4.117	0.343	0.192	1.12	0.741	0.705
1003	2596	4.117	0.343	0.15	1.12	0.741	0.720
1003	2506	1.999	0.167	0.15	1.12	0.741	0.820
1248	2506	1.999	0.167	1	0.548	1	0.709
1248	2506	1.999	0.167	0.421	0.548	1	0.793
1248	2506	1.465	0.122	0.309	0.548	1	0.849
1248	2596	4.117	0.343	1	0.548	1	0.401
1248	2596	4.117	0.343	0.421	0.548	1	0.569
1248	2836	2.67	0.223	0.231	0.548	1	0.742
1248	2836	2.67	0.223	0.192	0.548	1	0.749
1248	2857	3.22	0.268	0.309	0.548	1	0.664
1248	7516	1.001	0.083	0.231	0.548	1	0.870
1248	7516	1.001	0.083	0.192	0.548	1	0.873
1248	7516	1.001	0.083	0.15	0.548	1	0.877

The dual problem is:

$$\min_{\alpha,\alpha^*} \quad \frac{1}{2}(\vec{\alpha}-\overrightarrow{\alpha^*})Q(\vec{\alpha}-\overrightarrow{\alpha^*})+\varepsilon\sum_{i=1}^{l}(\alpha_i+\alpha_i^*)+z_i\sum_{i=1}^{l}(\alpha_i+\alpha_i^*) \tag{3}$$

Subject to $e^T(\vec{\alpha}-\overrightarrow{\alpha^*})=0$
$0\le\alpha_i,\alpha_i^*\le C, i=1,\dots,l,$

where $Q_{ij}=K(\overrightarrow{x_i},\overrightarrow{x_j})\equiv\phi(\overrightarrow{x_i})^T\phi(\overrightarrow{x_j})$.

Problem (3) can be solving, the approximate function can be got:

$$\sum_{i=1}^{l}(-\alpha+\alpha^*)K(\overrightarrow{x_i},\vec{x})+b \tag{4}$$

where $K(\overrightarrow{x_i},\overrightarrow{x_j})$ is the kernel function; α_i,α_i^* are Lagrange multipliers is the penalty factor.

In this SVM model, the output is $\overrightarrow{\alpha^*}-\vec{\alpha}$.

Dr. Lin Zhiren of Taiwan University has developed the SVM toolbox in different programming languages (Chang C.C., Lin C.J., 2011). This article solves problems using the LIBSVM, a tool from Lin's SVM toolbox.

4 RESULT AND ANALYSIS

Wall factors predicted by the SVM model can be show as Figures 1 and 2. The training model behaves well. Test set is also used based on this model.

As we can see, the predicted value is very close to the actual value, whether the test set or the training test. The actual-prediction points locate near the straight line, 45 degrees from x-axis.

Through the above, SVM model is proved to be a Sufficient accuracy model for predicting the wall factor, especially for the small sample case. This model has an important guiding significance for industrial applications.

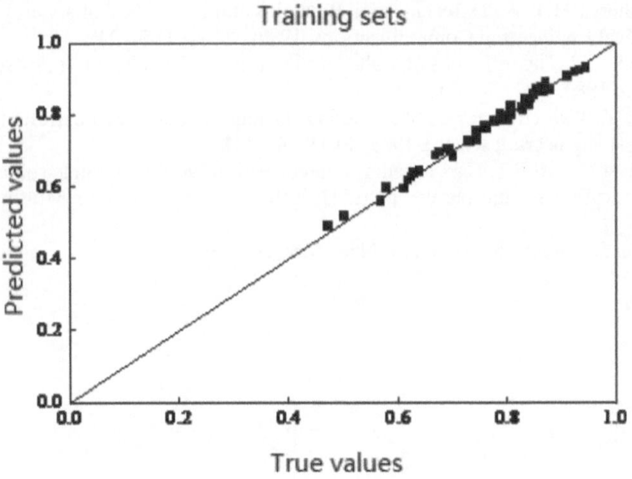

Figure 1. Wall factor predicted by training sets.

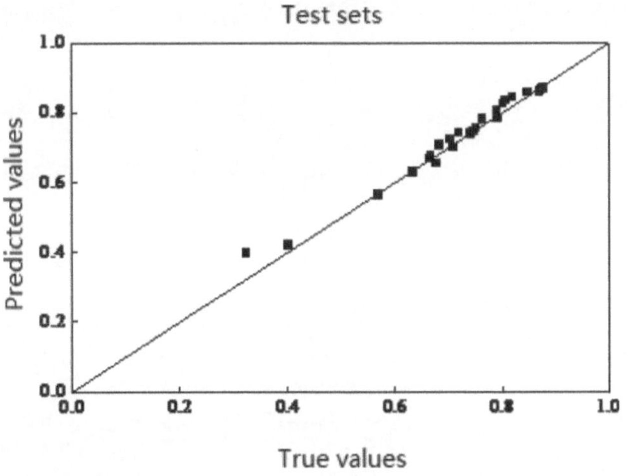

Figure 2. Wall factor predicted by test sets.

5 CONCLUSIONS

In this article, a new method is proposed to describe the influence of walls on particles setting velocity in any fluid flow of Rectangular channel, based on the SVM model.

According to the training and testing by 70 experimental dates, the following conclusion can be obtained.

1. SVM model has high accuracy. Its training set error is less than 5%. The error of the testing set' prediction is less than 5% for 90% calculated results.
2. The SVM model has a broader scope and can meet the needs of more complex projects.

REFERENCES

Chang C.C., Lin C.J. LIBSVM. a library for support vector machines [J]. ACM Transactions on Intelligent Systems and Technology (TIST), 2011, 2(3): 27.

Chhabra R.P., Uhlherr P.H.T. Wall effect for high Reynolds number motion of spheres in shear thinning fluids [J]. Chemical Engineering Communications, 1980, 5(1–4): 115–124.
Kawase, Y., Ulbrecht, J.J. The influence of walls on the motion of a sphere in non-Newtonian liquids. Rheologica Acta, 1983, 22(1): 27–33.
Machač I., Lecjaks Z. Wall effect for a sphere falling through a non-Newtonian fluid in a rectangular duct [J]. Chemical engineering science, 1995, 50(1): 143–148.
Miyamura A., Iwasaki S., Ishii T. Experimental wall correction factors of single solid spheres in triangular and square cylinders, and parallel plates [J]. International Journal of Multiphase Flow, 1981, 7(1): 41–46.
Vapnik V. Statistical Learning Theory. Wiley, New York, NY, 1998.

Modeling and Computation in Engineering III – Zhang & Xie (Eds)
© 2014 Taylor & Francis Group, London, ISBN 978-1-138-02680-3

Modern distance education technology and social protection for farmers: A different perspective

Jing Gong & Chun Yang
Institute of Agricultural Scientific and Technological Information, Beijing Academy of Agriculture and Forestry Sciences, Beijing, China

ABSTRACT: As a new public policy framework, social protection provides systematic and comprehensive measurement of the level of production and life security of the vulnerable groups including farmers. From the perspective of social protection, education and training is an effective way to help farmers to deal with shocks of risks. However, due to the impact of different factors, the current education and training programs for farmers are faced with many problems. Thanks to advanced information technology, modern distance education can overcome the obstacles caused by traditional factors such as time, space and costs, and significantly improve educational efficiency. Therefore, farmer's social protection can be strengthened through the application of modern distance education technology.

1 INTRODUCTION

Being an emerging public policy framework originated from 1990's, social protection represents a multichannel and multilayer intervention system, and aims at providing the vulnerable groups with systematic and comprehensive security of production and living. After entering 21st century, with growing attention being put to "the Problem of Agriculture, Countryside and Farmers", some researchers introduced the concept of social protection into China, and make it an approach to measure the level of production and living security of the biggest vulnerable group, namely farmers.

Among the interventions provided by this framework, education and training can strengthen farmer's initial resource endowment, and help them to deal with natural, economic and social risks. However, due to the vast territory and complex situation of rural areas in China, as well as scattered residences and high mobility of farmers, traditional education and training programs are largely left behind. Under such circumstances, it's necessary to make full use of the advantages of modern distance education technology, such as openness, flexibility, popularization and accessibility, so as to overcome the obstacles caused by time, space and costs as well as other traditional factors. As a result, farmers can be provided with more education and training opportunities, and their social protection situation can be improved significantly.

This article can be divided into five parts. The second part introduces the basic framework of social protection. The third part discusses the role of modern distance education technology in education and training programs for farmers. The fourth part analyzes the effect of the application of modern distance education technology in education and training programs for farmers, taking the modern distance education system for farmers in Beijing for example. The fifth part summarizes the article briefly.

2 BASIC FRAMEWORK OF SOCIAL PROTECTION

The term Social Protection mainly stems from Social Security. It has a western origination, and was put into practice in the 1990's (Zhang et al. 2011). However, there are still different

Table 1. Basic framework of social protection.

Types of intervention	Initial resource endowment			
	Individual and household	Group based	Public provided	Marked based
Promotional	(1) Preventive health practices (2) Migration (3) More secure income sources	(1) Collective action for infrastructure (2) Common property resource management	(1) Sound macroeconomic policy (2) Environmental policy (3) Education and training policy (4) Public health policy (5) Active labor market policies	
Preventive	(1) Crop and plot diversification (2) Income source diversification (3) Investment in physical and human capital (4) Self-insurance	(1) Occupational associations (2) Rotating savings and credit associations (3) Investment in social capital	(1) Agricultural extension (2) Protection of property rights (3) Social insurance	(1) Saving accounts (2) Microfinance (3) Commercial insurance
Protective	(1) Sale of assets (2) Loans (3) Child labor (4) Reduced food consumption (5) Seasonal or temporary migration	(1) Transfers from networks of mutual support	(1) Social assistance (2) Workfare (3) Subsidies (4) Social funds (5) Cash transfers	(1) Sale of financial assets (2) Loans

Source: Adapted from Holzmann and Jorgensen (2001).

opinions towards the definition of social protection in academic and political sectors. Holzmann and Jorgensen (2001) regard social protection as "public interventions to assist individuals, households, and communities better manage risk, and provide support to the critically poor". While Cook (2004) points out that "the term social protection is largely used to refer to longer-term policies that aim to protect and promote the economic and social security or well-being of the poor". Although these definitions are in different forms, their essences are in common, that is to say, they all emphasize the role of policies and interventions in safeguarding poor people's livelihood by means of resources allocation.

Social protection theory maintains that people can obtain resources through two different channels. One is initial endowment, namely resources that have already been obtained or easily to be obtained, such as household assets, individual labor and some kind of social rights. The other is market exchanging, namely resources obtained through trade or other financial activities. As for the vulnerable groups, shocks such as illness, accidental injuries and unemployment may exhaust their initial resource endowment, and cause market exchange failure. As a result, they can hardly obtain sufficient living resources. Therefore, social protection must be in place to increase their resources accumulation. Under the social protection framework, the problem of decrease of initial resources can be solved by individual, household, group and government interventions, while market exchange failure should be resolved through market interventions.

The interventions to be adopted can be divided into the following three types: promotional intervention, preventive intervention and protective intervention. Promotional intervention aims at reducing the probability of a shock or negative fluctuation, such as preventive health practices, collective infrastructure construction and common property resource management. Preventive intervention aims at reducing the impact of shocks, such as diversification of production and income, participation in social insurance and purchase of commercial insurance. Protective intervention aims at relieving the impact of a shock after it occurs, such as sale of assets, reduction of food consumption, mutual support and social relief. Table 1 lists some of the major interventions of different types.

3 MODERN DISTANCE EDUCATION FOR FARMERS UNDER SOCIAL PROTECTION PERSPECTIVE

The social protection framework indicates that education and training belongs to promotional interventions provided by public or government, and should be regarded as one of the most effective measures to improve abilities to tackle risks for vulnerable groups. That is because education and training can not only improve farmers' cultural quality, scientific and technological standards, and business management abilities, but enhance their risk awareness, and help them to learn methods to deal with risks in a scientific way, therefore, indirectly increase their resources for coping with risks. This is in accordance with the principle set by the central government. The No.1 central government document of 2012 highlights the role of education and training in boosting agricultural scientific and technological innovation. However, due to the restriction of political, economic, historic, social and cultural factors, traditional education and training for farmers still faces many problems. For example, there are still no laws and regulations relating to education and training for farmers at the national level, and relative standards and norms need to be streamlined. Education and training is unsystematic, and education resources dispersedly managed by different departments should be integrated. There is also imbalanced resource allocation between urban and rural areas, which leads to insufficient investment in rural education. The educational pattern lacks flexibility and can't meet the diversified and differential requirements of farmers. What's more, most farmers still do not emphasize education and training too much.

Modern distance education is a kind of education that emerges with the development of information science and technology. It integrates the advantage of broadcast, television and online education, and bond together different media such as text, audio and video information, which makes it possible to create a virtual classroom where teachers can bypass the obstacles of time and space, and teach students as well as communicate with students freely. Thanks to its unique edges, modern distance education has been accepted by many countries, and gradually becomes an important tool for universal education and lifetime education. Modern distance education technology was introduced into China in the 1970's. Since then, it has been developing rapidly, and was put into practice in farmers' education in the 1990's. During this progress, relying on its openness, flexibility, popularization and accessibility, modern distance education overcomes the restrictions posed by traditional factors such as time, space and costs, and made it a reality that farmers could receive education and training at home, which represented a significant boost to farmers' education and training work.

These results mentioned above have already been proved by many research works. Li (2007) upholds that openness, flexibility, popularization and lifetime accessibility as well as other advantages of modern distance education provide different group of people in rural areas with sufficient options and supportive services, and consequently share with farmers precious education and learning resources. Yang (2008) believes that the implementation and development of agricultural distance education as a new scientific mean and educational technology can secure equal education opportunities for farmers, and substantially promote educational efficiency. He points out that China has cultivated a huge amount of high quality new farmers by means of distance education, and laid down a strong foundation for agricultural and rural development in terms of talents and wisdom. Based on empirical studies, Sun et al. (2011) regards openness, flexibility, popularization and interactivity as the four major features of modern distance education, and maintains that distance education can play an important role in boosting education equality, farmers' lifetime education and socialist new countryside construction.

Considering the advantages it possesses, the central government issued in 2007 an important policy document, named "Proposal on Carrying out Modern Distance Education for Party Members and Officials in Rural Areas throughout the Country", and decided to make the trail aiming at "frequently educating party members and officials and permanently benefiting rural residents" a universal movement.

4 ANALYSIS ON EFFECTS OF MODERN DISTANCE EDUCATION FOR FARMERS IN BEIJING

Beijing municipality is, to some extent, the precursor of modern distance education for farmers. In 1999, Beijing municipality established a modern distance education system for farmers covering the whole Beijing region as well as some other places in the country, featuring an integration of different technologies such as satellite broadband, IP network and digital TV broadcasting. This system got recognition both domestically and internationally, and United Nations Educational, Scientific and Cultural Organization (UNESCO) named it the "Beijing Model". After the implementation of modern distance education project for rural party members and officials, this system has been upgraded with computer broadband network technology and interactive new media being introduced in, and Content Delivery Network (CDN) and Peer to Peer (P2P) technologies being innovatively used. As a result, a modern agricultural education and training platform featuring "two networks, two nodes and ten modules" came into being, complimented by a new training model which combines distance education, face to face training and supportive services together.

Targeting the special study requirements of farmers, the platform designed different modules with rich functions, and set up a supportive service system to monitor and guide users' learning activities. With these components in place, farmers can overcome difficulties encountered in their learning process through Application Forum, BBS, Technological Hotline and Curriculum Projection, and enjoy personalized study process through Learning Archives, Curriculum Recommendation and Expert Consultancy. There is also a far-reaching transmission system, with two sub platforms and two stations, namely municipal platform, county-level platform, township training station and village terminal station, playing their own roles. After the integration of resources from Culture-sharing Project, Digital Family

Project and Ainong Station, 5058 terminal reception stations have been established, covering all towns and villages in suburban areas. In order to meet the demand of different groups, totally 11919 hours training resources have been developed, relating to various fields such as culture, business, science and technology and agriculture. At the same time, a 5000-member terminal station management team comprises mainly of college-student-village-officials has been founded through classified training and demonstration.

Since starting its service in 2010, the platform has conducted as many as 3 million person time education and training activities, making distance education a sustainable and scaled way to train farmers. Among these training activities, the platform opened special columns such as Valley Economy, Food Security, Characterized Plantation, Ecological Breeding and so on, and organized face to face technological instruction with the help of rural grassroots technicians. Above all, 162 new varieties and 1732 suitable technologies relating to vegetable, flower, edible fungi, livestock and poultry have been applied. With their total application reaching 92.6 thousand hectors and 600 thousand units of animals, these new technological achievements can increase farmers' income by 434 million RMB, yielding a yearly benefit of 145 million RMB.

5 CONCLUSIONS

From the analysis given above, it can be found out that education and training is one of the major promotional interventions in social protection framework, and it should be given priority as a primary type of measure that helps farmers to cope with risks. However, due to the impact of many factors, the current education and training system for farmers is unsystematic, and is restrained by dispersed management, low standardization and normalization, insufficient funding as well as the lack of enthusiasm of farmers. Modern distance education is a kind of education that emerges with the development of information science and technology, openness, flexibility, popularization and accessibility are some of the advantages that make it an important way for farmers to receive education and training. The example of modern distance education for farmers in Beijng shows that relying on its advantages, modern distance education can overcome the restrictions posed by traditional factors such as time, location and costs, and make it a reality that farmers can receive education and training at home, thus providing a significant boost to farmers' education and training undertaking. Therefore, the application of modern distance education technology is conducive to the improvement of social protection for farmers as well as rural stability and prosperity.

REFERENCES

Cook S. 2004. Social protection in Asia. Har-Anand Publications Pvt Ltd.: 17–18.

Holzmann R. & Jørgensen S. 2001. Social risk management: a new conceptual framework for social protection and beyond. *International Tax and Public Finance*, (8): 529–556.

Li Z. 2007. Modern distance education: an approach to the construction of socialist new countryside. *Rural Economy*, (8): 121–122.

Sun S. et al. 2011. Practice exploration and efficiency analysis of modern distance education for farmers. China Agricultural Scientech Press.

Yang J. 2008. Boost the development of modern agricultural distance education—accelerate construction of new countryside. *Chinese Agricultural Science Bulletin*, (12): 576–579.

Zhang X. et al. 2011. Comparison of rural social protection in China and India. Social Sciences Academic Press: 15–16.

Modeling and Computation in Engineering III – Zhang & Xie (Eds)
© 2014 Taylor & Francis Group, London, ISBN 978-1-138-02680-3

Flood forecasting test based on multiple sources of rainfall data

Zhiyuan Yin
Hubei Key Laboratory for Heavy Rain Monitoring and Warning Research, Institute of Heavy Rain, China Meteorological Administration, Wuhan, China

Fang Yang
Meteorological Information and Technology Support Center of Hubei Province, Wuhan, China

Tao Peng & Junchao Wang
Hubei Key Laboratory for Heavy Rain Monitoring and Warning Research, Institute of Heavy Rain, China Meteorological Administration, Wuhan, China

ABSTRACT: The Shuibuya control watershed on upstream Qingjiang in Hubei Province was taken for example in this paper. Firstly, the grouped relationship of Z-I is used to bonding surface meteorological stations to calibrate the radar estimate precipitation, then the genetic algorithms and neural networks method were combined to establish a revised model of AREM forecasting precipitation, both of before and after revised AREM forecasting precipitation are inputted in Xinanjiang hydrological model to experiment flood forecasting at last. The results show that the revised AREM forecasting precipitation can significantly improve the forecasting accuracy of the process cumulative precipitation. The average relative error reduction rate exceeds more than 60%. The average flood forecasting experiment deterministic coefficient of AREM forecasting precipitation between before and after revised was improved from −32.6% to 64.38%, peak relative error decreases from 39% to 25.04%. The accuracy of overall flood forecasting have been improved to some extent.

1 INTRODUCTION

With the continuous development of modern meteorological technology and more rich observational data, making full using of these data in reservoir basin of having less hydrological stations or no hydrological stations can strengthen scientific and predictability of reservoir flood control operation. Currently, there are many experts input the rain gauge, Quantitative Precipitation Estimation (QPE), Quantitative Precipitation Forecasting (QPF) and other information into the hydrologic model for simulation and testing, and achieve better results. Peng Tao use grouped Z-I relationship and ground rainfall precipitation to calibrate the radar estimated precipitation. The precipitation was input into hydrological model for flood forecasting, results show that, the calibrated precipitation's precision of radar estimation of flood forecast is greatly improved. Zhang Yaping calibrate the radar estimated precipitation by selecting different number of rain gauge, and input the calibrated results to the TOPMODEL hydrological model for runoff simulation, satisfactory results are obtained. The results show that, considering the rainfall forecast period has obvious advantages on improving flood forecasting results than to without considering the forecast period precipitation. Wang Lili combined the GRAPES model and hydrological model for flood forecasting, the result shows that GRAPES meteorological hydrological model is obvious to prolong the flood forecast period and has higher precision for flood simulation. The application of the above results biased in favor of one or two class data for the study of fusion, three or more kinds of the information is rare. Therefore, this paper make the best of the observed precipitation meteorological station, radar estimated precipitation and forecast precipitation to calculate the area rainfall precipitation which used as input to hydrological model considering the reservoir basin without the reliable rainfall data. This work can supply scientific reference for reservoir flood control.

2 THE STUDY AREA AND RESEARCH METHODS

This paper selects Qingjiang Shuibuya control basin as the research object. Normal water level of the reservoir is 400 m, the storage capacity is 43.12×10^8 m^3. It is a reservoir having multi-year regulating performance and engineering in power generation, flood control, aquaculture, tourism, irrigation etc. Because only collect part of reservoir inflow data, we regard it as basin which have no rainfall stations.

Giving full play to the role of genetic algorithm and BP neural network and combining both of them is currently a very active research field. This paper will establish precipitation correcting model of fusing meteorological observing station, radar estimation and forecasting data according to genetic neural network algorithm (hereinafter referred to as the GA-BP algorithm). The model's output is taken as input of hydrological model for flood forecasting test in order to improve the precision of flood forecast. In terms of the characteristics of the GA-BP algorithm, BP algorithm training which inevitably exist the local extreme based on weight modification principle of error gradient descent. Genetic algorithm is good at global searching, but incapable of local precise searching. The combination

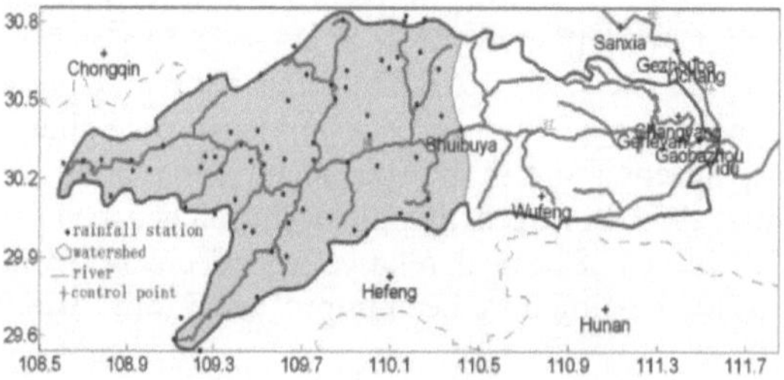

Figure 1. Overview diagram of the Qingjiang basin (the grey part is the research basin).

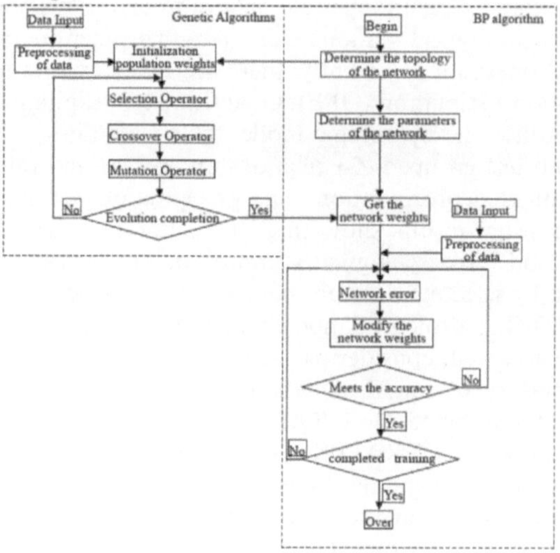

Figure 2. Flow chart of GA-BP algorithm.

of genetic algorithm and BP algorithm achieve the complementary advantages to better solving practical problems.

3 THE EXAMPLE ANALYSIS

In the evolving iterative process of genetic algorithm, set population size as $P_o = 500$, evolution algebra as $k = 1000$, crossover probability as $P_c = 0.6$, mutation probability as $P_m = 0.001$. Firstly, optimized the initial connection weights of three layers BP neural network using data of nine precipitation processes in statistics period, then calculate the optimal network connection weights through the BP algorithm and establish the revised AREM precipitation forecast model at last. Table 1 gives the results of both GA-BP algorithm and unrevised AREM forecasting accumulated precipitation. The results show that the mean absolute error and mean relative error of unrevised AREM accumulated precipitation are respectively 18.6 mm and 77.22% in statistics period, while the mean absolute error and mean relative error of revised by GA-BP algorithm are respectively 6.99 mm and 26.09%. According to the simulation results, both the absolute error and relative error of the revised forecasting precipitation are decreased more than 60% than the unrevised precipitation. In addition, unrevised forecasting accumulated precipitation has the worse stability. The minimum relative error is 0.57%, the maximum relative error is 235.15%. In contrast, the relative error of the forecasting accumulated precipitation revised by GA-BP algorithm range is only 20% of the former and the stability of the prediction effect is obviously superior to the former. When the cumulative precipitation exceed 20 mm, the average relative error is within 20% which greatly improve the precipitation forecasting accuracy.

From the Figure 3, unrevised AREM precipitation can accurate forecast the maximum hourly precipitation appeared in the process, the magnitude of error is within 10%, the forecasting error of time is less than 2 h. From the precipitation process, in addition to the data near by the maximum hourly precipitation close to the reality, the forecasting result is lower. The revised precipitation is worse than unrevised on maximum hourly precipitation, but the whole forecasting process is closer to the reality.

The results of Table 2 show that the average absolute error and relative of the unrevised AREM accumulated precipitation error were 19 mm and 31.25%. The average absolute error and relative of the revised by the GA-BP algorithm were 4.4 mm and 9.97%. From the statistics, the average relative error of the accumulated precipitation of GA-BP algorithm to process decreases in more than 20%, the maximum is decreased by 48%, its correction effect is obvious. In addition, the average relative error of the four accumulated precipitation

Table 1. The compare of GA-BP and unrevised accumulated precipitation in statistics period.

Processes	Time	Measured accumulated rainfall (mm)	Unrevised accumulated rainfall			Revised accumulated rainfall		
			F/mm	*A*/mm	*R*/%	*F*/mm	*A*/mm	*R*/%
1	2009.5.11	18.3	27.1	8.8	47.94	28.4	10.1	55.12
2	2009.6.19	35.0	34.9	0.1	0.57	45.0	10.0	28.43
3	2009.6.28	20.5	48.4	27.9	135.99	27.1	6.6	32.15
4	2010.6.06	45.6	33.6	12.0	26.48	48.1	2.5	5.30
5	2010.7.07	65.0	51.5	13.5	20.88	76.8	11.8	18.14
6	2010.8.14	34.2	70.6	36.4	106.35	43.9	9.7	28.32
7	2010.8.24	25.4	47.4	22.0	86.99	27.4	2.0	7.96
8	2011.7.25	19.9	26.8	6.9	34.73	21.3	1.4	7.02
9	2011.9.27	17.1	57.4	40.3	235.15	26.1	9.0	52.35
Average	/	31.3	44.2	18.6	77.22	38.2	6.99	26.09

Note: *F* represents the prediction results; *A*, *R* represent respectively the absolute error and relative error.

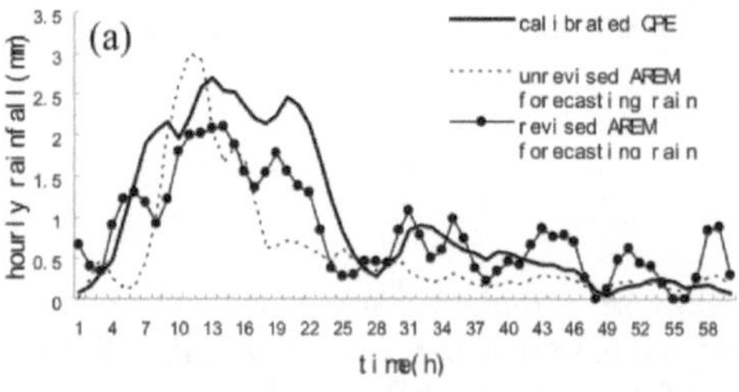

(a) precipitation process in Aug 3, 2011

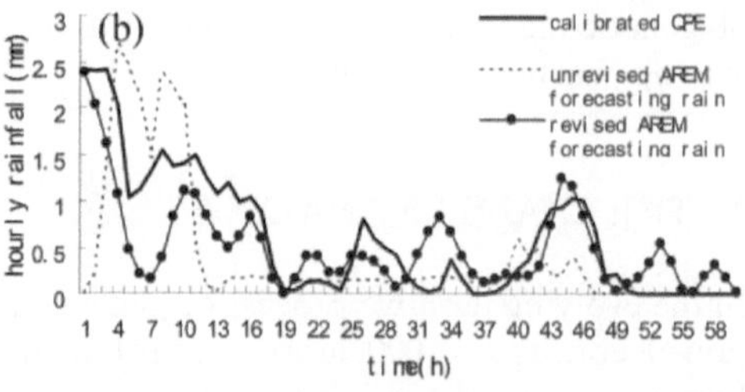

(b) precipitation process in Apr 10, 2012

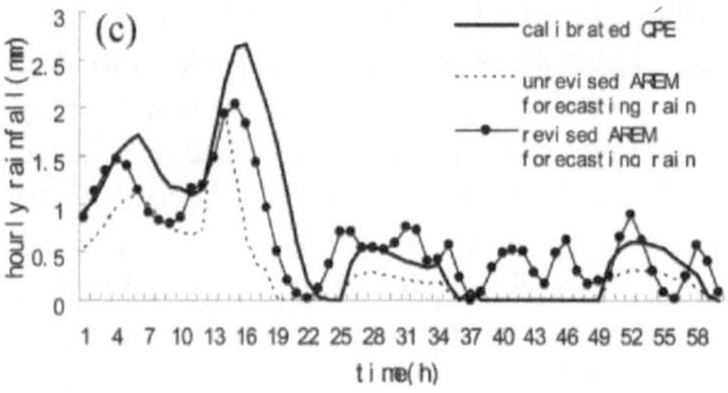

(c) precipitation process in May 25, 2012

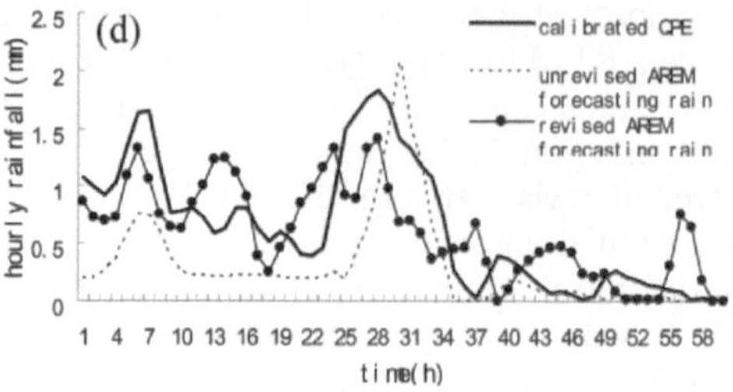

(d) precipitation process in July 7, 2012

Figure 3. Contrast of unrevised AREM forecasting precipitation, revised AREM forecasting precipitation and calibrated radar estimated precipitation in test period.

Table 2. The compare of GA-BP and unrevised accumulated rainfall in experiment period.

Processes	Time	Measured accumulated rainfall (mm)	Unrevised accumulated rainfall			Revised accumulated rainfall		
			F/mm	*A*/mm	*R*/%	*F*/mm	*A*/mm	*R*/%
1	2011.8.03	57.4	34.6	22.8	39.72	49.6	7.8	13.59
2	2012.4.10	35.7	23.6	12.1	33.89	29.9	5.8	16.25
3	2012.5.25	41.3	20.3	21.0	50.85	39.1	2.2	5.33
4	2012.7.07	38.1	18.0	20.1	52.76	36.3	1.8	4.70
Average	/	43.12	24.12	19.0	31.25	38.73	4.4	9.97

Note: *F* represents the prediction results; *A*, *R* represent respectively the absolute error and relative error.

processes which were above 30 mm is within 20% in test period, this also is consistent with the analysis conclusion of the statistics period.

Flow prediction results of four forecasting processes are given in Figure 4. Unrevised precipitation forecasting flow is better than revised only on April 10, 2012. This is not because the unrevised precipitation and live in good agreement, but the process of precipitation is mainly concentrated in the first 20 h, unrevised forecast precipitation mainly concentrated in the first 13 h. Although the latter time span is relatively short and hourly precipitation is relatively large, accumulated precipitation is close to each other between forecasting and reality in the first 13 h, the unrevised precipitation forecasting inflow is close to the reality. With the forecasting precipitation decreasing rapidly, the error become more and more big after 13 h. In contrast, the distribution of the revised forecasting precipitation and the reality is close to each other, but in the main precipitation process of in the first 13 h, the accumulated precipitation compared to the amount of precipitation is obviously lower, which leads to the revised flow process and the peak flow precipitation slants small, the amount of water is lower too.

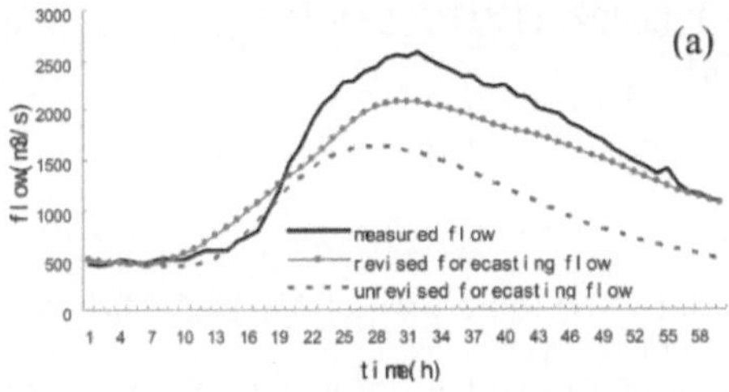

(a) precipitation process in Aug 3, 2011

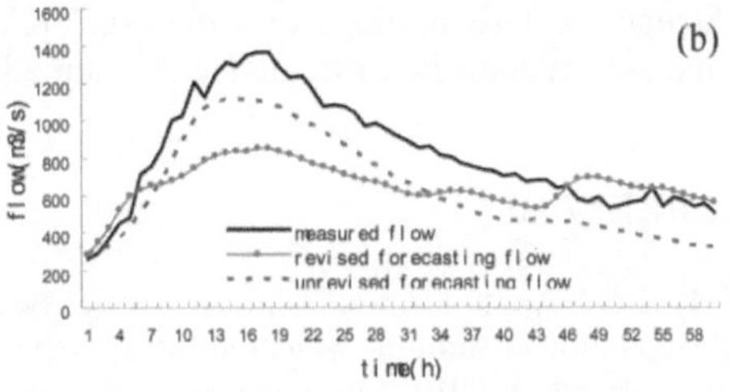

(b) precipitation process in Apr 10, 2012

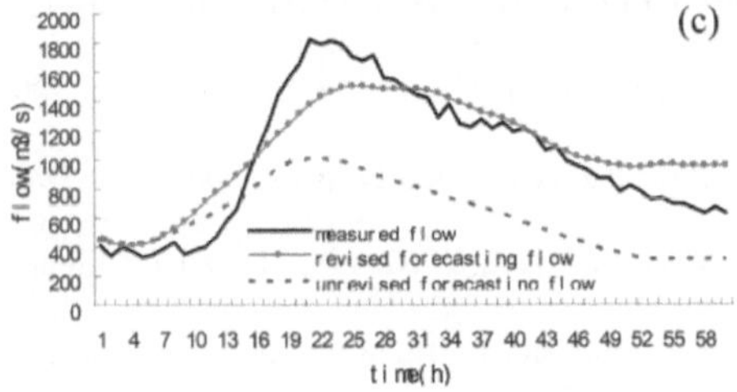

(c) precipitation process in May 25, 2012

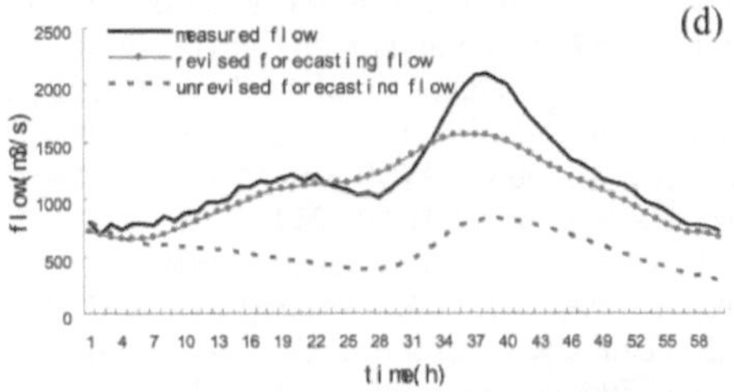

(d) precipitation process in July 7, 2012

Figure 4. Contrast of forecasting flow between unrevised and revised AREM precipitation.

In the other three processes, the revised forecasting precipitation fit well with the reality than unrevised no matter on accumulated precipitation or its distribution, so the prediction of the peak flow and runoff are closer to the reality.

4 CONCLUSIONS

1. The mean value of the calibrated radar estimated precipitation is lower than rainfall station's. The more accumulated precipitation is the bigger mean absolute error is. But the average relative errors were relatively stable, maintained at around 30%.
2. The revised AREM forecasting precipitation improved obviously in the accuracy of accumulated precipitation forecasting, especially for the accumulated precipitation over 20 mm, the average relative error can be controlled within 20%. The average absolute error of hourly forecasting precipitation and the average relative error is decreased by 41.56%, 38.16% respectively than unrevised. The effect is obvious, but there is still a certain gap compared with actual results.
3. The comparative test results show that: 1) unrevised AREM forecasting precipitation is the same as reality in the maximum hourly rainfall and the time of coming, but the process of precipitation forecasting is not ideal, the revised AREM forecasting precipitation is more accurate in the whole process of precipitation and the unrevised is more accurate in the maximum hourly precipitation; 2) when using the revised AREM forecasting precipitation for flood forecasting, the deterministic coefficient and peak relative error is improved obviously than unrevised. The prediction results of the deterministic coefficient have reached a level of Grade B operational forecasting. The forecasting accuracy of the peak relative error is close to the requirement, but needs further improvement.

ACKNOWLEDGMENT

This work has been supported by the National Natural Science Foundation of China (41205086, 51079099, 51379149); Special Research Project (Meteorology) of the Ministry of Science and Technology of China (GYHY201206028, GYHY201306056); Special basic business of Wuhan Institute of Heavy Rain (1014).

REFERENCES

Cai, R. 2007. Application of Artificial Neural Network (ANN) to Forecast of Nanjing's Short-term Precipitation in Autumn. *Bimonthly of Xinjiang Meteorology* 1(1):49–52.
Chen, D.H. et al. 2010. Interpretation and Application of BP Artificial Neural Network in MM5 Model Forecasting Gale for Coastal Regions of Fujian Province. T*orrential Rain And Disasters* 29(3):65–69.
Cui, C.G. et al. 2010. The Flood Forecast Test on QPF Coupling with Hydrological Model in Flood Season in Medium and Small Catchment. *Meteorological Monthly* 36(12):56–61.
Gong, Y. 2007. Evaluation and Analysis of the Rainfall Prediction of AREM in Flood Season of 2007. *Torrential Rain And Disasters* 26(4):372–380.
Gu, X.P. et al. 2006. GA-BP Ann Model For River Catchment Precipitation Forecast. *Journal of Tropical Meteorology* 22(3):248–252.
Jin, L. et al. 2005. Short-Term Climate Prediction Model of Neural Network Based on Genetic Algorithms. *Plateau Meteorology* 24(6):981–986.
Li, Z.Y. et al. 2003. Parameter Optimization of Heavy Rain Intensity Formula Based on Genetic Algorithm. *Plateau Meteorology* 22(6):637–639.
Liu, J. et al. 1999. Classified Z-I Relationship and Its Application to the Measurement of Rainfall by Weather Radar over the Huaihe River Basin. *Scientia Meteorologica Sinica* 19(2):213–220.
Liu, X.Y. et al. 2003. Run off simulation using radar and rain gauge data. *Adv Atmos Sci* 20(2):213–218.
Luk, K.C. et al. 2000. A study of optimal model lag and spatial inputs to artificial neural net-work for rainfall forecasting. *Journal of Hydrology* 227:56–65.
Peng, T. et al. 2010. Application of Radar QPE to Flood Forecast in Flood Season Hydrological Model. *Meteorological Monthly* 36(12):50–55.
Song, X.M. et al. 2010. Application of Xinanjiang Model Coupling with Artificial Neural Networks. *Bulletin of Soil and Water Conservation* 6(30):135–138.
Wang, L.L. et al. 2012. Application of GRAPES Meteorological and Hydrological Coupled Model to Flood Forecast. *Journal of Applied Meteorological Science* 23(3):274–284.
Yin, Z.Y. et al. 2012. Experiments of Bayesian probability flood forecasting based on the AREM model. *Torrential Rain And Disasters* 31(1):1–6.
Zhang, Y.P. et al. 2007. Application of Radar Rainfall Estimates to Runoff Simulation in Foziling Basin. *Journal of Applied Meteorological Science* 18(3):295–305.
Zhao, J.B. et al. 2011. Precipitation Forecast Test of Township Based on Genetic Algorithm Neural Networks. *Journal of Meteorological Research and Application* 32(1):29–33.

Modeling and Computation in Engineering III – Zhang & Xie (Eds)
 ISBN 978-1-138-02680-3

Tidal current numerical modeling of Chengmai Bay

Mengguo Li
Key Laboratory of Engineering Sediment of Ministry of Transport, Tianjin Research Institute for Water Transport Engineering, Tianjin, China

ABSTRACT: Chengmai Bay is in the north of Hainan Island and is adjacent to Qiongzhou Straits. In order to develop Dongshui deepwater harbor at Dongshui Lagoon in Chengmai Bay, it is necessary to understand tidal current field. In this paper, a numerical tidal current model of Chengmai Bay with irregular triangular grid is set up based on TK-2D software. The model is validated with the *in situ* data, and the results show that the calculated water levels at 5 stations and currents at 13 fixed gauging stations (see Fig. 1) are in good agreement with measured data in both magnitude and phase. The basic characteristics of tidal current are investigated in detail according to the numerical simulation results.

1 INTRODUCTION

Chengmai Bay is in the north of Hainan Island and is adjacent to Qiongzhou Straits. It starts from Tianweijiao in the east to Yubaojiao in the west (see Fig. 1). There are two lagoons in the bayhead, that is, Dongshui Lagoon and Huangchang Lagoon. In the two lagoons, Dongshui Harbor and Macun Harbor were constructed respectively. Dongshui Lagoon is 12 km long, 100 m ~1000 m wide and –2 m ~ –5 m deep. In the whole Chengmai Bay, the entrance of Dongshui Lagoon is only about 1 km away from –10 m isobath, hence, Dongshui Harbor possesses better conditions to be developed into a deepwater harbor of over 100000 DWT.

In order to develop Dongshui Harbor a deepwater harbor, it is necessary to understand tidal current field. In this paper, using TK-2D software, a numerical tidal current model with irregular triangular grid is set up. After the model is verified with in situ observation data, the model is applied to simulate the tidal current field and then the tide and current features are analyzed.

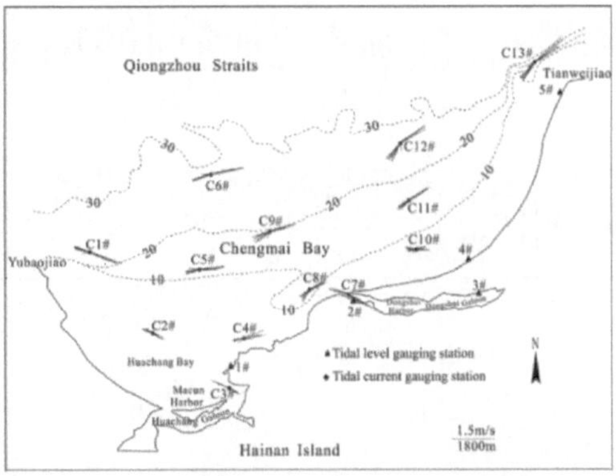

Figure 1. Sketch of Chengmai Bay sea area.

2 NUMERICAL MODEL OF 2-D TIDAL CURRENT

The TK-2D software developed by the author (Li, *et al*, 2006) is used to set up a numerical model of 2-D tidal current. Its governing equations consists of continuity equation and momentum equation. An explicit finite difference numerical method with irregular triangular grid is adopted to discretize the governing equations.

2.1 *Governing equations*

Continuity equation

$$\frac{\partial \zeta}{\partial t}+\frac{\partial[(h+\zeta)u]}{\partial x}+\frac{\partial[(h+\zeta)v]}{\partial y}=0 \tag{1}$$

Momentum equation

$$\frac{\partial u}{\partial t}+u\frac{\partial u}{\partial x}+v\frac{\partial u}{\partial y}-fv=-g\frac{\partial \zeta}{\partial x}-\frac{gu\sqrt{u^2+v^2}}{c^2(h+\zeta)}+\varepsilon\left(\frac{\partial^2 u}{\partial x^2}+\frac{\partial^2 u}{\partial y^2}\right) \tag{2}$$

$$\frac{\partial v}{\partial t}+u\frac{\partial v}{\partial x}+v\frac{\partial v}{\partial y}+fu=-g\frac{\partial \zeta}{\partial y}-\frac{gv\sqrt{u^2+v^2}}{c^2(h+\zeta)}+\varepsilon\left(\frac{\partial^2 v}{\partial x^2}+\frac{\partial^2 v}{\partial y^2}\right) \tag{3}$$

In equations (1) through (3), x and y are the coordinates in the Cartesian coordinate system, in which xy plane coincides with still sea surface, u and v are the depth-averaged velocity components in x and y directions respectively, g is the acceleration due to gravity, ε is the eddy viscosity coefficients of water in x and y directions respectively, t is time; f is the Coriolis parameter, ζ is the water surface displacement relative to the xy plane; h is the still water depth, c is Chezy coefficient, $c=(h+\zeta)^{1/6}/n$, n is the Manning roughness coefficient.

2.2 *Numerical method*

An explicit finite-difference numerical method with triangular grid is used here for Eqs. (1)–(3). The differential equations are as follows, respectively.

$$\zeta_M^{K+1}=\zeta_M^K-\Delta t\left\{\left[\frac{\partial(Hu)}{\partial x}\right]_M^K+\left[\frac{\partial(Hv)}{\partial y}\right]_M^K\right\} \tag{4}$$

$$u_M^{K+1}=\frac{u_M^K-\Delta t\left[\left(v\frac{\partial u}{\partial y}\right)_M^K+g\left(\frac{\partial \zeta}{\partial x}\right)_M^{K+1}-fv_M^K-\varepsilon\left(\frac{\partial^2 u}{\partial x^2}+\frac{\partial^2 u}{\partial y^2}\right)_M^K\right]}{1+\Delta t\left[\left(\frac{\partial u}{\partial x}\right)_M^K+g\frac{\sqrt{\left(u_M^K\right)^2+\left(v_M^K\right)^2}}{\left(c^2H\right)_M^{K+1}}\right]} \tag{5}$$

$$v_M^{K+1}=\frac{v_M^K-\Delta t\left[\left(u\frac{\partial v}{\partial x}\right)_M^K+g\left(\frac{\partial \zeta}{\partial y}\right)_M^{K+1}+fu_M^K-\varepsilon\left(\frac{\partial^2 v}{\partial x^2}+\frac{\partial^2 v}{\partial y^2}\right)_M^K\right]}{1+\Delta t\left[\left(\frac{\partial v}{\partial y}\right)_M^K+g\frac{\sqrt{\left(u_M^K\right)^2+\left(v_M^K\right)^2}}{\left(c^2H\right)_M^{K+1}}\right]} \tag{6}$$

where M is the number of nodal point, K is the time step index, Δt is the time step of integration, H is total water depth, $H=h+\zeta$.

2.3 *Computation domain and grid division*

The computation domain is shown in Fig. 1, the whole Chengmai Bay being included. The west boundary reaches 109°56.5'E, and the east boundary reaches 110°9.8'E. The north boundary reaches 20°5'N, and the south boundary reaches the head of Chengmai Bay. It is about 23.4 km long in the east-west direction and about 15.9 km wide in the south-north direction. Irregular triangular grid is adopted to divide the computation domain (see Fig. 2). In order to describe the features of complex coastline and bottom topography, fine grids are used in Dongshui Lagoon and nearby. There are 17324 triangular grid nodes and 33941 triangular elements in the domain. The spatial resolution ranges from 30 m to 200 m.

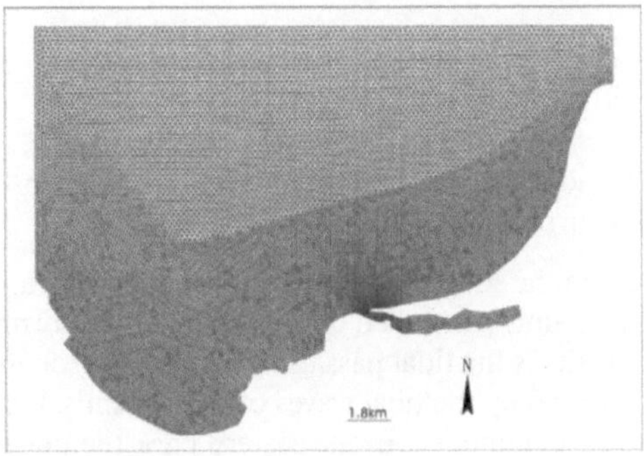

Figure 2. The computation grid.

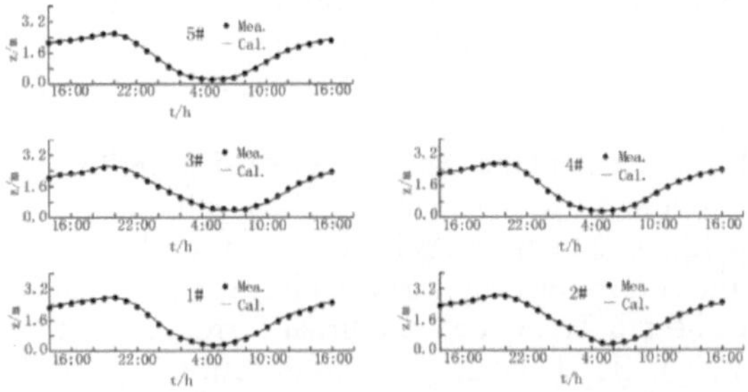

Figure 3. Comparisons of water level time series.

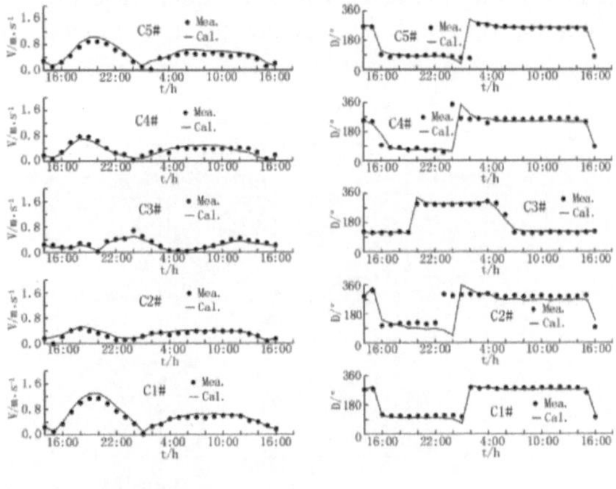

Figure 4. Comparisons of current speed and direction time series.

2.4 *Model verification*

There are three synchronous hydrometric surveys conducted in Chengmai Bay. The first (neap tide) was conducted from June 13 to 14, 2004, the second (moderate tide) from June 15 to 16, 2004, and the third (spring tide) from June 18 to 19, 2004. There are 5 tidal level observation stations and 13 tidal current gauging stations in each survey (see Fig. 1). To validate the model, three tidal simulations have been conducted for the same periods as the above mentioned. The simulated results are in good agreement with the observed data in both magnitude and phase. As the pages are limited, only verifications of tidal levels and tidal currents of spring tide are presented (Figs. 3 and 4). Others can be refered to Li *et al*, 2004. The calculated water levels at 5 stations and currents at 13 fixed gauging stations agree well with measured data in both magnitude and phase.

3 ANALYSIS OF THE SIMULATION RESULTS

According to the simulated tidal current, together with the in situ observation data, the characteristics of tide and tidal current are as follows.

1. Chengmai Bay is located on the south bank of Qiongzhou Straits, and the tide and tidal current in Chengmai Bay are mainly controlled by the tide and tidal current system in Qiongzhou Straits. As Qiongzhou Straits is the tidal passage for seawater exchange between Beibu Gulf and South China Sea, affected by the tidal waves of its two ends, the tidal wave propagation in Qiongzhou Straits is rather complex. In the eastern part, the tide is irregular semi-diurnal tide and is consistent with that in the north of South China Sea, in the western part, the tide is regular diurnal tide and is consistent with that in Beibu Gulf, while in the middle part, the tide is the transition between regular diurnal tide and irregular semi-diurnal tide. Chengmai Bay is located in the western part of Qiongzhou Straits, its tide belongs to regular diurnal tide (the tide coefficient F = 7.3 in Dongshui Harbor and F = 7.1 in Macun Harbor).
2. In Dongshui Harbor, the maximum high tide level is 3.07 m (Xiuying theoretical base, the same below), the minimum low tide level is –0.03 m, the mean high tide level is 2.43 m, the mean low tide level is 0.72 m, and the mean sea level is 1.60 m; the maximum tidal range is 3.08 m, the minimum tidal range is 0.05 m, and the mean tidal range is 1.70 m; the mean flooding duration is 12.3 h, and mean ebbing duration is 9.4 h.
3. In Chengmai Bay, the current of neap tide is mainly semi-diurnal current, while the currents of spring tide and moderate tide are diurnal currents. The tidal current is basically a to-and-fro current. The flood and ebb current direction is basically consistent with the isobaths in outer sea, while near the coastline and the entrances of the lagoons, it is basically consistent with the coastline (see Figs. 5 and 6).
4. The tidal wave in Chengmai Bay is generally progressive wave, that is, the turn time of tidal current occurs at middle tide level, while the tidal wave in Dongshui Lagoon and Huangchang Lagoon is standing wave, that is, the turn time of tidal current occurs at high tide and low tide.

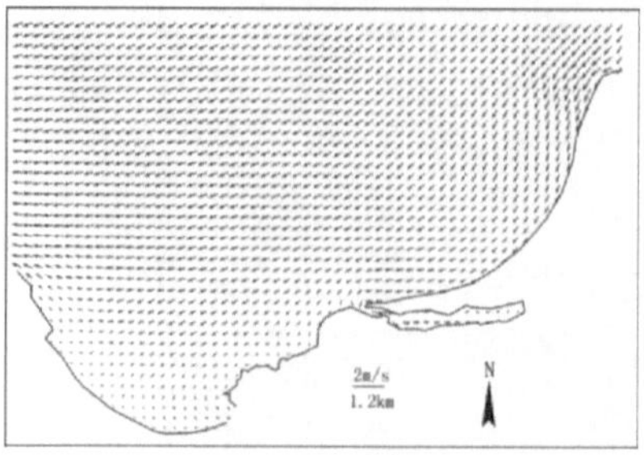

Figure 5. Current vectors flooding in Dongshui Lagoon and flowing westward in Chengmai Bay.

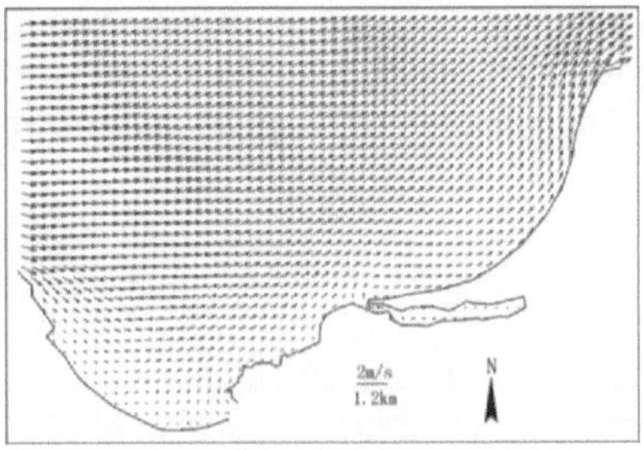

Figure 6. Current vectors ebbing in Dongshui Lagoon and flowing eastward in Chengmai Bay.

5. In Chengmai Bay, the water flows eastward above middle tide level and westward below middle tide level. Thus, the water current has features of flowing westward at flood, flowing eastward at flood, flowing westward at ebb and flowing eastward at ebb. The westward current is dominant at flood and eastward current is dominant at ebb. However, the tidal current in the two lagoons is different from that in outer sea, and there are obvious phase differences inside and outside the lagoons.
6. The phase of tidal current lags gradually from west to east. The turn time of west current near Yubaojiao lags behind that near Tianweijiao by 2.5 h, and the turn time of east current near Yubaojiao lags behind that near Tianweijiao by 0.5 h. The phase of tidal current also lags gradually from out sea to coast, that is, the turn time leads nearshore and lags offshore. According to statistical analysis of Macun Harbor section, Dongshui Harbor section, the lag time ranges from 0.5 h to 7.0 h.
7. In Chengmai Bay, the velocity of east current is greater than that of west current. The mean velocity of west current of spring tide, moderate tide and neap tide ranges from 0.09/s~0.62 m/s, while the mean velocity of east current ranges from 0.05/s~0.79 m/s. The velocity at the entrance of Dongshui Lagoon reaches 0.89 m/s.
8. The current velocity at Yubaojiao, Tianweijiao, the entrance of Dongshui Harbor and in the outer deep sea is obviously greater than that in shallow water area.
9. In Chemgmai Bay, the mean duration of west current is about 15.3 h, while the mean duration of east current is about 8.9 h. The mean duration of west current is longer than the mean duration of east current.

4 CONCLUSIONS

In this paper, TK-2D software is used to set up a mathematical model of tidal current in Chengmai Bay. On the basis of verifications of spring tide, moderate tide and neap tide, the model is applied to simulate the tidal current field in Chengmai Bay. The main conclusions are as follows.

1. The tidal wave in Chengmai Bay is generally progressive tide wave, while the tidal wave in two lagoons is standing wave.
2. In Dongshui Harbor, the maximum tidal range is 3.08 m, the minimum tidal range is 0.05 m, and the mean tidal range is 1.70 m.
3. In Chengmai Bay, the water flows eastward above middle tide level and westward below middle tide level. The westward current is dominant at flood and eastward current is dominant at ebb. There are obvious phase differences inside and outside the lagoons.
4. In Chengmai Bay, the current of neap tide is mainly semi-diurnal current, while the currents of spring tide and moderate tide are diurnal currents. The tidal current is basically a to-and-fro current.

5. The mean velocity of west current ranges from 0.09/s~0.62 m/s, while the mean velocity of east current ranges from 0.05/s~0.79 m/s. The velocity at the entrance of Dongshui Lagoon reaches 0.89 m/s. The velocity increases gradually from coastline to offshore.

REFERENCES

Li MG (2004). "Mathematical modeling of tidal current and sediment for Planning project schemes of Doushui Harbor". *Report of Tianjin Research Institute of Water Transport Engineering*. (in Chinese).

Li MG, Zhang HQ, and Chen HB (2006). "Study on multi-function mathematical model software package TK-2D and its application for coast and estuary". *Journal of Waterway and Harbor,* 27(1):pp51–56. (in Chinese).

Modeling and Computation in Engineering III – Zhang & Xie (Eds)
© 2014 Taylor & Francis Group, London, ISBN 978-1-138-02680-3

Brief and performance analysis of the fingerprint matching program based on GPU

Chunsheng Wu
Beijing Criminal Science Institution, Beijing, P.R. China
Computer Network Information Center, Chinese Academy of Sciences, Beijing, P.R. China
Graduate University of Chinese Academy of Sciences, Beijing, P.R. China

Caigang Feng
Beijing Criminal Science Institution, Beijing, P.R. China

Xuebin Chi
Computer Network Information Center, Chinese Academy of Sciences, Beijing, P.R. China

ABSTRACT: This paper introduces the basic concept of fingerprint matching and the development trend based on GPU computing; Discussed computation efficiency of fingerprint matching in traditional and the advantage of fingerprint matching based on GPU; The speedup ratio test between CPU and GPU fingerprint computing program, separately according to different database scale and different minutia point number. We analyzes the operation efficiency and time distribution of GPU dynamic library by using the "Profiler" software; achieving a comprehensive analysis on the bottleneck problem of GPU computing; putting forward the optimization schemes based on the existing problems.

1 INTRODUCTION

The fingerprint is one of the most important biology features for person. Since 60th twenty century, the Automatic Fingerprint Identification System (AFIS) was manufactured by several countries, for example USA, England and France (Jason, 2010). Because the application is more and more extensive, the capacity of fingerprint database become bigger and bigger. Since 1990, AFIS is used in China. In 2005, the capacity of fingerprint database is more than 30 millions in China. It is rapidly increase more than 20 percent each year. In 2010, the number is more than 70 millions. The rapid growth of fingerprint number for rapid computing of the fingerprint has brought new challenges.

Since 2003, the float calculation performance and memory bandwidth of GPU (Graphics Processing Units) superior to the CPU (Nvidia Corporation, 2009). Beside, The GPU has obvious advantages than CPU in power consumption and cost. GPU parallel computation ability has played a good speedup ratio in the field of graphics. Many algorithms based on GPU were successively put forward (Wang, 2006) (Gress, 2006) (John, 2007). With the augmentation of the GPU programmability, GPU gradually began to be used for general computing beyond graphics (Nvidia Corporation, 2009). General computing based on GPU become the focus in the field of high performance computing in recent years (Wu, 2004) (John, 2007). Since CUDA (Compute Unified Device Architecture) is put forward, the difficulty of GPU development is greatly reduced (Nvidia Corporation, 2009).

In this paper, the fingerprint matching technology based on GPU is introduced. Then the computing efficiency of fingerprint GPU matching program is tested fully. The I/O bottleneck problems of GPU computation is analyzed by test. In view of the I/O bottleneck problems, large memory mode is put forward for data storage and calculation. In this paper the performance of GPU fingerprints matching software is evaluated quantitatively. In view of the current existing problems, the specific optimization method is put forward.

1.1 *Basic concept of fingerprint matching algorithm*

The mainstream fingerprint computing method is positional similarity computation based on fingerprint feature points in domestic at present. In the case of CAFIS fingerprint system, after a complete fingerprint image is processed by feature extraction software, it can be 80–150 feature points are extracted. Every feature points include three parameters, namely: X coordinate, Y coordinates, and angle (D). These feature points show the uniqueness of the fingerprints. Through computation the feature point sets of the two fingerprints are consistent or atypism, we can judge whether the two are the same fingerprints.

For the number of feature points is n, we can analyze the complexity of two fingerprints matching. The basic method of fingerprints matching is to compare each feature points one by one in cycle, until find the right combination or the end of the cycle. We can conclude that the complexity is T (n) = O (n!). The amount of calculation was obviously too huge to practical application. To this end, the researchers have come up with a lot of way to reduce the complexity, such as: matching algorithm based on fingerprint center, point pattern matching algorithm, the matching algorithm based on local structure and algorithm based on line matching, etc. On the basis, we use a lot of optimization algorithm to make it suitable for practical application.

At work, our fingerprint system USES a triple matching algorithm based on feature point matching, including special point pair match, local structure match, global match (Feng, 2010). Because of the feature points computing is divided into level 3, have reduced the amount of calculation, its time complexity is O (n^2). Although the time of a fingerprint matching is very short, it can calculate about 2000 per second. The main job of automatic fingerprint identification system is to compute mass fingerprints. In Beijing, the number of fingerprint database is more than 2.5 million persons. A fingerprint matching time is about $2.5 * 10^7/2000 \approx 11000$ seconds. The efficiency is obviously hard to bear. Although some computing can distinguish fingerprint position and type, the computing time is still too long for real-time fingerprints computing and mass task. The CAFIS system matching algorithm basic flow is partitioned three levels, as shown in Figure 2.

1.2 *The fingerprint algorithm base on GPU*

We are used to parallel computing, such as multiple processes are run in multi-core processors at the same time and multi-node are structured by multiple computers. Parallel computing

Figure 1. Fingerprint feature points.

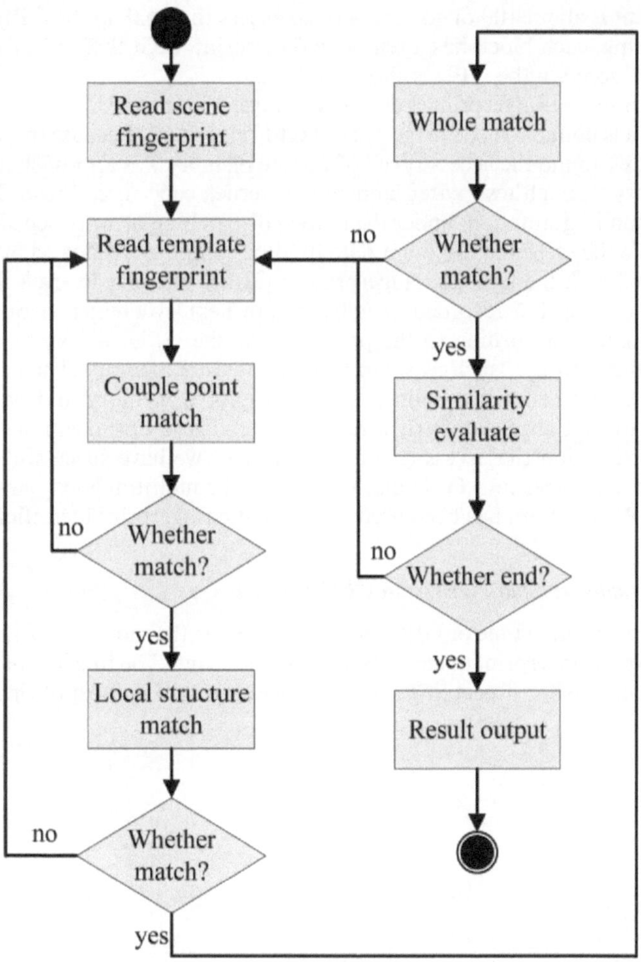

Figure 2. The fingerprint algorithm flow diagram.

of multiple processes is realized by way of MPI. But this mode can not improve the computing ability of single process, it need a plenty of hardware investment to improve the overall capacity. Aiming at the situation, the fingerprint calculation based on the GPU has be studied in the Beijing criminal science and technology institute since 2009.

The main method is to rebuild the first and second of three-level matching algorithm, make them suitable for GPU parallel computing. In many cases, the first and second level above accounted for 95% of all time. Due to the time proportion is very big, these steps need accelerate by GPU parallel computing. Tens to hundreds of Stream processors are included in the GPU. Threads are performed by single instruction multiple data in Stream processors. In fingerprint matching, each feature point matching computing is not associated with other feature points matching computing. That is to say with calculation of independence. In addition, because there is no connection between the fingerprint data, the computing can also be independent between multiple template fingerprints. So the high parallelism of fingerprint computing is suitable for the GPU computation.

Due to the limitation of GPU memory size, only the fingerprint data segment can be read into the global memory of GPU. In practice, about 390000 fingerprints are divided into a segment, around 400 MB. Each paragraph is committed to a kernel function to perform. The argument of kernel function is gridSize(16384, 1, 1), that is the block size. The argument of each block is blockSize(128, 1, 1), that is the thread size. Because the feature point number in each fingerprint is about 100 or so, it is defined as less than 128 in the GPU computation. In GPU computing, each block has a template fingerprint, each thread has a feature point. Fingerprint data stored in the GPU is shown in Figure 3.

Because the number of fingerprint in database is huge, it cannot be loaded in shared memory to call fleetly. So it is bound to texture memory in GPU algorithm. Because the texture memory is regular access, we found the access-speed of texture memory was approach the access-speed of shared memory through test. After launching a kernel, each thread read fingerprint data from the corresponding situation, under the unified dispatching of warp scheduler by parallel computing. While the scene feature point data need be computed, it is read in share memory. It will be matched with the template fingerprint in texture memory by cycle. For the feature points of similarity, it will be recorded, then be used by next level match algorithm.

In the development, according to the properties of the GPU, we use as more as possible proprietary data storage, such as share memory, texture memory. The access speed was increased greatly. Aim at memory non-coalesced in global memory and bank conflicts in share memory; we optimize data structure and code. These optimizations can effectively avoid the above situation the access latency. At present, we have successfully transplanted CPU serial algorithm ideas into GPU and implement the multithreaded parallel computing on GPU. The GPU program has been used in automatic fingerprint identification system.

1.3 *The performance compare GPU and CPU program*

Comparison the computing time of GPU and CPU based on the same data. Random scene fingerprints and suspect fingerprint database were used for testing. The fingerprint matching computation time is relate with suspect fingerprint database size and number of fingerprint feature

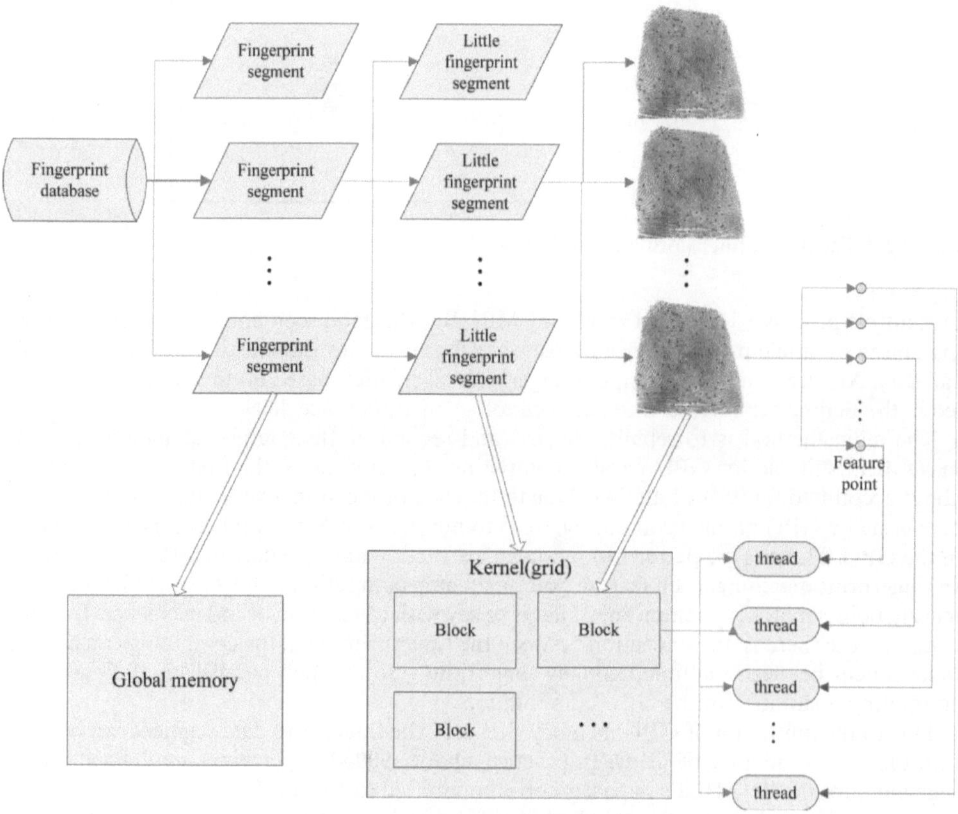

Figure 3. The fingerprint data mapping in CUDA model.

points, therefore, respectively according to the different suspect fingerprint database scale and scene fingerprint feature points number for testing. The test environment, CPU: Intel core i5 dual-core processor, 2.67 GHz, memory: 4 GB, GPU: GeForce GTX 580. This GPU is Fermi architecture, 1.5 GB memory, 16 SM (Stream Multiprocessor), 512 SP (Stream Processor).

1.3.1 *Test1: Different suspect fingerprint database scale (scene fingerprint point number is 15)*

Test result is shown in Table 1.

Through Table 1 data, we can obtain the speedup ratio of CPU, as shown in Figure 4.

It can be seen that with the increase of data size, speedup ratio gradually increased. It is linear after 390000. Because GPU memory capacity is limited, the template data is submitted circularly every 390000 in development. If the data cannot be divided into the entire length, it will be filled empty data.

1.3.2 *Test 2: Different scene fingerprint point number (suspect fingerprint database scale is 390000)*

Test result is shown in Table 2.

By the data in the table above, it is concluded that the speedup ratio using different number of feature point, as shown in Figure 5.

It can be seen that speedup ratio decrease gradually with the increase of feature points number. The number of feature points is between 10 and 30 usually, speedup ratio is steady in this area. Analyzing the main reason of speedup rate drop is: the CPU algorithm is a serial, in the process of coarse matching, when it find a target fingerprint is not in conformity with the

Table 1. The test result of different suspect fingerprint database scale (unit: second).

	Scale						
Type	1000	10000	50000	200000	390000	780000	1170000
CPU	0.640	5.531	26.688	108.844	216.687	432.906	645.859
GTX580	0.865	2.229	4.282	8.336	11.900	24.036	35.762

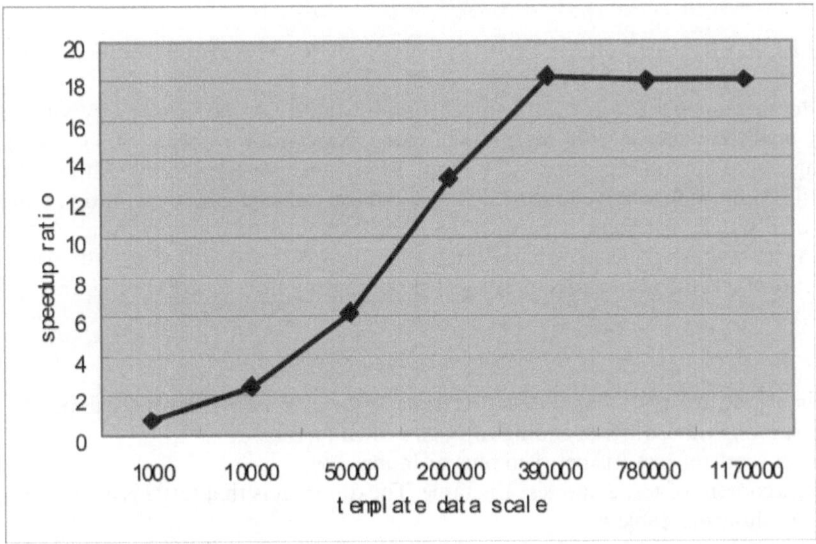

Figure 4. Test 1 speedup ratio.

Table 2. The test result of different scene fingerprint point number (unit: second).

	Feature point number						
Type	6	10	15	20	30	50	70
CPU	205.390	212.062	216.687	277.343	301.078	274.937	384.343
GTX580	10.009	10.65	12.072	14.665	17.056	20.837	29.572

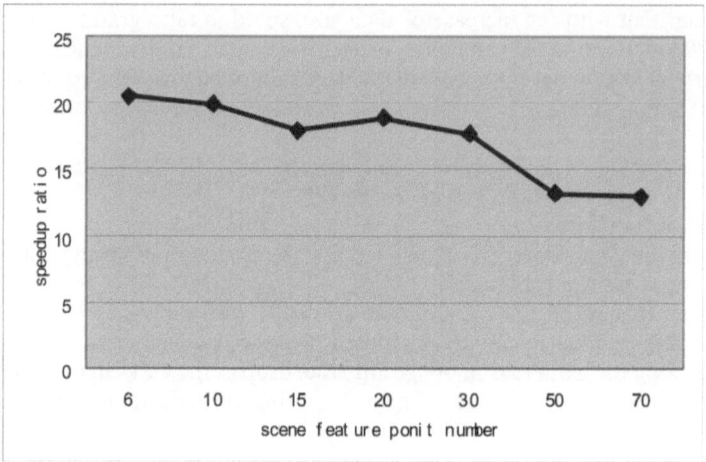

Figure 5. Test 2 speedup ratio.

conditions, it will no longer compute next steps. The anticipation will increase when feature point increase. But GPU parallel computing is a target for all the fingerprint feature points, cannot achieve the condition judgment between threads. So the GPU computing time significantly longer with the increase of feature point, it cause speedup ratio decrease.

Although a single test data have certain limitations, the above test data can quantitatively reflect the GPU computation advantages compared with the traditional CPU computation.

2 THE FINGERPRINT GPU COMPUTING PROGRAM EFFICIENCY ANALYSIS

GTX580 core frequency 855 m, the theoretical floating point calculation capacity is 437 Gflops; Memory read frequency is 4100 M, a 384-bit wide, bandwidth is about 190 G/SEC. Intel i5 CPU single core computing capacity is about 10 Gflops. We can see that although fingerprint GPU computing relative to the CPU can achieve excellent speedup ratio, it still have a number of distance that comparing with theoretical calculation capacity. Analysis the GPU programs run main steps, as shown in Figure 6.

The following is the test results of every step computing time of GPU program.

2.1 *The distribution of the GPU fingerprint computing time*

Test 1 time used mainly include data I/O time from disk into memory, calling the GPU program running time, other spending time, as shown in Table 3.

The error scope of test 1 is between plus or minus 3%.

The base content of test 2 and test 1 is same. The different is that test 2 is alternation point number. As shown in Table 4.

Because template data scales are the same in test 2, the same disk I/O and others time is used in Table 4.

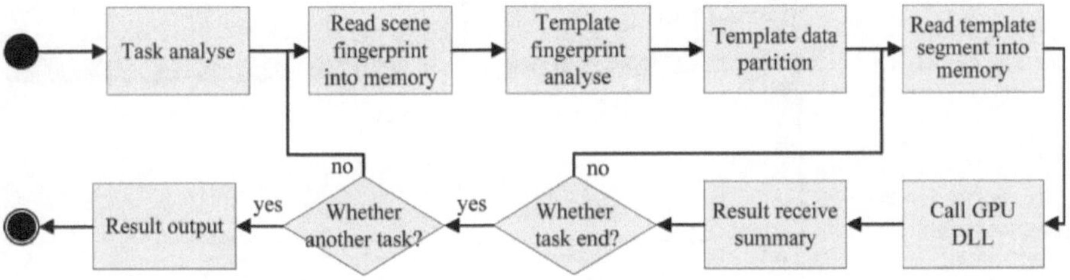

Figure 6. GPU program main process.

Table 3. The time distribution of different suspect fingerprint database scale (unit: second).

	Scale						
Type	1000	10000	50000	200000	390000	780000	1170000
Disk I/O	0.231	1.094	1.507	2.890	5.297	10.423	15.469
GPU running	0.234	0.735	2.375	5.046	6.203	12.813	19.093
Others	≈0.4	≈0.4	≈0.4	≈0.4	≈0.4	≈0.8	≈1.2

Table 4. The time distribution of different scene fingerprint point number (unit: second).

	Feature point number						
Type	6	10	15	20	30	50	70
Disk I/O	≈5.297	≈5.297	≈5.297	≈5.297	≈5.297	≈5.297	≈5.297
GPU running	4.312	4.953	6.375	8.968	11.359	15.140	23.875
Others	≈0.4	≈0.4	≈0.4	≈0.4	≈0.4	≈0.4	≈0.4

2.2 *The GPU time distribution inside the equipment*

Tables 3 and 4 GPU devices running time is refers to execute fingerprint GPU computing dynamic library time. Not only the time is the key to evaluate the GPU program performance, but also the foundation of GPU program optimization. Within GPU code, its running time can be estimated by Profiler software provided by NVIDIA Company for evaluation. Figure 6 is a Profiler time assessment. It analyzes the run time in this condition that the feature points is 15, template data is 390000.

In Figure 7, the percentage is GPU run-time proportion, instead of all run time. The main parameters and the key data of influence performance in the kernel function aradj_sortBlending_Texture2D() are shown in Tables 5 and 6.

2.3 *Performance analysis*

Analyzing schedule from Tables 3 and 4, we can see the main time-consuming of fingerprint GPU program is hard disk data I/O and dynamic library computing.

Hard disk access time is directly related to the template data size, subject to the hard disk I/O speed. Due to adopting the tactics of every 390000 fingerprints is divided into a segment, when the data is less than 390000, with the data increase, the read speed is on the rise, but the data after more than 390000, the data access time is linear segment. Every segment is about 350 MB; read speed is about 70 MB per second.

From Figure 6, Tables 5 and 6, we can see that aradj_sortBlending_Texture2D(), the kernel function of fingerprint matching, takes up 93.76% of the GPU running time in the GPU

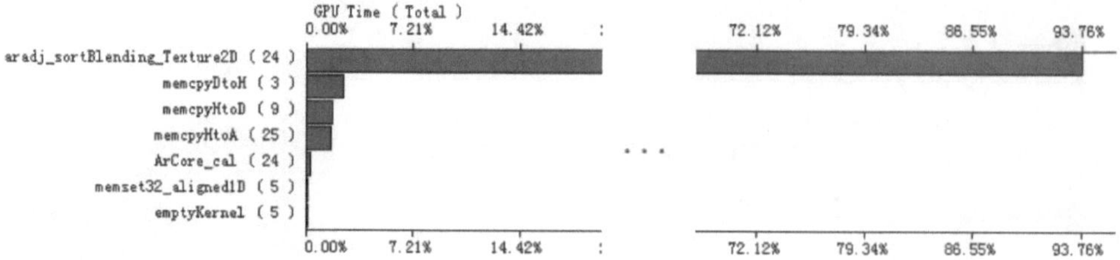

Figure 7. GPU program performance analyze.

Table 5. The main arguments.

Kernel function	Grid size	Block size	Share memory per block	Register per thread
aradj_sortBlending_Texture2D()	16384	128	2820 byte	37

Table 6. The key performance data.

Kernel function	Average run time	Average bank conflict	Average percent texture memory hit	Branch ambiguity	SM active percent
aradj_sortBlending_ Texture2D()	76.3 ms	2.89%	97.28%	11.36%	99.75%

working process, and the copy time between memory and GPU memory is 5.92%. Though the number of fingerprint data is huge, but fingerprint recognition still can be regarded as a compute-intensive applications. Further analysis the time compose of the kernel function, we can see while the function aradj_sortBlending_Texture2D() read data from Shared memory, bank conflicts is only 2.89%, hit ratio of texture memory is 97.28%, global memory is used very little. So the performance has been optimized in memory input and output. Stream multiprocessor active rate is 99.75%, no significant latency. We can see that the GPU performance can be brought into play fully. Branch ambiguity is 11.36%, it illustrate that the internal of algorithm has conditional branch is not split. It will cause the thread group in warp performed useless instruction.

Analyzing GPU programs run on various data, we can deduce:

A. comparing with other time, the transmission time between memory and GPU memory can be ignored;

B. GPU code does not appear obvious memory fusion and bank conflict problems;

C. Computing time occupies a large part in GPU code. This part needs algorithm optimization. We do not discuss it in this paper.

D. Serial computing still exist in the GPU code. This change also refer to the adjust of algorithm. It will not be discussed.

Therefore, rapidly improving fingerprint matching efficiency method is to optimize the way of fingerprint data read from hard disk.

3 THE OPTIMIZING SCHEME

Hard disk read speed is determined by the hardware features of hard disk. the mainstream of the current hard disk is 7200 revolutions per second, part of the server's hard disk to 10000–15000 revolutions per second. cost performance of this method is unsatisfactory. The following scheme of several can effectively avoid the hard disk be read.

3.1 *Large memory storage*

At present, the Beijing fingerprint database size has nearly 20 GB, and now the fingerprint data of Beijing has exceeded 100 GB. Through the matching application server is installed the large memory, the fingerprint file can be read into memory at the beginning of computing. Host program only need to submit memory address to GPU program.

This way can hold the existing pattern maximize unchanged, with minimal changes to get greater gains. But the shortcomings are obvious: one is increase the large memory is needed on each node, it will cause rising costs; another is limited by the upgrade capacity of server, such as: hardware does not support large memory, the operating system is 32-bit.

3.2 *Multi node memory storage*

On the basis of the large memory storage, we can divide the fingerprint data into sections. Each section is stored in memory of a independent server node. Through scattering data, we can reduce requirement of memory space. Suppose there are n nodes, each node memory only store 1/n fingerprint data. For matching task, scheduler distributes it to nodes. After each node complete the subtask, result will be returned the scheduler. Scheduling will summarize these result to the final result. Dividing data can be sequence segmentation, each file on average, classification according to the date type.

This way can effectively use the existing memory resources, greatly reduce the memory acquisition cost, no need to change the fingerprint core algorithm. But this way need to adjust the existing fingerprint system task distribution strategy. If a fingerprint only compute in one of notes, it has possibility occurring a single point of failure. There also is hidden danger in terms of load balancing. Need to select reasonable data allocation strategy based on test.

3.3 *Multi node GPU memory storage*

On the basis of multi nodes memory allocation, we consider that we can put data in GPU memory further. No longer read data files from CPU memory in computing process.

This way can further increase the computing efficiency than the second way, but increase is limited. No longer rely on the size of CPU memory. But still have existing the problems of the second way.

3.4 *The way of dynamic storage*

On the basis of absorbing the previous ways advantages, data dynamic storage way is put forward. All of fingerprint data are stored in local hard disk of each node. In advance part of data is read into memory. According to the size of the GPU memory, part of memory data is read into GPU global memory in advance. According to the date in GPU memory of each node, the Scheduler allocates computing tasks. When a node failure or efficiency is slower, Scheduler can automatically allocate the subtask of the nodes to another based on allocation strategy. The allocation strategy is: a. the nodes with the same data in GPU memory, b. the nodes with the same data nodes in memory, c. other idle nodes. The node of accepted task will be according to actual condition read data from the hard disk, memory to GPU memory. Matching nodes receive subtask under the unified scheduling strategy of Scheduler, ensure that every subtasks can be completed in time and be returned scheduler to summary.

This solution under the condition of the node number of enough, do not need to add new hardware cost, and effective to avoid the single point of failure and load imbalance.

Summarize above four kinds of solutions, as shown in Table 7.

3.5 *The test of large memory*

In view of the first Solution author has taken the simple test. On the premise of not modify the program, we use the method that memory is regarded as virtual hard disk. The fingerprint data can be read in the virtual hard disk read. With 390000 template fingerprints and

Table 7. The summary of solutions.

	Efficiency improve	Cost control	Easy transform	Load balancing	Fault freedom
Solution 1	★★	★	★★★	★★	★★★
Solution 2	★★	★★★	★★	★★	★
Solution 3	★★★	★★★	★★	★★	★
Solution 4	★★	★★★	★	★★★	★★★

15 feature points test the I/O performance. We came at a conclusion that template data read time just is 0.2 seconds (the original time is 5.297 seconds), the computing time is about 6.2 seconds. The total time is reduced to 55.65%, efficiency has been improved significantly.

The above way can effectively improve the efficiency of the GPU computation, under the premise of without changing the basic algorithm, improve the speedup ratio. Concrete plans need to choose according to actual situation in the practical work. Due to the limitation of paper length, we has only carried on the quantitative test of large memory way. Other ways will be discussed further after practical application.

4 CONCLUSION

Fingerprint feature points matching algorithm is suitable for application of GPU technology. through the technological transformation of the original algorithm, we can effectively improve the speedup program operation. But the data I/O, especially the hard disk data read and other auxiliary computing functions also have some problems which need to be improved. These problems need further research.

ACKNOWLEDGEMENT

Fortunately, This research funded by Beijing Municipal Science and Technology Project (No: Z121100000312099). Here and now, I would like to extend my sincere thanks to all those who have helped me make this thesis possible and better.

REFERENCES

Feng Cai-Gang. The National Eleventh-Five Year Research Program of China the core technology research of ten millions automatic fingerprint identification system and heterogeneous platform [R]. Beijing: Beijing Criminal Science Institution, 2010.

Gress A, Guthe M, Klein R. GPU-based collision detection for deformable parameterized surfaces [J]. Computer Graphics Forum, 2006, 25(3): 497–506.

Jason Sanders, Edward Kandrot. CUDA by Example: An Introduction to General-Purpose GPU Programming [M]. Beijing: Tsinghua university press, 2010.

John D Owens, David L, Naga G, et al. A survey of general-purpose computation on graphics hardware [J]. Computer Graphics Forum, 2007, 26(1): 80–113.

Liu Bao-Quan, Liu Xue-Hui, Wu En-Hua. Real-Time Rendering Depth Images on GPU by Forward Warping [J]. Journal of software, 2007, 18(6): 1531–1542.

NVIDIA Corporation. NVIDIA CUDA Programming Guide Version 2.3.1 [M/OL]. 2009. [2010-05-13], http://www.nvidia.cn/object/cuda_get_cn.html.

Wang Jing, Wang Lili, Li Shuai. Pre-Computed Radiance Transport All-Frequency Shadows Algorithm on GPU [J]. Journal of Computer Research and Development, 2006, 43(9): 1505–1510.

Wu En-Hua. State of the Art and Future Challenge on General Purpose Computation by Graphics Processing Unit [J]. Journal of software, 2004, 15(10): 1493–1504.

Modeling and Computation in Engineering III – Zhang & Xie (Eds)
© 2014 Taylor & Francis Group, London, ISBN 978-1-138-02680-3

Water quality assessment and control countermeasures of rural non-point source pollution: A case study on Xichong River

Li Zhou
State Key Laboratory of Hydraulic and Mountain River Engineering, College of Water Resources and Hydropower, Sichuan University, Chengdu, China

Tianjiao Han
School of Sciences, China University of Mining and Technology, Beijing, China

Lingling Wu
Chengdu Bureau of Hydrology and Water Resources Survey of Sichuan Province, Chengdu, China

Min Ke, Hongbo Zhang, Xiaodong Li & Tianqi Ao
State Key Laboratory of Hydraulic and Mountain River Engineering, College of Water Resources and Hydropower, Sichuan University, Chengdu, China

ABSTRACT: The arithmetic average water quality index method and Nemerow index method of quality evaluation index methods were applied to the water quality assessment in two sections which were Yanjiaxiang and Xiyangsi Power Station. Then it put forward some appropriate recommendations for the control of non-point source pollution. At last, it come to four conclusions: (1) The water pollution in Xichong River (Xichong's territory) is very serious and the data show that the water quality is substandard almost all the year round. (2) Nemerow index method has a better reflection of water quality than arithmetic average water quality index method. (3) The situation of water pollution in Hongxi River is more serious than Longtan River. (4) Non-point source pollution is the main reason of water pollution and wet months have an obvious effect on diluting pollutants.

1 INTRODUCTION

Water quality evaluation is a method, which is according to the evaluation target, choosing the corresponding water quality parameters, water quality standards and evaluation methods, to evaluate the use of water quality and the water processing requirement. Since 1960s, the concept of water quality assessment has been derived by Horton, et al, both at home and aboard has carried out a series of important study (Wei Wang, 2012). Zuxin Xu (Zuxin Xu, 2005) put forward a new method called water quality evaluation index method. The feature is both qualitative evaluation and quantitative evaluation. Weijun Lu and Tao Zhang (Weijun Lu, 2009) claimed that the combination of qualitative and quantitative evaluation method is more useful and practical. Pollution index method is based on an evaluation method of water environmental function areas. The basic idea is to get a comprehensive index to synthetically evaluate the water quality, towards the measured data, which compares the corresponding water quality of water environment function category. Generally, non-point source pollution refers to the runoff pollution after a rainfall (especially heavy rain) or snowmelt carrying through erosion and natural process, and eventually depoits into the rivers, lakes, wetlands, coastal and groundwater (Zaohong Zhou, 2010). Statistically (Fanghua Hao, 2006), among the pollutants in the rivers and lakes, more than 50% derives from non-point source pollution, where countryside accounts for 1/3. Therefore, non-point source pollution, especially in the countryside, is one of the main reasons of water pollution.

2 STUDY AREA

Xichong County is located in the north of the Sichuan basin, whose latitude and longitude coordinates are 104°4′E~105°36′E and 30°52′N~31°15′N. The altitude is from 282 m to 569 m and the high terrain of the northwest, southeast low, lying northwest to southeast tilt. The mountain range runs north and south, and the north is on the high side. Xichong River is the largest one in Xichong County, flowing through Jincheng town, Taiping down and Yixing down etc, including Hongxi water system and Longtan water system. The Hongxi River is 42.5 km long, the drainage area is 289.87 km^2. And Longtan River is 39.8 km long, the drainage area is 189.73 km^2. The main stream runs along the Jialing district and Shunqing district into the Jialing River. The total population of the Xichong County is 659 thousands in 2012, where agricultural population is 524 thousands and non-agricultural population is 135 thousands. Xichong River (Xichong's territory) basin involves the population of 30 thousands, accounting for almost half the county. Xichong River has been a pollutant river since 1980s (Nianfu Yin, 2012). Recently, the ecological environment is getting worse, which has a bad effect on people's lives. So in order to improve people's living environment, it's significant to give Xichong River (Xichong's territory) water quality assessment and analysis of water quality change rule and water pollution causes, besides give the corresponding water pollution prevention and control suggestions.

3 METHODOLOGY

Arithmetic average water quality index is derived by health institutes of University of Ferrara, which will make pollution index added to obtain the arithmetic mean (Yali Ma, 2012).

$$\bar{F} = \frac{1}{n}\sum_{i=1}^{n}\frac{C_i}{C_{oi}} \tag{1}$$

where, $\bar{F}$—arithmetic average index; C_i—measured value of sampling point of evaluation index i, mg/L; C_{oi}—standard values of evaluation index i, mg/L; n—the number of evaluation index. On the basis of arithmetic average index, Nemerow considered the most serious polluted factor and make the mathematical process concise and physical concept clear (Bo Wang, 2005). Although the method highlights the biggest factor so much that ignores the weight of other factors, it is still used widely.

$$F = \sqrt{\frac{F_{max}^2 + \bar{F}^2}{2}} \tag{2}$$

where, F_{max}—maximum of the single set value.

Two 2011 monitoring data of water quality in Yanjiaxiang and Xiyangsi are selected, including COD, NH_3-N, BOD, TN and $KMNO_4$ index, to evaluate the water quality. And Table 1 shows the data of monitoring section, X represents Xiyangsi Power Station monitoring, Y represents Yanjiaxiang monitoring.

Standard GB 3838-2002 addresses the: Surface water quality assessment should be based on water area functional category to select the corresponding standard (GB3838-2002). Xichong River is required to reach the III water quality standard. And Table 2 shows the quality standards of surface water.

Yanjiaxiang monitoring section (30°53′07″N, 105°47′10″E) is located in Longtan River system in Xichong, which has a strong representation of the water assessment. Xiyangsi Power Station monitoring section (30°53′23″N, 105°52′06″E) is lying in the confluence of the Longtan River and Hongxi River. And this monitoring section represents water conditions of the Hongxi River and the entire Xichong River system strongly.

Table 1. The data of monitoring section.

Index	COD		NH_3-N		BOD		TN		$KMNO_4$	
Month	Y	X	Y	X	Y	X	Y	X	Y	X
1	18.1	22.2	1.7	0.4	2.0	4.2	1.4	1.6	2.3	4.5
2	18.3	22.3	0.7	1.2	2.0	4.0	1.2	1.3	2.2	4.3
3	19.1	26.6	1.1	5.9	2.9	4.9	1.5	7.6	3.4	5.4
4	50.2	53.8	0.9	5.7	14.5	15.6	1.3	7.5	15.7	17.0
5	37.2	41.1	2.6	1.1	8.2	9.8	3.4	1.9	15.0	16.0
6	35.5	38.7	1.9	1.1	4.4	6.9	3.9	2.3	14.4	15.4
7	36.0	39.2	0.8	1.0	9.6	10.0	2.4	4.1	14.9	15.8
8	19.4	26.4	0.9	1.3	2.6	3.5	1.5	2.1	5.4	6.0
9	15.6	16.7	0.3	0.6	3.2	4.7	1.5	2.6	4.5	4.8
10	16.9	21.3	0.4	0.8	3.1	4.8	1.0	1.2	3.6	5.3
11	16.4	21.1	0.4	0.8	2.8	4.0	1.1	1.3	3.3	5.1
12	16.9	19.3	0.6	0.8	1.7	1.9	1.1	1.4	3.7	4.9

Table 2. The quality standards of surface water (GB 3838-2002).

Number	Index	I	II	III	IV	V
1	COD≤	15	15	20	30	40
2	NH_3-N≤	0.15	0.5	1	1.5	2
3	BOD_5≤	3	3	4	6	10
4	TN≤	0.2	0.5	1	1.5	2
5	$KMNO_4$≤	2	4	6	10	15

4 RESULTS

The arithmetic average water quality index method and Nemerow index method of quality evaluation index methods are applied to to the water quality assessment in two sections which are Yanjiaxiang and Xiyangsi. The results are shown in Table 3.

Due to the Nemerow index method's consideration of the most serious polluted factor, Nemerow index is larger than arithmetic average water quality index which are shown in Table 3 and lead to more persuasion. Two methods have the same trend, but arithmetic average index is generally low and closer to III. This inconformity states Nemerow index with the most serious polluted factor reflects the water pollution better. The whole year of Xichong River's water quality condition is unoptimistic. All the indexes exceed except October, November and December of Yanjiaxiang. Form the year, Nemerow index surges into extreme point from February to April and descends from June to July, which is different from the results of literature (Xuemin Gao, 2000) that the water quality of the dry season is better than the wet season's.

Xiyangsi Power Station monitoring section is located in the confluence of the Longtan River and Hongxi River, so the data not evaluate the Hongxi River's water quality but also the whole Xichong River's. As the Table 3 shows, except May and June, Xiyangsi Power Station monitoring section is worse than Yanjiaxiang's, which means the pollution of Hongxi River system is more serious than Longtan River's. According to the data of livestock breeding and population distribution supplied by Xichong county, the result, that the most serious non-point source pollution towns, Jincheng and Changlin in Hongxi River system, is coincide with the water quality monitoring.

Table 3. The water quality index of Yanjiaxiang and Xiyangsi section.

	Yanjiaxiang		Xiyangsi	
Monitoring month	Average index	Nemerow index	Average index	Nemerow index
1	0.97	1.39	0.96	1.29
2	0.73	1.00	1.08	1.21
3	0.97	1.23	3.39	5.86
4	2.20	3.00	4.53	6.20
5	2.49	3.00	2.03	2.37
6	2.21	3.17	1.93	2.27
7	1.96	2.24	2.44	3.38
8	0.99	1.26	1.32	1.76
9	0.82	1.20	1.21	2.05
10	0.74	0.90	1.03	1.12
11	0.70	0.89	1.00	1.13
12	0.73	0.95	0.90	1.19

5 DISCUSSION

5.1 *Non-point source pollution's influence on the water quality*

Xichong River water quality caused by severe pollution of non-point source factor mainly has the following three aspects (1) large Population density. Human activities are the dominant factors of agricultural non-point source pollution in rural area, mainly including farmland production, livestock farming and large population density, whose life is bound to cause pollution to the environment. Based on the analysis, Xichong's population density in 2011 is 612 people/km^2, far more than the population of the world average of 33 people/km^2 and Chinese average density value of 130 people/km^2, where the largest value of Jincheng is 2301 people/km^2. (2) Improper fertilization. Studies have shown that improper fertilization is the main source of water pollution and eutrophication. Xichong River's total fertilization reached 8699 tons in 2011, including 4211 tons of nitrogen fertilizer. (3) Livestock and poultry breeding pollution. Xichong agricultural population and livestock breeding have developed rapidly. The lack of drainage pipe with sewage treatment system aggravate the pollution. In 2012, according to the data provided by Xichong Animal Husbandry Bureau, Xichong River has amounted to 316516, where the free-range is 127582. Centralized livestock farming means having a certain scale farming, and equipped with facilities for sewage treatment plants and little contribution to water pollution. The free-range is different. Livestock and poultry dung wanton emissions, directly into the river causing the water pollution seriously. The range of livestock and poultry breeding in Xichong River accounts for as much as 40.3% of the total livestock farming, which is one of the main reasons of Xichong River pollution.

5.2 *Effect of dry season, normal season and wet season on water pollution*

Xichong River, in a humid subtropical monsoon climate zone, has mild climate, abundant rain, appropriate light, and is clear four seasons. Annual average temperature is 17.4°C and annual average rainfall is 1045.8 mm, especially focused on July to September, accounting for 52.1% of the annual precipitation. The runoff is given priority to with water recharge, perennial peak flow velocity of 3.12 m/s, flood peak flow of 2500 m^3/s. Years of low water flow velocity is 0.42 m/s and low water flow rate is 1.89 m^3/s. Nemerow index changes as Table 3 shows, mainly due to the month before flood season approaching, gradually increased rainfall and increased surface runoff to cause heavy surface runoff non-point source pollution. Before flood season, because of the big river runoff and obvious dilution effect of pollutants, the Nemerow index fells. After flood season, the rainy season continues, but the river runoff was much smaller and agricultural non-point source pollution appeared. Point source pollution plays a main part in dry season.

For there is only one year of water quality monitoring data, the interannual analysis the water quality changes also does not reach the designated position, but the results also verify the effect of the dry, normal and wet season on water pollution.

6 RECOMMENDATIONS FOR CONTROL COUNTERMEASURES OF RURAL NON-POINT SOURCE POLLUTION

6.1 *Giving overall consideration and innovating work process*

Since rural non-point source pollution related to the peasants' interests and economical sustainable development closely. So there're 3 relationships to be dealt with: (1) the relationship between rural economy development and environmental protection. (2) the relationship between urban environmental protection and the rural. (3) the relationship between and prevention and management.

6.2 *Overall promotion and highlighting the key points*

Rural environmental protection is a wide range of project, requiring overall promotion and the distribution. In view of Xichong's actual situation and towards the centralized large-scale modern agricultural farming road, Paying special attention to Livestock breeding pollution prevention and specifying range of livestock breeding is the key point.

6.3 *Adaptation to local condition and classification guidance*

Combining the reality and adhering to the people-centered, according to the natural ecological environment conditions and the level of economic and social development, Xichong should make corresponding rural environment protection countermeasures and measures, emphasizing the pertinence and effectiveness. Xichong has poor surface water and small water storage, belonging to the river of rectum type. Xichong, affected by both natural factors and human factors, caused a large amount of soil erosion in the basin, which leads to large water loss and soil erosion. At last, the erosion destroyed the benign ecological system, forming the vicious ecological environment.

6.4 *Combination of engineering measures and management measures*

On one hand, government needs to take corresponding engineering measures for pollution sources, to ensure the discharging standard and cut pollution load requirements, on the other hand, government needs to pay attention to improve the related environmental management measures, including village regulations, reducing the pollutants from the source, meanwhile it ensure that pollution control facilities for long-term operation. Since 2013, Xichong has newly built 7 sewage treatment stations, greatly improving the state of water pollution.

6.5 *Government leading and social participation*

Give full play to the government leading role and implement the responsibility of the government to protect the rural environment. And strengthen farmers' environmental education, establish the public participation mechanism, encourage and guide farmers and social forces to participate in and support rural environmental protection. For Xichong as a large agricultural county, it should be more solid popularize environmental protection education.

7 CONCLUSIONS

1. Water quality pollution of Xichong River is very serious. According The Nemerow index calculation of Xiyangsi station monitoring section water quality throughout the year isn't up to standard, and Yanjiaxiang's is also not optimistic.
2. The arithmetic average water quality index method and Nemerow index method are both reflect the water quality pollution objectively. But the Nemerow index with the most serious pollution factor gets a better result.
3. Hongxi River's pollution is worse than Longtan River's, so it's urgent to manage the serious pollution of villages and towns.
4. Xichong, as a large agricultural county, whose non-point source pollution is very large. Water quality index of dry season, normal season and wet season showed a regular change: rainfall gradually increased in February, and accumulated with the runoff of non-point source pollutants runs into the river in dry season causing serious pollution. In June and July, after the flood season, rainfall concentrated, significant increase in water, whose dilution effect of pollution is more than flush pollution of non-point source pollutants into the river. So water quality pollution index descends. Since there is only one year data, Water pollution change of dry season and wet season still needs further research.

ACKNOWLEDGEMENTS

The research work was supported by International S&T Cooperation Projects, the Ministry of Science and Technology of P. R. China under Grant No. 2012DFG21780, S & T Project of Sichuan Environmental Protection under Grant No. 11HBT-01 and Start Fund for Young Teachers's Scientific Research of Sichuan University under Grant No. 2012SCU11065.

REFERENCES

Bo Wang, He Han. 2005. The Application and Defect of Nemerow Method in Water Quality Evaluation. *Chinese Journal of Urban and Rural Industrial Hygiene*, (6), pp.16–17.

Fanghua Hao, Hongguang Cheng. 2006. *Non-point Source Pollution Models—Theory and Application.* Beiijing: China Environmental Science Press.

GB3838-2002, The National Standard of the People's Republic of China-Surface Water Environment Quality Standards.

Nianfu Yin, Tiesong Li, etc. 2012. Study of Water Environment Capacity of Xichong River in Sichuan Province Based on Blind Number Theory. *Bulletin of Soil and Water Conservation*, (1), pp.233–237.

Wei Wang, Mei Ji & Yanan Su. 2012. Reviews of the Progress in the Research of Water Quality Evaluation and the Methods for Water Quality Evaluation. *Sci-Tech Information Development & Economy*, 2(13), pp.129–131.

Weijun Lu, Tao Zhang. 2009. Comparison and Analysis of Several Appraisal Methods for River Water Quality. *Environmental Science and Management*, (6), pp.174–176.

Xuemin Gao. 2000. Water Quality Assessment and Modeling of Rivers and Lakes along the Yangtze River. Peking University.

Yali Ma, Tianqi Ao. 2012. Water Quality Assessment and Treatment Advices for Luxian Section of Laixi River. *Sichuan Environment*, (5), pp.49–53.

Zaohong Zhou. 2010. Agricultural Non-point Source Pollution Empirical Analysis and Policy Choice. Beijing: China Meteorological Press.

Zuxin Xu. 2005. Comprehensive Water Quality Identification Index for Environmental Quality Assessment of Surface Water. *Journal of Tongji University (Natural Science)*, (4), pp.482–488.

Modeling and Computation in Engineering III – Zhang & Xie (Eds)
 ISBN 978-1-138-02680-3

Water environment management at home and abroad research situation and enlightenment

Xiuju Zhang, Wenrong Zhao, Ning Qiu, Julius Lissa Tommy & Huan An
College of Hydrology and Water Resources, Hohai University, Nanjing, Jiangsu, Jiangsu, P.R. China

ABSTRACT: Water environment problem is one of the most important factors that restrict the development of national economy. With the growth of population and the improvement of urbanization, it is even more urgent to solve the water pollution and improve water environment. In recent decades, a lot of research has been done in order to control the water environment pollution and many governance measures have been carried out for some major lakes and urban rivers. The paper summarizes the successful experience of water environment management at abroad, such as the representative management cases analysis of the Rhine River, the Thames, the Biwa Lake. The paper analyzes the successful and not so successful cases in some main lakes in China and concludes their water environment governance measures, the present situation and existing problems. The paper provides reference for the future reasonable water resource utilization and protection in China.

1 INTRODUCTION

Water resource is a basic natural resource and strategic economic resource. With rapid development of the social economy, the per capita possession of water resources is shrinking, at the same time; the problem of water resources shortage and water quality deterioration has become increasing prominent. The problem of water environment seriously affects and restricts national economic development, water environment protection and management has risen to the height of national strategic development, at the same time, corresponding theory and technical support is urgently needed. Since the 1950s, many countries in the world have carried on the comprehensive treatment and development to major rivers and lakes. In the early 70s, developed industrial countries in Western Europe and North America became conscious of the water environment pollution and its harm mechanism research, made it the main research direction of their studies, a series of research about water pollution monitoring methods, evaluation methods and the laws of transformation began to be carried out (Vander Weijden CH. et al, 1989). In the late 70s to the early 80s, researchers started their research field on the pollution control and the behavior characteristics of organic matter in water (Perret D et al, 1994). In recent decades, a lot of research has been done by some countries, and government has spent huge sums to carry a large range management on some major lakes and urban river. Japan, the United States, Germany, Switzerland and other developed countries have started to reflect on their previous ideas of water environment management in recent years. They have put forward the new concept of ecological water conservancy, paying attention to the natural ecology and the recovery and protection of natural environment, making the comprehensive service functions of rivers and lakes show better. From the view of the domestic research situation, the domestic research on water management has developed fast in recent years. A large-scale prevention and control of water pollution in key river valleys of China was in full swing, such as "three rivers and three lakes" which means Huai River, Hai River, Liao River, Taihu Lake, Chao Lake and Dianchi Lake, etc. ((SEPA), 2000). After several years of effort, some phased results have been achieved and the water quality of Parts of the Rivers has improved. However, due to historical reasons, water environment problem is more complex in China and the start time of water environment management in China is relatively late and is subject to the impact of some aspects, such as economic development and the difference in local management system, management efficiency, the division of labor is unknown and so on. The results obtained are not particularly significant. Therefore, under the condition of existing economy and technology, solving the problem of water environment also needs to go through a slow process. The paper summarizes the successful experience of water environment management in some foreign countries and analyzes the current water environment governance measures, the present situation and existing problems in some main lakes of our country. The paper sums up the successful cases on the comprehensive improvement of water environment, such as Biwa Lake, Rhine River, and Thames River, and on that basis puts forward some considerations for water environment comprehensive management strategy and provides reference for the future water environment protection and remediation in China.

2 FOREIGN WATER ENVIRONMENT MANAGEMENT

2.1 *The European water environment management*

The research of European water environment management has changed from traditional single basin water management to sustainable and comprehensive management which aims to keep the quality of life and sustainable development.

2.1.1 *The management of Rhine River*

1. The previous pollution of Rhine River

The Rhine River originated in the Alps of Switzerland, flowing through France, Germany, the Netherlands, and finally runs into the North Sea (Grijp Nvander et al, 2003). The Rhine River passing through the economy developed areas of Western Europe, and its downstream flowing through the center of Ruhr. Ruhr is the most concentrated area in West Germany with iron and steel, coal, machinery, military and other important industries. A great deal of wastewater and pollutants were dumped into the river. Since the middle of the 20th century, it is estimated that more than 1000 species of harmful substances was discharged into the Rhine River. (Hurkmans R, 2008) Industrial wastewater and domestic sewage was poured into the river on the scale that 5000–6000 m^3/d, until 1965, COD of the river reached 30–130 mg/l, BOD reached 55 mg/l, DO < 1 mg/l (Hongzhi Liu et al, 2001). Therefore, the Rhine once became "the largest sewer of Europe".

2. The management measures of the Rhine River

The management of Rhine River began in the middle of the 19th century and the shipping management institutions were set up. In July 1950, "the international committee on protection of the Rhine" was established in Basel, Switzerland for completely handling the Rhine river watershed protection problems and exploring solutions. Their work involved in the field of Rhine-related groundwater, aquatic and terrestrial ecosystem, the pollution and flood control engineering, etc (Havings. A.J, 2000). In order to strengthen sewage treatment of the Rhine River, the relevant countries along the Rhine River have taken some practical and effective methods to control pollution of the river that is based on international cooperation. The first was to develop a total amount of pollutant emissions according to the population proportion, allocated the total amount to the several countries along the way, strictly enforced water quality monitoring, constructed a number of water quality monitoring stations. Sewage discharged into the river had been treated rigorously and the river oxygenated on a regular basis. In addition, the corresponding laws and regulations have been formulated; industrial wastewater that discharged into the river must have been forced harmless treatment.

3. The governance effect of the Rhine River

Through decades of efforts and huge capital investments, the Rhine River finally returned to its original natural appearance. From the late 1980s, the number of salmons in the Rhine gradually increased in 1995, the fish species of the Rhine River in Germany has reached 45 (Gangyan Zhu, 2007). The Rhine has basically become clear in 2002 and the water discharged into the Rhine has reached the emission standards. The successful management of the Rhine River is closely related to the close cooperation between the relevant countries, and the operation system of the Rhine river commission (ICPR).

2.1.2 *The management of the Thames*

1. The previous pollution of the Thames

Thames River is a "Mother River" in Britain which is originated in the Cotswold Hills of southwest England, flowing through London and ten more cities, finally running into the North Sea. After the British industrial revolution, the population gradually concentrated and a large number of municipal sewage and industrial wastewater directly discharged into the river without treatment. At the same time, a large accumulation of garbage dirt was cumulated along the coast. In addition, the river is also affected by the tides. During the period of rising tide, back flow of sewage was generated suddenly under the tidal force, caused a situation in which sewage and dirty flows to the stinking streets.

2. The management measures of The Thames

In the process of governance and management of the river Thames, Britain tried a bold reform and scientific management that was called "a revolution of water industry management". It merged more than 200 pipe water units of the whole basin and built a new water authority- the water management agency of the river Thames. The administration unified managed various kinds of businesses such as water treatment, aquaculture, irrigation, animal husbandry, shipping, etc. Firstly, a comprehensive development was carried out from the sewage treatment to fish, planting, irrigation, navigation and water ecological environment monitoring etc. that fully mobilized the enthusiasm of each department (Jamieson D, et al, 1996). Secondly, the water quality of the Thames in London was monitored and the sewage treatment was strengthened. At the beginning of the 20th century, hundreds of sewage treatment plants were conducted in London, formed a complete urban sewage treatment system and started sewage chlorination. Thirdly, the agency managed the river through legislation. The industrial wastewater and domestic sewage that directly discharged into the river has been strictly formulated. Fourthly, in the late 20th century, the industry of water supply and management became privatized in Britain. The governance of the River Thames, uses the conventional measures such as, intercepting pollution, biological oxidation, aeration oxygenated, and microbial activated sludge. However, a bold structural reform and scientific management method was conducted on administration (Eum HI et al, 2010).

3. The governance effect of the river Thames

After about 150 years of governance, the content of dissolved oxygen has obviously increased, while the content of ammonia nitrogen, toxic and harmful substance contents and Biochemical Oxygen Demand (BOD) has decreased significantly, in the 1980s, there are more than 100 kinds of fishes and 350 species of invertebrates back thriving there and the river Thames has already become one of the most clean city rivers again. In the aspect of biological communities, the benthic animals and fish communities have obviously changed with the improvement of the water environment of the Thames (Hendry K et al, 2003).

2.2 *The water environment management of Japan*

Research on water environment management and protection mechanism in Japan, gives priority to establish ecological concepts. Advanced technology is just an auxiliary. The most successful example all over the world on lake management is the governance of the Biwa Lake in Japan. Its pollution situation is very similar to the lake pollution in China. Their tips might be worth hearing.

2.2.1 *The previous pollution of the Biwa Lake*

Biwa Lake is the largest freshwater lake in Japan, and an important water resource of people living and industry production in Kyoto, Osaka and Kobe (Xiaojuan Hu, 2007), but with the industrial development and population growth of the Shiga, water quality of the lake had become worsened. In the early 1970s, the water pollution of the Biwa Lake reached its peak. From 1977 to 1992 (except 1986), red tides emerged in the Biwa Lake every year.

2.2.2 *The government measures of the Biwa Lake*

In order to improve the water quality of Biwa Lake, the Japanese government has taken a number of water quality protection measures. The success of its governance can be contributed to the following four aspects: (1) Comprehensive management of the environment in accordance with the law. A series of rules and regulations has been formulated by the government of the Shiga, the emissions of domestic sewage and industrial wastewater around Lake Biwa area were clearly defined. (2) A comprehensive plan for maintaining ecological balance. According to the different characteristics of the different regions, different tactics were developed. (3) To increase the awareness of the participation of all citizens. The Government of Shiga attached importance to the mass participation and education work, encouraged them to participate in the protection of the environment of the Biwa Lake. (4) Paid attention to exchanges with other countries.

2.2.3 *The government effect of the Biwa Lake*

As a result of the application of the above measures, the pollution of Lake Biwa has been effectively controlled, cyanobacterial blooms has disappeared and water quality become better. Water quality is equivalent to the surface water quality of II standard in China, transparency reached above to six meters. The successful experience of the management of the Biwa Lake gives us an important enlightenment that the management of lake pollution must start from the source of lake, rather than form the trail of the river.

3 THE DOMESTIC RESEARCH ON WATER ENVIRONMENT MANAGEMENT

3.1 *The successful cases of Qinhuai River water environment management*

The United Nations human settlement bureau (UN-Habitat) thinks that the experience of water management in Qinhuai River of Nanjing is worth reference (RAO Zhengfu, 2009). Its successful experience is worth us studying and thinking about.

3.1.1 *The general situation of Qinhuai River basin*

Qinhuai River is the mother river of Nanjing. Since reforming and opening in 1978, with the accelerated pace of industrialization and urbanization in Nanjing and rapid urban population growth, the emissions of industrial wastewater and domestic sewage has increased rapidly, sewage is been discharged into the river causing water pollution.

3.1.2 *The water environment management of Qinhuai River*

Since 2002, the government invested large sums of money and adopted a series of comprehensive measures. The first was to manage the downstream of Qinhuai River which flows across the main urban areas of Nanjing. The project legal person responsible system of the Qinhuai River Engineering has been implemented by the government of Nanjing. In July 2003, the Nanjing municipal government authorized the establishment of construction and development Co., Ltd of the Qinhuai River which is a coalition of investment and financing, construction, management and operation brought together. Subsequently, Nanjing skillful used policies that opened the door of market-based financing successfully: financing though the development of 3,000 acres of land that are along the river within 200 m; urban sewage treatment fee of water charges rose to 0.5 Yuan per degree, 75 million per year for the governance of Qinhuai River for 20 years. According to the characteristics of the Qinhuai River, the governance should stand on the height of watershed and pay more attention to the source water (Jianlan Hua et al, 2010), make use of clean water resources that is I~III class water to rinse, dilute and replace sewage which above V class water. Now the sewage of Qinhuai River has been rinsed by getting water from the Shijiu Lake through Tianshengqiao River. For the sewage of polder and ditch, to take water released from the reservoir or before took water from the Shijiu Lake to flush waste, opened the culvert immediately, clean water was imported to restore the water level of dike ditch.

3.1.3 *The management effect of Qinhuai River*

With the sewage interception along the Qinhuai River and a large area of afforestation on both sides of the river, not only constructed a green landscape corridor in the middle of the city and in a certain extent, improved the air quality along the river, but also made the water quality of the river meets the requirements of the landscape water, produced a positive impact on the health of urban populations. The water quality change of Qinhuai River after treatment (see Table 1), the water quality of Qinhuai River after management is better than IV class water and conforms with the water quality requirement, made the natural ecological environment and urban living environment fused together and created a green system that harmonized man, nature and city.

3.2 *Less obvious cases on management effect of domestic water environment*

3.2.1 *The water environment management of the Taihu Lake*

1. The water pollution situation of Taihu Lake

Taihu Lake is the third largest freshwater lake of China, the area of Taihu Lake Basin is 36500 km^2. Since the beginning of the 1980s, the average water quality of Taihu basin changed from the original II class water dominated to III and IV class water dominated, local areas of Taihu lake has been already in V class water or inferior to V class water. Since the 1990s, the phenomenon of cyanobacteria has bloomed frequently (Qin B, 2007). At the end of May 2007, due to the large quantity of the pollutant emissions, the outbreak of cyanobacteria in Taihu Lake and other reasons, resulted in water quality pollution of water source field in Wuxi, seriously affected the normal life of nearly one million local people.

2. The water pollution control measures of Taihu Lake

The water environment management of Taihu Lake mainly used some means, such as adjusted industrial structure, comprehensive management, efforts to prevent water pollution (Guang bo Cui et al, 2004). Concrete measures are: ① to adjust the industrial structure, and transformed the mode of economic development, reduced the pollution sources effectively. ② The use of advanced technology to promote the comprehensive water environment management. The relevant departments and local has implemented water pollution control and the demonstration of water body restoration technology, agricultural non-point source pollution control and aquatic vegetation restoration. ③ The use of economic levers to reduce the amount of sewage discharge. To raise water price, including water resource fee, sewage

Table 1. The water quality changing conditions of Qinhuai River after treatment.

Water body	The changing situation of COD (unit: mg/L)				
Inner-Qinhuai River	1999 53.32	2001 36.59	2004 27.18	2008 20.46	2012 17.467
	The changing situation of main pollution evaluate index				
	Year	COD/mg · L^{-1}	TN/mg · L^{-1}	TP/mg · L^{-1}	
External-Qinhuai River	1999 2012	50.61 26.131	0.85 0.305	0.179 0.153	

treatment fee and sewage charges, which can play an important role in water conservation. ④ Worked together to prevent water pollution and improve the ability of regional coordination. Regional water pollution control was included in the conference content at the joint session of mayors of the Yangtze River Delta, the establishment of consultative mechanism have achieved certain results.

3. The present situation on water environment management of Taihu Lake
First of all, after a period of management, the pollution sources have been controlled to some extent. While rapid economic development, some pollutant indices were stable or improved in local water of Taihu Lake (Zhu Xi, 2011). To get the water quality of different monitoring points through analyzing the water quality indices in 33 monitoring points of Taihu lake from 2007 to 2009 (see Fig. 1), from Figure 1 we can see, the water quality in 33 conventional monitoring points all exceed the standard of III class water. The worst water quality appeared in non-flood season of 2007 and gradually improved after 2007; the percentage of monitoring points of IV class water accounts for the total monitoring points in flood season of 2008, 2009 respectively are 18.2%, 21.2%, increased to18.2%, 30.3% in non-flood season separately. The quantity of V class and inferior V class monitoring points basically decreased step by step. This shows a series of management measures was adopted after water crisis in 2007 that made the water quality of Taihu Lake to be improved in 2008 and 2009. Secondly, the control of industrial point source pollution has achieved significant progress. Thirdly, the management of urban domestic pollution sources has achieved certain results. There is an improving trend on the water environment of Taihu Lake. To protect the safety of drinking water, the construction of portable water sources and water supply pipe network has been organized, and a urban drinking water security system has been basically formed of the Taihu Lake, which has effectively improved the water supply security capacities of the basin.

3.2.2 *The water environment management of Dianchi watershed*

1. Water environment pollution situation and main causes
Dianchi Lake is located in the downstream of the main urban districts of Kunming, and it is the lowest point of the watershed. The main rivers of the basin running through the densely populated towns, accepted the sewage of industrial, agricultural production and urban living along the way, and carried large amounts of pollutants flows into the Dianchi Lake. For a long time, most of the pollutants have remained in the lake, the water quality of the environment deteriorated sharply. In addition, the replacement cycle of Dianchi Lake is long and stream flow is slow which has evolved into a semi-enclosed lake (Jianqiang Yang, 2000). Due to the long residence time of the water body, the slowly flowing velocity and significant reuse, led to the imbalance of substances in and out of the lake which accelerated water pollution and the accumulation of pollutants in the lake. Furthermore, the pipe network construction was lagging behind and the main and sub-main sewers didn't match, the sewage treatment capacity of sewage treatment plants and processing load rate of TN, TP, and COD usually can not meet the design capacity. From Figures 2 and 3, it can be seen since 1960, there was

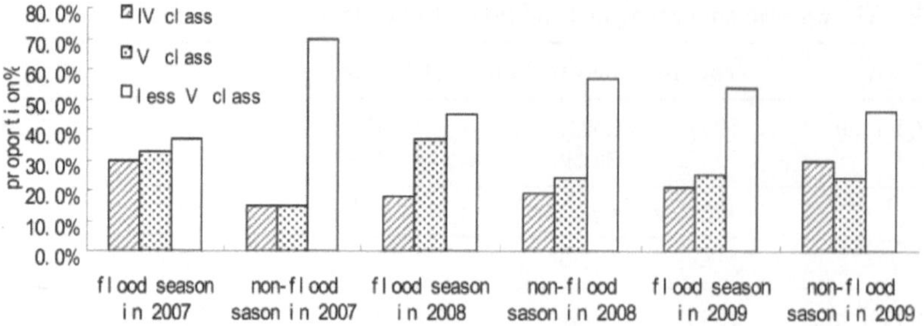

Figure 1. Water quality in 33 monitoring points of Taihu Lake from 2007 to 2009.

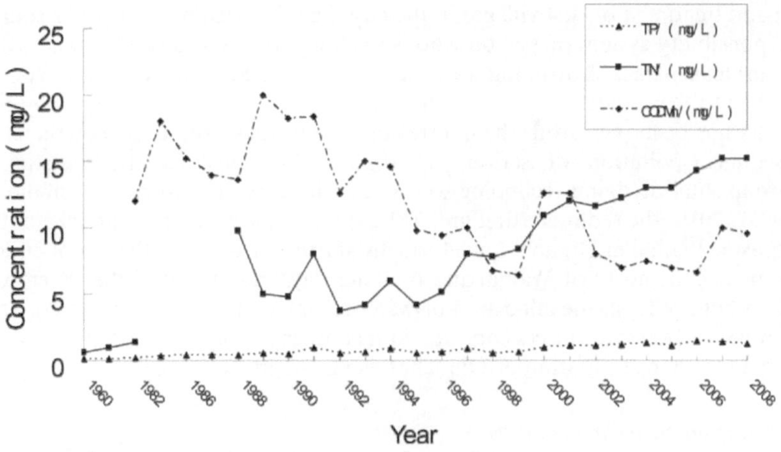

Figure 2. TN, TP, COD_{Mn} change in Caohai of Dianchi Lake.

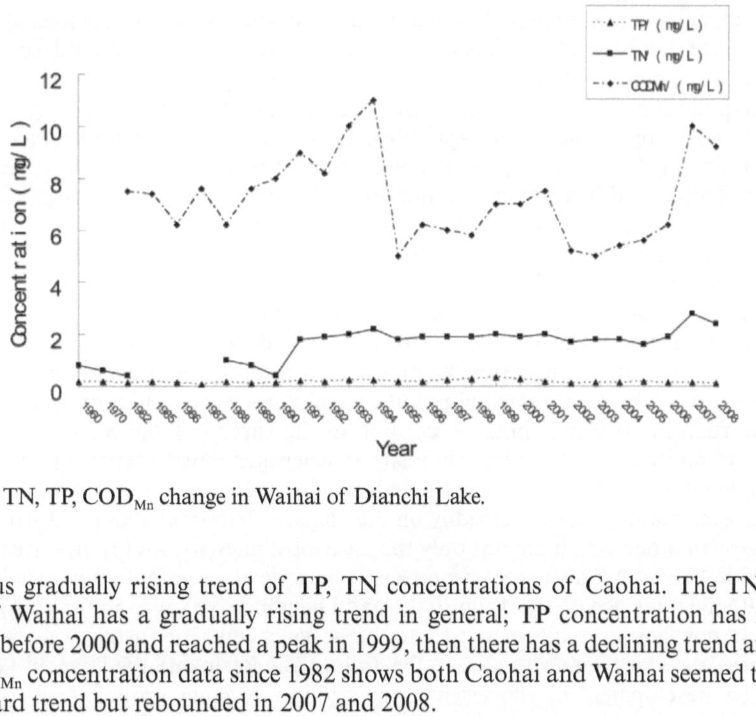

Figure 3. TN, TP, COD_{Mn} change in Waihai of Dianchi Lake.

an obvious gradually rising trend of TP, TN concentrations of Caohai. The TN concentration of Waihai has a gradually rising trend in general; TP concentration has remained increased before 2000 and reached a peak in 1999, then there has a declining trend after 2000. The COD_{Mn} concentration data since 1982 shows both Caohai and Waihai seemed to present a downward trend but rebounded in 2007 and 2008.

2. The management measures and current situation of Dianchi Lake

Yunnan Province and Kunming have taken a series of measures to comprehensively promote water pollution control of the Dianchi Lake (Yi Ding et al, 2011). Including, strengthening the supervision and management of polluting enterprises, proceeded standard assessment of emissions in each enterprise, rectified outfalls and the implementation of sewage discharge permit system, and gradually promoted and standardized the clean production to reduce the pollution load of Dianchi Lake to a great extent. Afforestation, the control of water and soil erosion, ecological restoration; strengthen rubbish management and harmless disposal to prevent environmental and water pollution. Sewage interception around the lake to further strengthens the prevention and control of urban life pollution sources. In 2011, the sewage interception project of Dianchi Lake was fully completed and accomplished the sewage collection and treatment of 364 villages in the city. The establishment of water sources protection responsibility system of person who is in charge of reservoirs and ponds, the water quality of urban centralized drinking water sources such as Songhua dam and Yunlong reservoir reached to the standard of 100%. The use of environmental dredging technology, the initial results has been seen from the control of endogenous pollution. For the characteristics of sediment pollution was serious in Caohai of Dianchi Lake, the first-stage project used environmental dredging technology, cleared sediments of Caohai 4.24 million m^3. In December 31, 2010, the sediment dredging (Phase III) project of the main lake outlets and the key areas of Waihai in Dianchi Lake formally started, cleared off the major endogenous pollution such as the north of Waihai and Baofeng Bay, and improved the water quality of Waihai. Last but not least, the allocation of water resources and the utilization of reclaimed water improved the water resources contradiction between supply and demand. Table 2 shows the water quality change situation of Dianchi Lake Basin in recent years.

3.2.3 *Existing problems and analysis*

Although the management work of Taihu Lake and Dianchi Lake has been carried out for many years, the achievement of them is far less obvious than the management effect of Qinhuai River in Nanjing. After investigation and studies, they still have several problems still need to be urgently solved.

First of all, non-point source pollution and land use patterns are closely related. With the increasing usage of chemical fertilizers and pesticides, as well as the rapid development of rural aquaculture, non-point source pollutants have become a major factor of nitrogen and phosphorus pollution of water body in China. According to the survey data of Xinhua.net in 2009 shows that the proportion of non-point source pollution from agriculture has accounted for 30% to 60% in the various types of environmental pollution, which has exceeded the industrial pollution and has become an important feature of China's water pollution. The eutrophication of lakes and reservoirs has not been fundamentally resolved due to rural non-point source pollution always lacking of effective control, which is also a major cause of bad governance in Taihu, Dianchi Lake and so on.

Secondly, the industrial structure is not reasonable. From the result of the development in recent years, the pattern of economic growth mainly depends on the secondary industry changes a little that further increases the pressure on resources and environment. The economic losses caused by environmental pollution including water pollution were more than 500 billion Yuan in 2010 of China. According to the survey of the State Environmental Protection Administration, the current industrial wastewater emissions per year of China are more than 300 million tons, especially the industrial wastewater emissions which were carried by the seven major rivers are steadily on the increase. Irrational industrial structure and extensive growth mode which are not only the basic obstinacy without radical treatment for a long time in China, but also the reason caused by serious water pollution. The industrial structure at an early stage of the Taihu Lake and Dianchi Lake was dominated by the light industry and now has gradually evolved into the heavy chemical industry as the main body. Overall, there is a widespread situation in the Taihu Lake Basin and the Dianchi Lake Basin that the slow development of the service industry and a higher proportion of secondary industry, the intensive heavy industry seriously restrict the intensity of the water environment governance.

Thirdly, the management system is not perfect. The remediation work of Qinhuai River is solely responsible for by the People's Government of Nanjing. However, the phenomenon of

Table 2. The water quality change situation of Dianchi watershed.

Watershed	2002	2003	2004	2005	2006	2008	2009	2010	2011	2012
Caohai	Less V	Less V	Less V	Less V	Less V	Less V	V	Less V	V	Less V
Waihai	V	Less V	V	V	Less V	Less V	V	V	V	V

multi-management is generally in Taihu Lake and Dianchi Lake remains to be improved. The current management of Taihu Lake mainly including the management of the water administration departments on dikes, hydraulic structures, water intake, outfall etc. and the governance of the fisheries management department on an aquaculture license management etc., with the economic development and the increasing development levels of the exploitation and utilization of Taihu Lake. The various departments fulfilled their respective management responsibilities in accordance with the relevant laws and regulations, existed a phenomenon of overlapping management and multiple management, so in practice, there will be an unfavorable situation that sometimes nobody supervises and sometimes many people manage. In the process of management, the community often concerns about the management of watershed exploration function, but ignores the management and protection of the watershed welfare function. The management system has not yet formed an effectively integrated management department, so it needs to be improved and strengthened (Guihua Lu et al, 2009).

Fourthly, the environmental infrastructural construction of Dianchi Lake and Taihu basin is relatively lagging behind and the pollution of urban life has become the primary factor. Urban sewage collection network system makes the rain and sewage interflow, the ability of sewage treatment facilities is insufficient, part of the sewage directly into the sewage treatment plant through the river channel. In addition, although the lake areas of Dianchi Lake and Taihu Lake has been monitored by the relevant departments, monitoring indicators are only limited to water quality, monitoring sites are less, and some ecological indicators such as plankton, aquatic plants, aquatic animals etc. are not incorporated into the monitoring system, and it is also lack of effective means of positive measures when the sewage of the upstream centralized discharged into the lake or the forecast monitoring before the outbreak of blue-green algae. It is still far from enough only relying on artificial means to supervise.

Finally, the regulation of government is insufficient. In recent years, the treatment effect of water environment in China is not obvious due to government regulation is lax to a certain extent. The phenomenon of underlying discharges and under drainage exists in some enterprises. There is no monitoring. The main pollution sources that cause pollution are the illegal discharge of industrial enterprises along the river. Although some of these enterprises have installed sewage treatment equipment, but due to various reasons, very little sewage treatment equipment could normal operation everyday. Nevertheless, some local government departments still turn a blind eye to this because although these enterprises are large polluters, but it is also the main force of local economic development and the taxpayers, it is essential for local government. So, because of the huge capital investment of the state, on the other hand, the local government turns a blind eye to local pollution enterprises. More local governments chose the former between the local economic development and environmental protection. Therefore, there will be also an ending of more management to more pollution.

4 CONCLUSION

Through the analysis of overseas typical river management and regulation experience, we can find that water environmental pollution can be governed, but this kind of management is often time-consuming and much costly. China realizes the rapid economic development under the condition of relative lagging behind in economy and technology. Large population base, less per capita resources, large amounts of pollution emissions and different pollution conditions in different places leading to the prevention work of environmental pollution and ecological damage will be a long-term strategic task. Therefore, we must on that basis of conscientiously summing up the lessons on prevention and control of water pollution, drawing on the successful experience of the world, combined with the specific circumstances of our own country, continuing to strengthen policy innovation, institutional innovation and technological innovation, and gradually set out on a path of water pollution prevention with Chinese characteristics. In order to ensure the sustainable development of the water environment in each region, facing the present situation of water environment and existing problem in China, we must increase the intensity of the water environmental governance, strengthen the environment protection functions of the government, and strengthen environmental supervision and management. Specifics can start from the following several aspects.

1. In view of problems that heavy industry enterprises are intensive in some areas, pollutant emissions are large and so on, in the industry development planning, we should fully consider the carrying capacity of the water environment, actively adjust the industrial structure step by step, promote the clean production vigorously, and change the water body into virtuous circle. We should be determined to shut, stop, merge and turn the enterprises which are high water consumption, heavy pollution and pollution treatment costly much, develop the circular economy vigorously, improve the comprehensive utilization rate of water resources, and reduce pollutant emissions, accelerate the special rectification of the heavily polluting industrial enterprises. To strictly control the agricultural non-point source pollution, promote the use of organic fertilizers and bio-pesticides; reduce the use density of chemical fertilizers, pesticides, and mitigate the water environmental impact that caused by the loss of insecticide and fertilizer. Make rational layout to grow plant varieties in different regions, reduce and control pollution from the source water through scientific prevention and the development of ecological agriculture.
2. To strengthen the government supervision and improve the water resources management system. The phenomenon of underlying discharge and under drainage exists in some enterprises, and there is no monitoring. Therefore, the implementation of more rational water resources management system, strengthen supervision and functional distribution of the government on the water environment governance, and actively promote water reuse technology and the process of urban sewage treatment and wastewater reuse, and strengthen the efforts to water pollution management in key watersheds to prevent the deterioration of water quality. Join forces of different departments to control pollution. Water environment management will be done poorly by a solely department, only the water conservancy department, environmental protection department, hydrological departments and other departments united in together, unceasingly strengthen the communication and sharing among different departments which can be more conducive to carry out the work of water environment governance.
3. To speed up the engineering construction of centralized sewage treatment facilities. The urban centralized sewage treatment is an effective way to improve the urban water environment. We can also through the river network to intercept pollution, dredging, increase the carrying capacity and self-purification capacity of the water body, improve the water quality of river channel, restore the ecological functions and remediation on both sides of river, dredging the river, eliminate internal pollution; dredging the river to greatly reduce the content of contaminants and nutrients in the sediment, mitigate water pollution caused by the river channel sediment and improve the water quality condition.
4. Vigorously to carry out water-saving activities and take effective measures to reduce water consumption. Water-saving and efficient agricultural irrigation technology should be promoted organized. Vigorously develop the use of water closed-circuit circulation and minimize the amount of wastewater emissions. Pay close attention to the development of environmental economic policies which is favorable to the environment protection; further strengthen environment economic means under the system of market economy. As soon as possible to improve the standard of sewage charge, make it higher than the cost of pollution control; to develop the relevant policies on water pollution prevention and establish the compensation mechanism of resources updates and fully realize the principle of "polluter pays", increase the sewage treatment charge in water charge generally to be used for the operation cost of the city sewage treatment plants; the preferential treatment of tax reduction or exemption should be given to the production of renewable resources and products that utilize renewably.

REFERENCES

Eum HI, Sredojevic D, Simonovic SP. 2010. Engineering procedure for the climate change flood risk assessment in the Upper Thames River Basin. *Journal of Hydrologic Engineering*, 16(7): 608–612.

Gangyan Zhu. 2007. The Rhine River Basin Management Experience and Enlightenment. *Water Conservancy and Hydropower Letters*, 2007(3):28–29.

Grijp N vander, Olsthoorn X. 2003. Institutional Framework for the Management of the Rivers Rhine and Meuse in the Netherlands: An Overview. *Institute for Environmental Studies*. Amsterdam.

Guang bo Cui. 2004. A new concept for comprehensive management of water environment in Taihu Basin—Thoughts aroused by water division experiment of "Diverting Water From Yangtze to Taihu". *China Water Resources*, (6):43–44.

Guihua Lu, Qian Ma. 2009. Analysis on the causes of forming black water cluster in Taihu Lake. *Advances in Water Science*, 20 (3):438–442.

Havinga HSmits AJM. 2000. River Management along the Rhine: a Retrospective View. *New Approaches to River Management*, 15–32.

Hendry K, Cragg HD. 2003. Ecology of the Atlantic Salmon. Conserving Natural 2000 Rivers. *Ecology Series* No. 7.

Hongzhi Liu, Xueyun Lu. 2001. The comparison of Chinese and foreign water pollution prevention. *World Environment*, (4):27–30.

Hurkmans R, De Moel H, Aerts J, et al. 2008. Water balance versus land surface model in the simulation of Rhine river discharges. *Water Resources Research*, 44(1).

Jamieson D, Fedra K. 1996. The "WaterWare" Decision Support System for River-Basin Planning. *Journal of Hydrology*, 117(3):163–175.

Jianlan Hua, Wei Wang et al. 2010. The review of Qinhuai river comprehensive treatment engineering. *Urban Road and Bridge and Flood Control*. (9):182–184.

Jianqiang Yang. 2000. Dianchi Lake pollution control and ecological protection. Journal of Hydraulic Engineering, (5):17–21.

Nam-choon K. 2005. Ecological restoration and revegetation works in Korea. L*andscape and Ecological Engneering*, (1):77–83.

Perret D, Newman ME, Negre JC, et al. 1994. Submicron particles in the Rhine River-I. Physico-chemical characterization. *Water Research*, 28(1):91–106.

Qin B, Xu P, Wu Q, et al. 2007. Environmental issues of lake Taihu, China. Hydrobiologia, 581(1): 3–14.

Rao Zhengfu, Potter B, Webb D. 2009. Sustainable Water Resources Management: River Basin Modelling And Decision Support Framework. *Wuhai University Journal of Natural Sciences*, 14(6):543–551.

State Environmental Protection Administration (SEPA), 2000. Water Protection in the Three Rivers and Three Lakes of China. *Chinese Environmental Press*, Beijing: 151–175.

Vander Weijden CH, Middelburg JJ. 1989. Hydrogeochemistry of the river Rhine: long term and seasonal variability, elemental budgets, base levels and pollution. *Water Research*, 23(10):1247–1266.

Xiaojuan Hu. 2007. Biwa Lake Management Reference. *Environment*, (8):134–135.

Yi Ding, Lai Kui. 2011. Dianchi Protection governance experience and enlightenment. *Southwest University for Nationalities*, (7):111–115.

Zhu Xi. 2011. Taihu Lake Water Environmental Effects and Its Implications. *Eutrophication control and water special issue of Ecological Restoration Technology of China*. Water Resources Development Center, (2).

Modeling and Computation in Engineering III – Zhang & Xie (Eds)
© 2014 Taylor & Francis Group, London, ISBN 978-1-138-02680-3

The study of high resolution implicit algorithm in unstructured grids and its application

Chuanteng Lu, Yufang Han & Xiaofeng Luo
Nanjing Hydraulic Research Institute, Nanjing, Jiangsu, China

ABSTRACT: The implicit algorithm theory is studied less than the explicit algorithm in FVM with unstructured grids. In this paper, N-S equations are discreted using LU-SGS implicit algorithm during the solving process ultra-relaxation coefficient is introduced. According to numerical flume test, the value of relaxation coefficient can improve the convergence rate of calculations, which ranges from 1.0 to 1.3. Based on LU-SGS implicit algorithm, 2D high resolution mathematical model in DanDong port is established to simulate the tidal current and suspended sediment transportation. The results show that LU-SGS implicit algorithm is of high precision and stable to simulate hydro-dynamic environment in the wide open sea.

1 INTRODUCTION

In recent years, with the rapid development of computer technology, FVM models have been widely used, in which the calculating area tends to large scale and the computing girds tend to high resolution. Explicit scheme of FVM has been studied and applied more than implicit scheme. Liu studied the influence to current fields which has been caused by spur-dikes in WENO scheme of FVM. The flow of ShengLi oil field seas was simulated by Zhang in Roe scheme of FVM. The water movement in Penglai Beach was simulated by Wang in Roe scheme. The flood in red river in Van-Coc was studied by Dan K. Nguyen whose model was built through advection discretization, time advancement and velocity correction. The tidal wave movement of Hooghly River was simulated by Soumendra Nath Kuiry. Francisco J. M. Simões, E. Bladél and Lorenzo Begnudellil have done some researches to the discrete scheme. FVCOM and the water power module of MIKE are also based on FVM. All above studies on FVM are explicit scheme.

On the other hand, it is applied less in the implicit scheme of FVM. Casulli have raised semi-implicit scheme which was applied well by Cui-Ping Kuang, HU De-chao. Lu have simulated the flow of Yangtze river near Nanjing by the implicit scheme in FVM. Jameson have raised LU-SGS algorithm which was applied more in aerodynamics than in hydrodynamics.

Based on the previous research, the application of LU-SGS algorithm has been studied deeply, using ultra-relaxation coefficient to accelerate the convergence rate of calculations. The model is applied to simulate the movement of tidal current and sediment in Dandong Port.

2 GOVERNING EQUATIONS

Flow and sediment movement equations of vector form can be written as:

$$\frac{\partial U}{\partial t} + \nabla E = M + \nabla E^d \tag{1}$$

where: $U = (H, Hu, Hv, HS)^T$

$$E = (F, G), \text{ where: } F = \begin{pmatrix} Hu \\ Hu^2 + gH^2/2 \\ Huv \\ HuS \end{pmatrix}, \quad G = \begin{pmatrix} Hv \\ Huv \\ Hv^2 + gH^2/2 \\ HvS \end{pmatrix}$$

The turbulence diffusion and sediment diffusion in Eq.(1) can be expressed as:

$$E^d = (F^d, G^d), \text{ where: } F^d = \begin{pmatrix} 0 \\ N_x H\partial u/\partial x \\ N_x H\partial v/\partial x \\ D_x H\partial S/\partial x \end{pmatrix}, \quad G^d = \begin{pmatrix} 0 \\ N_y H\partial u/\partial y \\ N_y H\partial v/\partial y \\ D_y H\partial S/\partial y \end{pmatrix}$$

The source term can be written as: $M = M_0 + M_f = \begin{pmatrix} 0 \\ gH(M_{0x} + M_{fx}) + fv \\ gH(M_{0y} + M_{fy}) - fu \\ F_s \end{pmatrix}$

where:
H—Depth (m);
u, v—The x, y components of velocity V (m/s);
t—Time (s);
f—Coriolis coefficient ($f = 2w\sin\varphi$, w—the earth spins, φ—latitude);
g—Gravitational acceleration (m/s^2);
N_x, N_y—Turbulence viscosity coefficient in x, y direction (m^2/s).
S—Sediment concentration (kg/m^3);
D_x, D_y—suspend sediment diffusion coefficient in x, y direction (m^2/s).
F_s—Source term of sediment (kg/(m^2·s));
M_{ox}, M_{oy}—Variation of bed bottom elevation in x, y direction;
M_{fx}, M_{fy}—Bottom friction in x, y direction.

3 NUMERICAL METHOD

3.1 *LU-SGS algorithm*

In this paper, Eq.(1) is dispersed based on triangular meshes where physical variables u, v, S are arranged in grid cell center, and H is in the nodes of the grid. The element of i can be denoted as Ω_i where the control equations can be integrated, and surface integral would be translated into line integral through Green formula:

$$\frac{\partial}{\partial t}\int_{\Omega_i} U d\Omega_i + \oint_{\partial\Omega_i} E \bullet \bar{n}_i dl = \int_{\Omega_i} S d\Omega_i - \oint_{\partial\Omega_i} E^d \bullet \bar{n}_i dl \tag{2}$$

where: $d\Omega_i$—Infinitesimal of surface integral, dl—Infinitesimal of line integral; $\overline{n_i} = (n_{ix}, n_{iy}) = (\cos\theta, \sin\theta)$.

The integral results of convection and turbulence terms can be expressed as $F(U)$, $G(U)$, then Eq.(2) reduced to the follows:

$$A_i \frac{U_i^{n+1} - U_i^n}{\Delta t} + \sum_{j=1}^{3} F(U^{n+1}) L_j = AS(U_i^n) - \sum_{j=1}^{3} G(U^n) L_j \tag{3}$$

While LU-SGS approximate factorization, the numerical flux $F(U)$ approximation solution for the Riemann problem in Roe scheme can be expressed as:

$$F(U) = F(U_L, U_R) = \frac{1}{2}[F(U_L) + F(U_R) - \rho_{ij}(U_R - U_L)] \tag{4}$$

where: U_L, U_R—The numerical flux on both sides of interface, ρ_{ij}—The spectral radius of Jacobian matrix.

A first-order approximation loads at the left second term of Eq.(3):

$$F(U^{n+1})-F(U^n)=[F(U_i^{n+1},U_j^{n+1})-F(U_i^n,U_j^{n+1})]+[F(U_i^n,U_j^{n+1})-F(U_i^n,U_j^n)]$$
$$=\frac{\partial F(U)}{\partial U_i}\Delta U_i+\frac{\partial F(U)}{\partial U_j}\Delta U_j \tag{5}$$

Using Eqs.(3) and (5) reduce to the following expression:

$$A_i\frac{U_i^{n+1}-U_i^n}{\Delta t}+\sum_{j=1}^{3}\left\{\frac{1}{2}\left[\frac{\partial F(U_i)}{\partial U_i}+\rho_{i,j}I\right]\Delta U_i^n+\frac{1}{2}\left[\frac{\partial F(U_j)}{\partial U_j}-\rho_{i,j}I\right]\Delta U_j^n\right\}$$
$$L_j=A_iS(U_i^n)-\sum_{j=1}^{3}F(U_j^n)L_j-\sum_{j=1}^{3}G(U^n)L_j \tag{6}$$

For each control volume is closed, so:

$$\sum_{j=1}^{3}\frac{\partial F(U_i^n)}{\partial U_i}L_j=0$$

Then, Eq.(6) can be expressed as:

$$\frac{A_i}{\Delta t}\Delta U_i^n+\sum_{j=1}^{3}\frac{1}{2}\rho_{i,j}I\Delta U_i^nL_j=-\frac{1}{2}\sum_{j=1}^{3}\left[\frac{\partial F(U_j)}{\partial U_j}-\rho_{i,j}I\right]\Delta U_j^nL_j+A_iS(U_i^n)$$
$$-\sum_{j=1}^{3}F(U_j^n)L_j-\sum_{j=1}^{3}G(U^{n+1})L_j \tag{7}$$

For $D=A_i/\Delta t+\sum_{j=1}^{3}1/2\rho_{i,j}L_j$, $\quad \mathrm{Re}=A_iS(U_i^n)-\sum_{j=1}^{3}F(U_j^n)L_j-\sum_{j=1}^{3}G(U^{n+1})$,

Consequently, Eq.(7) is found to be:

$$D\Delta U_i^n=-\frac{1}{2}\sum_{j=1}^{3}\left[\frac{\partial F(U_j)}{\partial U_j}-\rho_{i,j}I\right]\Delta U_j^nL_j+\mathrm{Re} \tag{8}$$

Solving Eq.(8) needs two steps:

1. Push ahead:

$$\Delta U_i^*=D^{-1}\left\{-\frac{1}{2}\sum_{j=1}^{3}\left[\frac{\partial F(U_j)}{\partial U_j}-\rho_{i,j}I\right]\Delta U_j^nL_j+\mathrm{Re}\right\} \tag{9}$$

2. Push back:

$$\Delta U_i=\Delta U_i^*-\frac{1}{2}\sum_{j=1}^{3}\left[\frac{\partial F(U_j)}{\partial U_j}-\rho_{i,j}I\right]\Delta U_j^nL_j \tag{10}$$

$$\mathrm{So}: U_i^{n+1}=U_i^n+w\cdot\Delta U_i \tag{11}$$

where: w—Relaxation coefficient.

3 BOUNDARY CONDITIONS

3.1 *Open boundary of water level*

According to the compatibility relation:

$$U_R + 2c_R = U_L + 2c_L \tag{12}$$

where: c_L, c_R—wave propagation velocity of element around.

$$U_R = U_L + 2\sqrt{gh_L} - 2\sqrt{g(Z_R - Z_d)} \tag{13}$$

where: Z_d—Border bottom elevation of flux integral points.

3.2 *Wall boundary*

An interface between two cells is a wall boundary if there is no current flow across that interface. The boundary condition specified, therefore, includes the normal velocity component being set as zero and $H_R = H_L$.

3.3 *Initial conditions*

$\zeta(x,y,t) = \zeta_*(x,y,t)$, $u(x, y, t) = u_*(x, y, t)$, $u(x, y, t) = u_*(x, y, t)$, $S(x,y,t) = S_*(x,y,t)$, $_*$—Initial numerical values.

4 A TEST CASE OF NUMERICAL FLUME

To test the computational performance of LU-SGS algorithm, a 30 km in length and 10 km in width numerical flume is built. In this model, the average space step is 400 m, and the upper and lower boundaris give a constant water level. There is sampling point A in the center of numerical flume, as shown in Figure 1. Using explicit scheme and implicit scheme simulate the flow of flume respectively.

In the condition of without ultra-relaxation coefficient ($w = 1$), the computing results are shown in Figure 2. It indicates that the time step of implicit algorithm is much longer than the explicit algorithm, so the implicit algorithm could save enormous time. When the current velocity of sampling point A becomes stable, the cost time of explicit scheme is the shortest, and the cost time of implicit scheme increases as the time step increasing. It suggests that the convergence rate of computing results decreases rapidly though the LU-SGS algorithm could take much larger time step. For unsteady flow numerical calculation, this will seriously affect the accuracy.

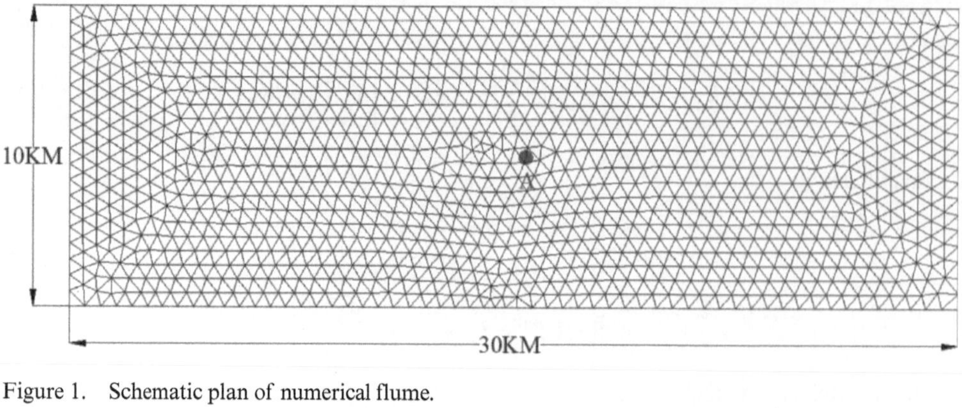

Figure 1. Schematic plan of numerical flume.

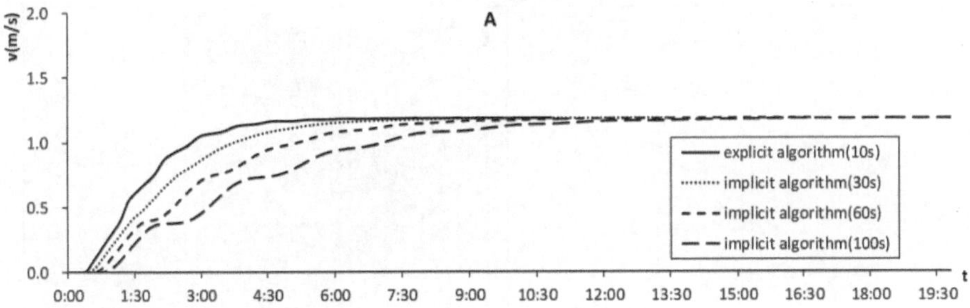

Figure 2. The velocity lines of sampling point A.

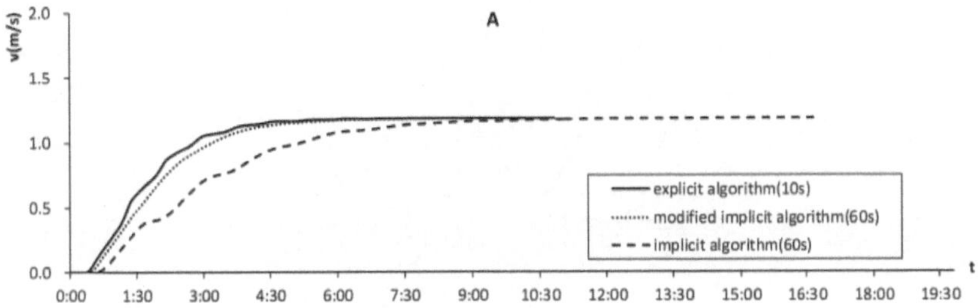

Figure 3. The velocity lines of sampling point A.

When the relaxation coefficient is 1.2 ($w = 1.2$), the computing results are shown in Figure 3. It suggests that the modified LU-SGS algorithm could accelerate calculating convergence rate obviously. When the time step becomes 60 s, the convergence rate of implicit algorithm approaches to explicit algorithm which indicates that using relaxation coefficient in the process of calculation could accelerate the convergence rate significantly. The value of relaxation coefficient ranges from 1.0 to 1.3 generally as the time step varies by the trial.

5 THE APPLICATION OF LU-SGS ALGORITHM IN DANDONG PORT

5.1 *The establishment of mathematical model*

Dandong Port, which is an international trade port in north China, is located in the northeast of Liaodong peninsula and the west bank of the Yalu estuary.

5.1.1 *The range and parameters of mathematical model*

In this model with 107179 meshes, upper boundary lies near Langtou port, the open sea boundary is to −50 m contour, the western boundary is to Shicheng Island and the eastern boundary is to Dahe island in North Korea. The length of the model is about 150 km, and the width is about 110 km. The meshes in the project area (Fig. 4) have been refined, and the minimum length of grid is 35 m while the maximum length of grid is about 2 km. For the LU-SGS algorithm, the time step of this model is 12 s while the explicit algorithm takes 1.2 s only. In this paper, the relaxation coefficient is 1.1.

5.1.2 *Key issues*

1. Moving Boundary

For many shoals in simulated area would expose or inundate as the change of tide level, moving boundary is necessary to reflect the shoals' dry/wet characteristic correctly. In this paper, dry-wet spinning is used to check whether the bottomland is exposed or not by the mesh water depth. If the water depth is lower than a controlled depth, this mesh will be expressed as land,

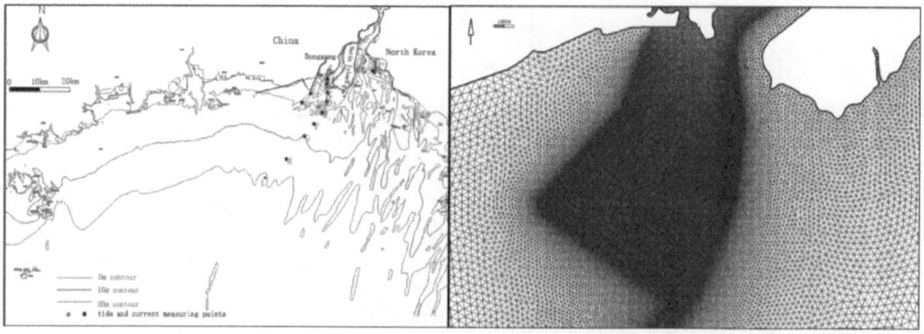

Figure 4. The range of mathematical model and mesh refinement in the project area.

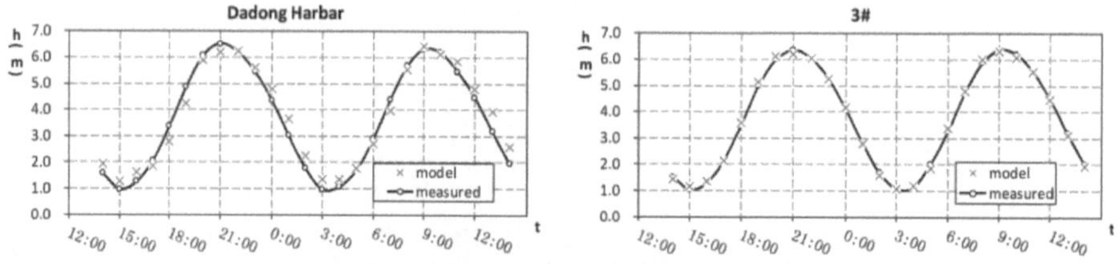

Figure 5. Sea level verification (24th 14:00~25th 14:00 Sep. 2010).

otherwise if the water depth is higher than a controlled depth, it will be calculated again. In this model, the depth of moving boundary is 0.02 m to ensure the water balance as far as possible.

2. Roughness
Roughness is one of the main parameters in mathematical model, which reflects the resistance characteristics in the tidal process. It is a comprehensive parameter related with the bed surface, water depth, and topography. Selecting a correct roughness could get more accurate results. In this paper, additional roughness formula is used as follows, considering the roughness response to the water depth:

$$n = n_0 + n' = 0.013 + 0.012/H \tag{14}$$

where: n_0—general roughness, n'—additional roughness.

3. The sediment source and sink function F_s
The sediment source and sink function F_s can be expressed as:

$$F_s = -\alpha w(S - \beta S_*) \tag{15}$$

where: α—Sediment settling probability; w—Sediment settling velocity (Median diameter is 0.03 mm); S_*—sediment carrying capacity;

$$\beta = \begin{cases} 1, S > S_* \\ 1, S \le S_* \quad V > V_c \\ S/S_*, S \le S_* \quad V \le V_c \end{cases} \tag{16}$$

where: V_c—sediment incipient velocity expressed as mean vertical velocity.

5.2 *Model verification*

The model verification data select spring tide simultaneous observation data from September 24th 2010 to September 25th 2010, and the verification points are shown in Figure 4.

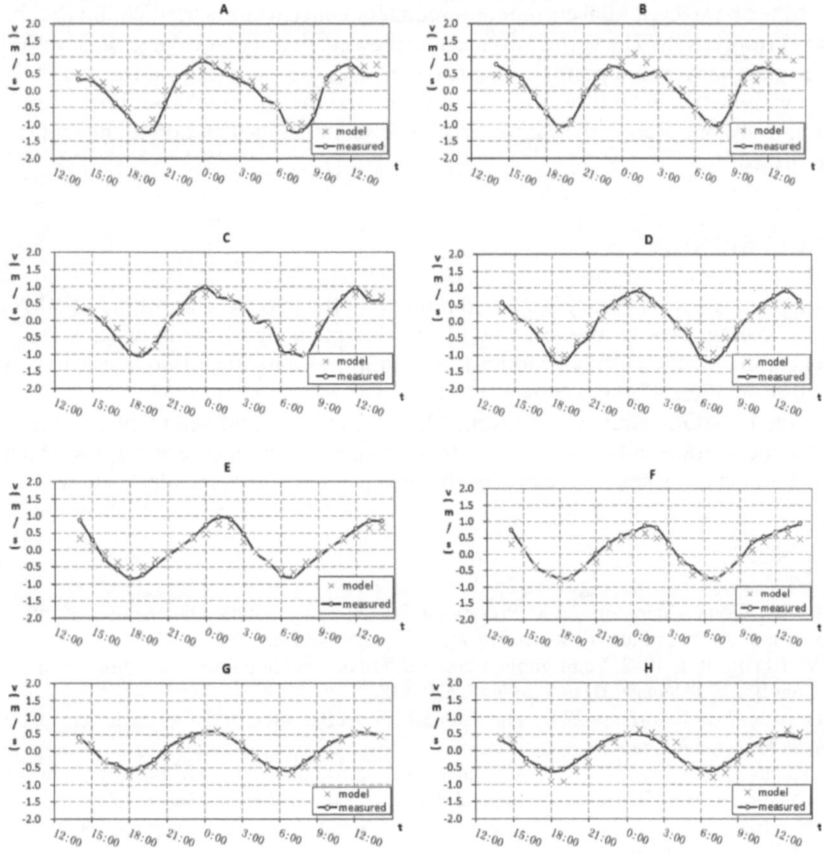

Figure 6. Tidal current verification.

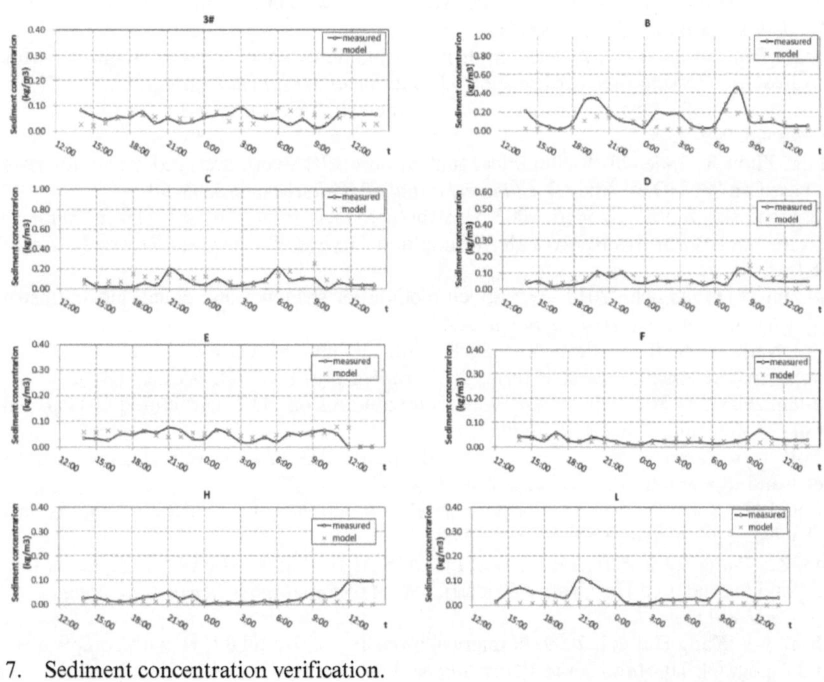

Figure 7. Sediment concentration verification.

The results of sea level, tidal current and sediment concentration verification are shown in Figures 5–7 respectively. The verification results indicate that the calculated sea level and tidal current of most verification points coincide better with the measured values, and the relative errors are within 10%. As uniform sediment in simulation, the bias of sediment verification is large relatively, but overall, the characteristics of sediment distribution in the DaDong seas can be reflected.

6 CONCLUSIONS

1. The explicit algorithm theory is studied more than the implicit algorithm in FVM. In this paper, N-S equations are discreted by LU-SGS implicit algorithm in which solving process relaxation coefficient is introduced. According to numerical flume test, the value of relaxation coefficient ranges from 1.0 to 1.3.
2. Based on LU-SGS implicit algorithm, 2D tidal current and sediment high resolution mathematical model in DanDong Port is established with good verifications which show that the model can simulate the process of tidal current and suspended sediment.

REFERENCES

Bladél E., M. Gómez-Valentín. 2008. Preserving Steady-State in One-Dimensional Finite-Volume Computations of River Flow[J]. *Journal of Hydraulic Engineering*: 1343–1347.

Casulli, V., Cheng, R.T. 1992. Semi-implicit finite difference methods for three dimensional shallow water flow[J]. *Int. J. Numer. Meth. Fluids 15*: 629–648.

Chen Zhi-chang, Luo Xiao-feng. 2007. The Physical Model Research in Dandong harbor[R]. *Nanjing Hydraulic Research Institute*.

Cui-Ping Kuang, Si-Yu Chen. 2011. A two-dimensional morphological model based on a next generation circulation solver I: Formulation and validation[J]. *Coastal Engineering*: 1–13.

Dan K. Nguyen, Yu-E Shi. 2006. 2D Shallow-Water Model Using Unstructured Finite—Volumes Methods[J]. *Journal of Hydraulic Engineering*: 258–269.

Dou Guo-ren, Dou Xi-ping. 2003. Sediment transport under the action of tidal currents and waves and its numerical simulation[C]. *The Dou Guoren papers*: 824–836.

Francisco J.M. Simões. 2011. Finite Volume Model for Two-Dimensional Shallow Environmental Flow[J]. *Journal of Hydraulic Engineering*: 173–182.

Hu De-chao, Fan Bei-lin. 2011. A Semi-implicit 3-D Numerical Model Using Sigma-coordinate for Non-hydrostatic Pressure Free-surface Flows[J]. *Journal of Hydraulic Engineering*: 212–223.

Jameson A, Yoon S. 1987. Lower—upper implicit schemes with multiple grids for the Euler equations [J]. *AIAA Journal*: 929–935.

Liu Yu-ling, Zhou Xiao-de. 2010. Numerical simulation of 2D depth-averaged model for rivers with groins based on WENO schemes[J]. *Chinese Journal of Hydrodynamics*: 93–98.

Lorenzo Begnudellil, Brett F. Sanders. 2006. Unstructured Grid Finite-Volume Algorithm for Shallow-Water Flow and Scalar Transport with Wetting and Drying[J]. *Journal of Hydraulic Engineering*: 371–384.

Lu Kang-ming, Li Guang-chi. 2010. An implicit method for shallow water equations on unstructured grids[J]. *Chinese Journal of Hydrodynamics*: 247–253.

Soumendra Nath Kuiry, Kiran Pramanik. 2008. Finite Volume Model for Shallow Water Equations with Improved Treatment of Source Terms[J]. *Journal of Hydraulic Engineering*: 231–242.

Wang Bo-lan. 2003. LU-SGS Scheme for Euler Computations on 2D Unstructured Grids[D]. *Nanjing University of Aeronautics and Astronautics*.

Wang Zhi-li, Gen Yan-fen. 2005. Numerical modeling of 2-D shallow water flow withcomplicated geometry and topography[J]. *ShuiLi XueBao*: 439–444.

Xin Wen-jie. 1993. Some Problems in the Numerical Computation of 2-D Tidal flow in Estuaries and Bays[J]. *Chinese Journal of Hydrodynamic*: 348–354.

Yuzhi Sun, Z.J. Wang. 2007. Efficient Implicit LU-SGS Algorithm for High-Order Spectral Difference Method on Unstructured Hexahedral Grids[J]. *45th AIAA Aerospace Sciences Meeting and Exhibit, Reno, Nevada*: 8–11.

Zhang Hong-wei, Kang Hai-gui. 2009. Numerical modeling of 2D tidal flow with complicated geometry and topography[J]. *Advances in Water Source*: 118–123.

Modeling and Computation in Engineering III – Zhang & Xie (Eds)
© 2014 Taylor & Francis Group, London, ISBN 978-1-138-02680-3

Numerical study on cold water emission for CNOOC Tianjin floating LNG project

S.S. Yao, M.X. Xie & M.G. Li
Tianjin Research Institute of Water Transport Engineering, Ministry of Communications, Tianjin, China

ABSTRACT: Taking CNOOC Tianjin floating LNG project as background, a 3-D numerical model based on the hydrodynamics and advection-diffusion equation is used to simulate the tidal current field and temperature diffusion caused by the cold water emission, the distribution of temperature diffusion is predicted both at surface layer and bottom layer, as well as the envelope area of the maximum temperature reduction and reduction value at the water intake. Calculating results show that: (1) tidal current shows a typical reciprocating characteristic in the project area, it flows toward SW during flooding tide and NE during ebb tide. Besides, current direction at the water outlet pointing to the open sea during the full tide, while it flows along the dyke sometimes at the surface layer, reflecting the 3-D current characteristics; (2) temperature diffusion distributed in a narrow shape, mainly along E-W direction, the diffusion area at surface layer is smaller than bottom layer, the envelop area of temperature reduction more than 0.5 degree is within 4 km^2; (3) the maximum temperature reduction value at the water intake is between 0.5 degree and 1.0 degree, and the difference between surface layer and bottom layer is only within 0.1 degree.

1 INTRODUCTION

As the first floating Liquefied Natural Gas (LNG) project in China, CNOOC Tianjin floating LNG terminal wharf project is a pilot key project of China using clean energy and floating technology. The completion of this project will effectively improve the situation of Tianjin natural gas shortage and the emergency supplies shortage, it will also provide guarantees for energy supplies in Beijing and even the whole area of North China, marking a new mode of domestic LNG terminal project, and playing an exemplary role for the subsequent floating LNG project construction. In this paper, based on the water intake and drainage project for supporting LNG wharf, using numerical model, the characteristic of current field and temperature diffusion due to cold water emission were simulated, and then the impact on surrounding water environment was analyzed, the conclusions can provide technical support to the local government for decision-making.

2 PROJECT CONDITION

Tianjin floating LNG wharf is located in Lingang Industrial Area, the southeastern part of Nanjiang harbor district in Tianjin port, lying at the north area of Dagusha channel. Tidal characteristics in project area belongs to irregular semidiurnal tide with mean tidal range 2.28 m, tidal current is mainly reciprocating flows, and current velocity is relatively weak, the average velocity for the whole tide is less than 0.3 m/s.[1–2] Dominant waves toward E, maily wind waves and varied seasonally. The beach surface sediment in the project area is clayey silt,[3] sediment concentration under normal weather is generally low, average sediment concentration for all seasons ranged from 0.05 kg/m^3 to 0.15 kg/m^3.

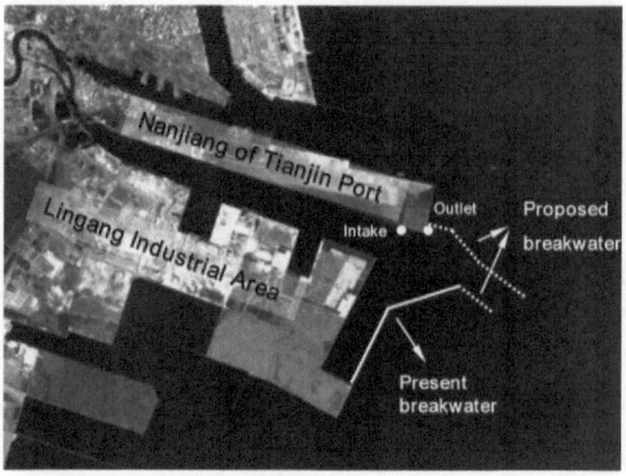

Figure 1. Study site of Tianjin LNG wharf project.

Because the north breakwater in Lingang Industrial Area is being designed, therefore the breakwater both in south and north should be considered with our project, the planar layout after the completation of the whole project is showed in Figure 1. Among them, the northern breakwater is composed of three parts from the root to tip of the breakwater, the first and second part are none water-crossing breakwater, the crest elevation of the breakwater is +5.5 m (under the theoretical base datum, the same below), the third part is a submerged breakwater with crest elevation +2.5 m and 2350 m long; while in the south, only a submerged breakwater is built with 1500 m long, the crest evelation is the same as the north breakwater. In order to meet with the water supplies for LNG terminal, the water intake and drainage project are built as supporting facilites, the layout of the intake and outlet can be seen in Figure 1, bottom emission is adopted, and the discharge of the intake and outlet are the same, in fact, there are two kinds of water use including short-term and long-term. In this paper, only long-term condition (discharge of 100000 m^3/h) was studied, then temperature diffusion was simulated, and the influence on surrounding water areas due to cold water emission was finally analyzed.

3 SIMULATION FOR HYDRODYNAMICS AND TEMPERATURE DIFFUSION

3.1 *Model set-up*

Hydrodynamic model is based on the conservative 3-D tidal current movement equations, unstructured triangular mesh and σ coordinates are also used. Temperature diffusion followed the 3-D Eulerian advection and diffusion equation, details related about the model theory can be seen in [4].

In this study, tidal current model was using the large scale and local scale model for nested calculation, the large model covers the Bohai sea, and the local model covers the Bohai Bay. Open boundary is controlled by the tide level, the mesh number of local model is up to 22000, and the minimum space step is even 10 m, the mesh generation is showed in Figure 2.

The 3-D numerical model was verified according to the hydrological measured data in both July, 2006 and August, 2008, readers of interest could see Zhang [5] for more details.

3.2 *Temperature diffusion simulation*

Tidal current is the main driving force affecting the temperature diffusion, so before calculating the temperature diffusion, tidal current field was simulated firstly. Taking spring tide as example, large scale current field at surface layer and local current field at surface and bottom

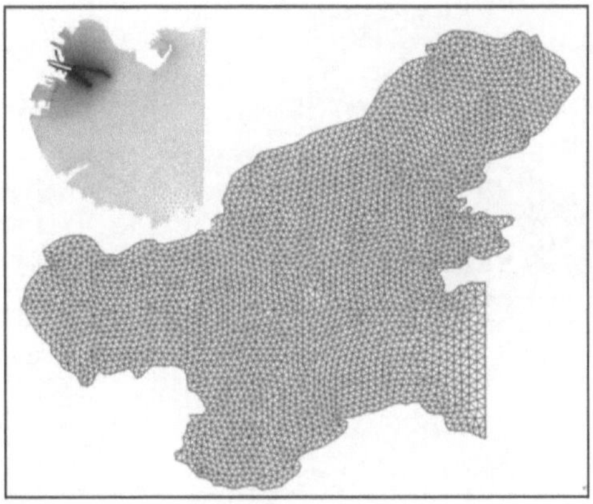

Figure 2. Mesh generation of numerical model.

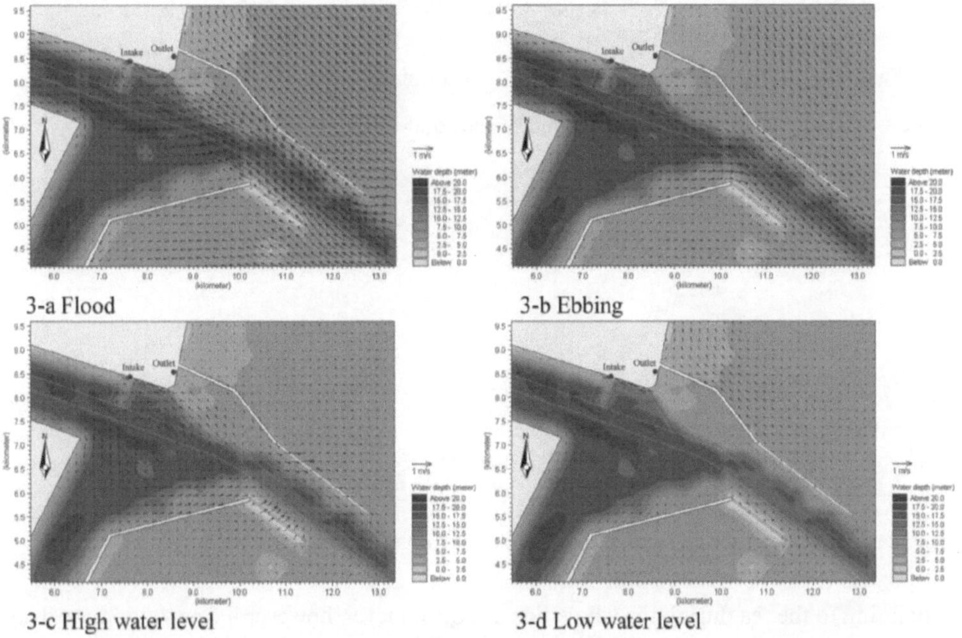

Figure 3. Large scale tidal current field at surface layer (Spring tide).

layer were showed in Figure 3~Figure 4. Through analysis, some conclusions are made as follows:

1. Due to the direction of the breakwater and tidal current are basically the same, current field is generally smooth. During flood tide, circulation current is produced at the inner area of the south breakwater, and it reaches the largest range at high water level, while in ebb tide, restrained by the submerged breakwater, tidal current flow out of the entrance parallelling to the channel.
2. There is obvious inflow and outflow at the intake and outlet, and many small circulation currents are produced around them, but the velocity is below 0.2 m/s. Because both the intake and outlet adopted bottom emission, so the impact on surrounding flows at bottom layer is more distinct than surface layer. Especially at the outlet, current direction is

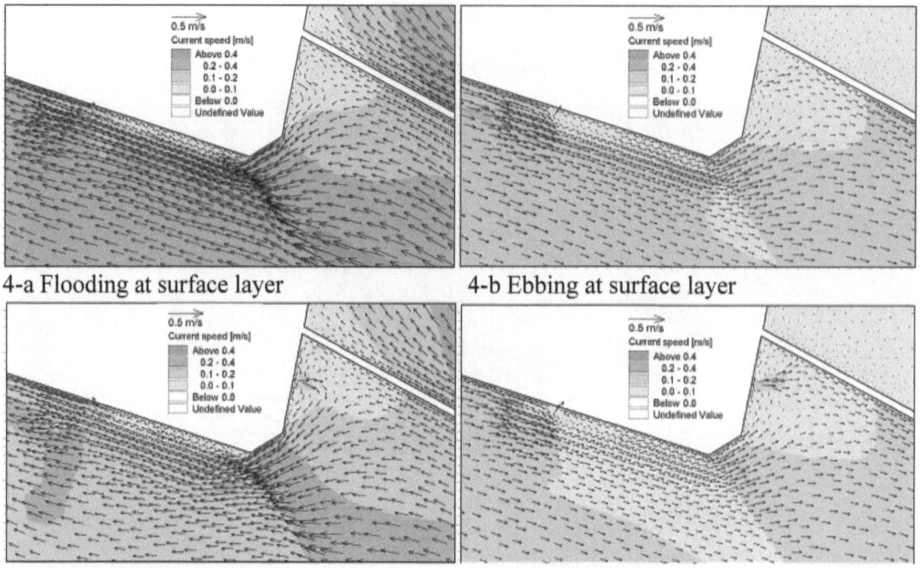

Figure 4. Local tidal current field around intake ande outlet (Spring tide).

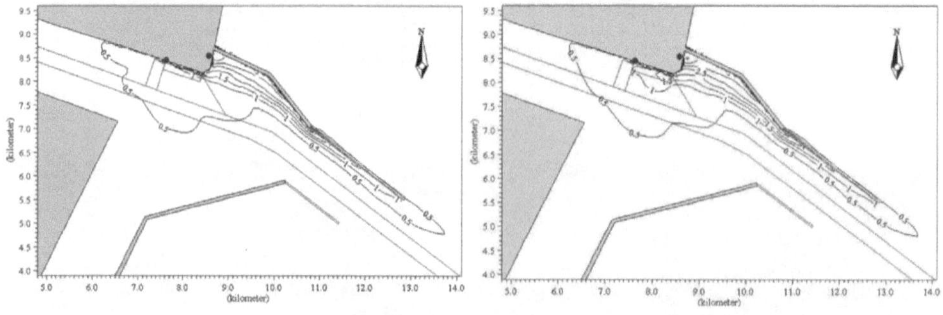

Figure 5. Envelope of the maximum temperature reduction (Left: Surface; Right: Bottom).

pointing to the sea during the whole tide at bottom layer, however it flows along the dyke sometimes at surface layer, reflecting significant three-dimensional features.

3. Overall, it is found that the contruction of the water intake and drainage project has nothing obvious impact on surrounding flow state, the influence area is only around the water intake and outlet.

3.3 *Temperature diffusion simulation*

Based on the hydrodynamic simulation, temperature diffusion on the surrounding water environment due to cold water emission was further analyzed. Considering the cold water required a certain period to reach the stable state since the beginnng of emission, so the simulation period was 15 d including the spring tide, middle tide and neap tide. The envelopes of the maximum temperature reduction both at surface layer and bottom layer were showed in Figure 5 and Table 1~Table 2. Through analysis, some conclusions are made as follows:

1. After the construction of the water intake and drainage project, temperature diffusion is distributed in a narrow shape, mainly advected and diffused from west to east, and the

Table 1. Statistics for the envelops area of temperature diffusion.

Surface layer				Bottom layer			
Discharge (m^3/h)	Time (d)	Temperature variation (°C)	Area (km^2)	Discharge (m^3/h)	Time (d)	Temperature variation (°C)	Area (km^2)
100000	15	≥0.5	2.54	100000	15	≥0.5	3.40
		≥1.0	1.07			≥1.0	1.32
		≥2.0	0.67			≥2.0	0.75
		≥3.0	0.29			≥3.0	0.34
		≥4.0	0.03			≥4.0	0.05
		≥5.0	/			≥5.0	/

Table 2. The maximum temperature reduction at the water intake.

Discharge (m^3/h)	Temperature reduction at the intake (°C)	
	Surface layer (°C)	Bottom layer (°C)
100000	0.84	0.89

west diffusion length is smaller than east length. The reason is that affecting by LNG pile group, cold water was held up in pile group area, thus it can not diffuse to the intake easily during flood tide. When in ebb tide, cold water was taken to the tip of submerged breakwater by tidal current.

2. Due to the intake and outlet are both bottom emission, so the diffusion at bottom layer is stronger than surface layer, no matter the temperature variation at the intake, or the maximum temperature envelops, three-dimensional characteristics is reflected obviously, but the difference is little. Through calculation, the maximum temperature reduction at the outlet was 0.5°C~1.0°C, but the difference between surface layer and bottom layer was only within 0.1°C. Temperature reduction more than 0.5°C both at the surface layer and bottom layer were respectively 2.5 km^2 and 3.4 km^2, while the reduction more than 1.0°C were respectively 0.03 km^2 and 0.05 km^2. It is worth mentioning that, the construction of submerged breakwaters will cause the water exchange not smooth between harbour and outside water, cold water is relatively difficult to diffuse out of the entrance. However, in general, the temperature diffusion caused by cold emission was only concentrated around the project area, and it has little influence on surrounding water environment.

4 CONCLUSIONS

1. After the project, restrained by the south and north breakwaters, the whole tidal current is relatively smooth. Circulation current is produced at the inner area of the south breakwater during flood tide, and it reaches the largest range at high water level. While in ebb tide, current flows out of the entrance parallelling to the channel. There is obvious inflow and outflow at the intake and outlet, and many small circulation currents are produced around them, but the velocity is below 0.2 m/s. The impact on surrounding flows at bottom layer is more distinct than surface layer, reflecting obvious three-dimensional features. From the view of tidal current, it is found that the contruction of the water intake and drainage project has nothing obvious impact on surrounding flow state, the influence area is only around the wate intake and outlet.

2. Temperature diffusion caused by the cold water emission is distributed in a narrow shape, mainly advected and diffused from west to east, and the west diffusion length is smaller than east length, it is mainly because cold water was held up by LNG pile group. Due to the intake and outlet are both bottom emission, so the diffusion at bottom layer is stronger than surface layer, no matter the temperature variation at the intake, or the maximum temperature envelops, three-dimensional characteristics are reflected obviously, but the difference between surface and bottom is little. Temperature reduction more than 0.5°C both at the surface layer and bottom layer were respectively 2.5 km^2 and 3.4 km^2, while the reduction more than 1.0°C were respectively 0.03 km^2 and 0.05 km^2. It is worth mentioning that, the construction of submerged breakwaters will cause the water exchange not smooth between harbour and outside water, cold water is relatively difficult to diffuse out of the entrance. However, in general, the temperature diffusion caused by cold emission was only concentrated around the project area, and it has little influence on surrounding water environment.

REFERENCES

Remote sensing investigation of peninsula coast in eastern Zhejiang, Land and Resources Remote Sensing, China, 2004, pp. 122–128.

L.C. Sun, N. Zhang, *Hydrodynamic, sediment and channel feasibility research for Nangang Industrial Research of Tianjin Port,* Technical Report of Tianjin Research Institute for Water Transport Engineering, M.O.T., Tanggu, China.

L.C. Sun, N. Zhang, *Key technique research on hydrodynamics and sediment of Tianjin Port,* Technical Report of Tianjin Research Institute for Water Transport Engineering, M.O.T., Tanggu, China.

M.X. Xie, S.S. Yao, *Numerical study on cold water emission of Tianjin floating LNG project,* Technical Report of Tianjin Research Institute for Water Transport Engineering, M.O.T., Tanggu, China.

H. Yang, Y.X. Wu, H.B. Zhao, Hydrology, sediment environment research and evolution analysis for Caofeidian sea area, Journal of waterway and harbor, China, 2005, 6(3):130–133.

N. Zhang, *Numerical study on tidal current of Tianjin floating LNG project,* Technical Report of Tianjin Research Institute for Water Transport Engineering, M.O.T., Tanggu, China.

Modeling and Computation in Engineering III – Zhang & Xie (Eds)

Research on sediment transport and deposition in the intake open channel under the typhoon

Na Zhang, Mengguo Li & Wendan Li
Tianjin Research Institute of Water Transport Engineering, Ministry of Communications, Tianjin, China

ABSTRACT: The Tianwan nuclear power plant is located at the Gaogong Island, south of Lianyungang Port in China. The ocean area near the Tianwan nuclear power plant is frequently struck by typhoon and strong wind of the cold wave. The sediment concentration near the bottom will become larger under the strong wind weather. In the paper, the three-dimensional model for the tidal current and sediment transport is used to predict the maximum sediment concentration and deposition depth in intake open channel. The results show that sediment concentration is high in outer segments and low in the breakwater; the maximum vertically averaged sediment concentration near the head of the breakwater and at front end is respectively about 1.27 kg/m^3 and 0.745 kg/m^3; under the action of "Wipha" typhoon, maximum deposition depth in the intake open channel is 0.11 m. So after the implementation of the project, it didn't emerge obvious sudden deposition.

1 INTRODUCTION

The Tianwan nuclear power plant is located at the Gaogong Island, south of Lianyungang Port in China, which is constructed under the cooperation agreement between the Chinese and Russian government. The ocean area near the Tianwan nuclear power plant is frequently struck by typhoon and strong wind of the cold wave. The sediment concentration near the bottom will become larger under the strong wind weather. Because of increasing of sediment concentration on windy days, back silting will be produced in water intake channel of Tianwan nuclear power plant. In order to obtain the design parameters related to the sediment for the water system in the power station, the sediment transport and back silting in the intake open channel is numerically investigated in the present study.

The intake project includes open channel and intake tunnel. The intake open channel for the 3rd and 4th units contains a North and a South training dyke which have the eastward extension of 1.5 km on the basis of the first intake open channel. The North training dyke is extended 500 m to the northeast, the South training dyke is extended 2600 m to the northeast, and the dike centerline spacing is about 320 m. The intake open channel is used by the planned capacity of Unit 1–8. For the single training dyke, the intake open channel is formed through excavation, where the bottom width and the elevation of the open channel are 190 m and –6.6 respectively. The intake flow for 1st–4th units are 51 m^3/s. The intake flow for 5th–8th units is 57.82 m^3/s.

A combination of culverts + drainage jetty + excavation open channel drainage way is used by drainage. An outfall is shared by 3#~6# units, which is located on the south of the first phase of outlet, which has been built. The drainages by open channel of 1 #~8 # units are formed by excavation, in which the excavation length is about 1300 m, the bottom width is 60~80 m, and the bottom elevation is –3.5 m. The outlet is located at about –2.0 m depth contour.

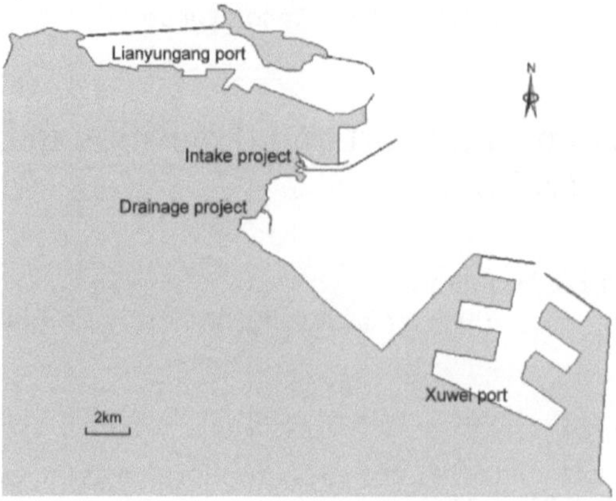

Figure 1. Position of the project.

2 METHOD

2.1 *Numerical model*

In the paper, typhoon wind-field model, wave model and the three-dimensional model for the tidal current and sediment transport is used to predict the maximum sediment concentration and back silting at the front end of the intake open channel of the 3rd and 4th reactor units under the influence of strong wind like typhoon Wipha. The calculation frame of the models is shown on Figure 2.

A large number of researches indicate, Fujita. T. (1952) and Takahashi (1939) formula can simulate pressure distribution under the typhoon well and has been applied to engineering practice.

The SWAN model can represent the wind wave generation and its evolution in the nearshore region. This model is used to simulate the wave under the typhoon condition.

The three-dimension MOHID model is used to simulate tidal current, sediment transport and deposition depth in intake open channel under typhoon Wipha.

2.2 *Model scope*

The size of the computational domain of wind field is 1590 × 2390 km. It is from the south of Taiwan Island to the Liaodong Bay, and eastern end is 132°E. The rectangular cell is adopted and the mesh size is 5 km × 5 km.

The nesting scheme is used to simulate the waves under typhoon conditions. Four level models are adopted, i.e. China sea model, Haizhou Bay model, Lianyungang model and project region model. The computational domain and model setup for the China sea model is the same to the typhoon model. The domain size of the Haizhou Bay model is 295 × 435 km with mesh size of 1 × 1 km. The southern, northern and eastern ends are 132°S, 37°N and 122°E. The domain size of the Lianyungang model is 74.4 × 61.6 km, including the ocean area near Lianyungang. Its southern end is near the estuary of Guan River, the northern end is in the northern of Lianyungang port, and the water depth at the eastern end is about 23 m.

The nesting scheme is used for the three-dimensional model. The larger region model is about 68 km × 55 km, and the northern and southern ends are at 34°55'N and 34°25'N, respectively. The grid of the smaller region model is refined in the area of interest, such as the main channel of Lianyungang port and the area near the power station. The maximum and minimum mesh sizes are 300 m and 10 m.

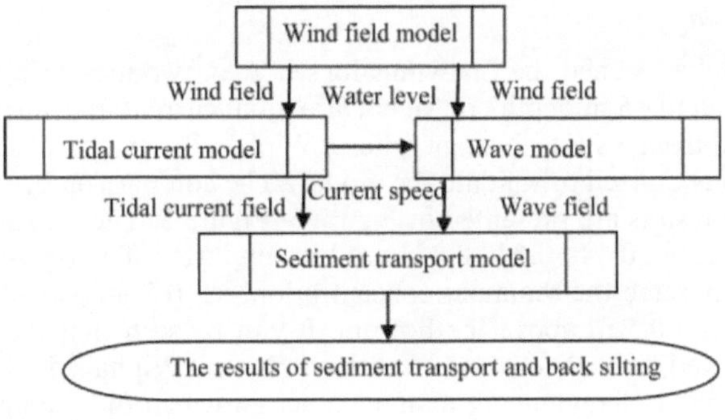

Figure 2. The calculation frame of the model.

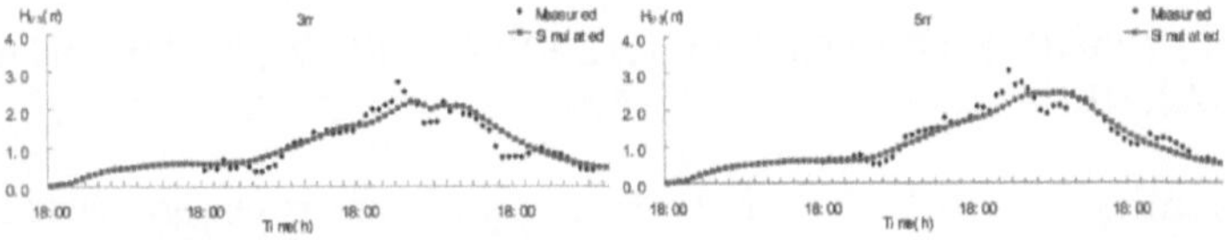

Figure 3. The compared result between computed wave height and measured wave height (The beginning time is 18:00 September 17, 2007).

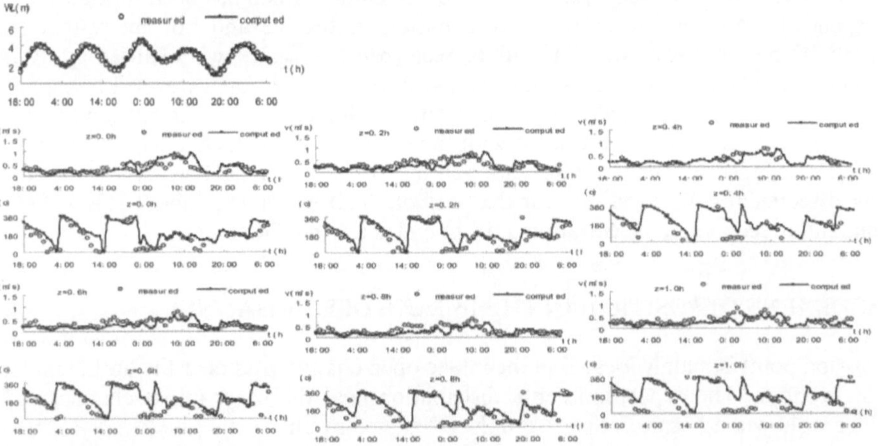

Figure 4. Verification of water level, flow velocity and direction process during the typhoon at –5 m station (The beginning time is 18:00 September 17, 2007).

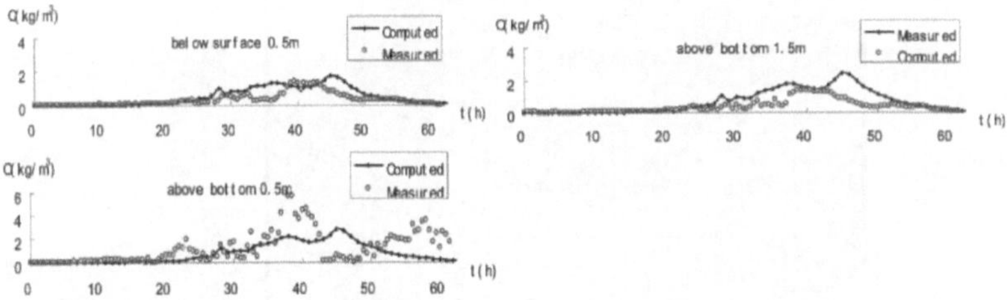

Figure 5. Verification of sediment concentration process during the typhoon at –5 m station (The beginning time is 18:00 September 17, 2007).

2.3 *Model validation*

Before typhoon Wipha reached the Lianyungang sea area, two measuring-stations were set respectively at –3 m and –5 m depth contour. Wave, tidal current and suspend load concentration are observed from 2 spots during typhoon Wipha. The in-situ observation data during typhoon Wipha is utilized to validate the models. The comparison between the observed data and calculated results are presented in Figure 3~Figure 5. The velocities in 6 layers are shown, i.e. surface, 0.2 h, 0.4 h, 0.6 h, 0.8 h and bottom layers. Three positions in the vertical are adopted to observe the sediment concentration, i.e. 0.5 m below the surface, 1.5 m above the bottom and 0.5 m above the bottom. It can be seen that the simulated results agree with the observed data. Before and after the typhoon "Wipha", 5.6 km of water depth measurement are done in lianyungang main channel, in which the average silting intensity is about 0.5 m. The computed average silting intensity is 0.45 m (the dry density 200 kg/m^3), which is basically identical with measured data.

3 SEDIMENT CONCENTRATION AT THE FRONT END OF THE INTAKE OPEN CHANNEL

Sediment concentration distribution of the project sea areas is closely related to weather conditions. Sediment concentration is relatively low on normal weather, however, on windy days, especially in typhoon effect, high-concentration layer at the near-shore will be occurred, in which the waves play a major role on this phenomenon. The sediment suspensions in intake open channel after construction under the "Wipha" Typhoon are calculated based on the tidal current and sediment transport numerical model, which has been verified.

Suspended sediment concentration distribution in surface and bottom waters during the typhoon period are shown in Figure 6. Seen from the figure, wave caused by "Wipha" typhoon is very large which can wash beach, sediment induced by waves will transport and diffuse in the tidal current. Sediment concentration is high in outer segments and low in the breakwater. Different drainage volume with the same intake open channel scheme has little effect on suspended sediment concentration in the project sea area. The maximum vertically averaged sediment concentration near the head of the breakwater is about 1.27 kg/m^3, and the maximum sediment concentration at front end is about 0.745 kg/m^3.

4 SEDIMENT DEPOSITION IN THE INTAKE OPEN CHANNEL

Deposition point is mainly located in the intake open channel and near the outfall under the Wipha typhoon. The deposition depth distribution near the water intake channel and the outfall are shown in Figure 7. Seen from the calculated results.

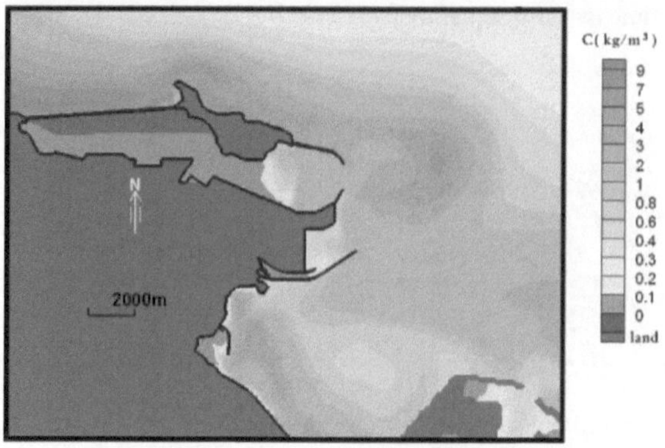

Figure 6. The distribution of bottom sediment concentration at 3:00 am on September 20, 2007.

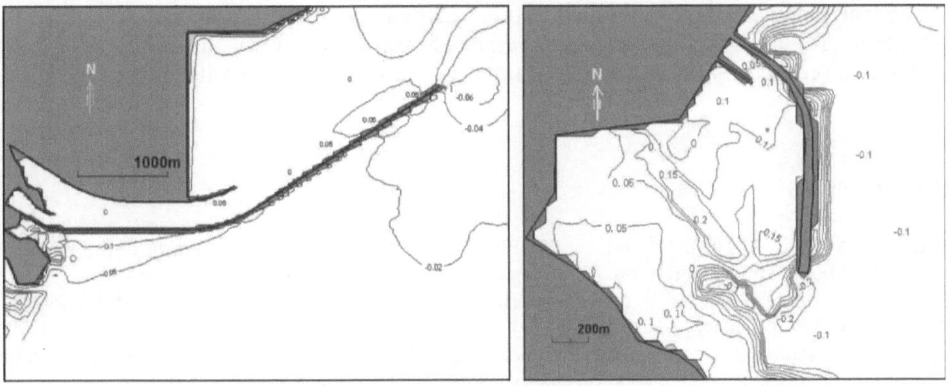

Figure 7. Distribution of deposition and erosion depths in the intake and drainage channel after the Typhoon (Unit:m).

Near the intake open channel: The erosion point is located in the outside and south of the south breakwater head, its deposition point is located in open channel, between the north breakwater and Qitai port and the south of the north breakwater. The distribution trend of deposition intensity in the intake open channel is slightly from the entrance, and then to increase first, then decrease gradually, near the water intake, basically not silt.

Near drainage channel: Erosion point is located in the entrance and outside of drainage channel and deposition point is located in the entrance of the breakwater. The deposition in the drainage channel is mainly caused by the existence of the drainage channel breakwaters. A sheltered harbor is formed near the outfall, the flow results show, many circulation is formed in the harbor, sediment entered from the entrance, is mostly settled down.

The deposition depth in the intake open channel is below 0.11 m and the maximum deposition depth near the outfall is 0.20 m under the Wipha typhoon.

5 CONCLUSIONS

1. In the paper, typhoon wind-field model, wave model and a 3D tidal current and sediment transport model are employed to study the sediment concentration and back silting in the intake open channel of the Tianwan nuclear power plant under the influence of typhoon like Wipha. Verification of wave height, tidal currents and suspended load concentration during the "Wipha" typhoon shows that the calculated values are in good agreement with the measured ones. In addition, the deposition in the main channel of Lianyungang during the "Wipha" typhoon are verified, and the calculated depths of deposition as well as their distribution are closed to the measured ones, indicating that the model generalization are right and the models can be employed for test on the intake open channel of the Tianwan nuclear power plant.
2. The sediment transport calculated results of mathematical model show that sediment concentration is relatively low in normal weather conditions and on windy days, especially during the typhoon, high-concentration layer at the near-shore will be occurred. Sediment concentration is high in outer segments and low in the breakwater. Different drainage volume with the same intake open channel scheme has little effect on suspended sediment concentration in the project sea area. The maximum vertically averaged sediment concentration near the head of the breakwater is about 1.27 kg/m^3, and the maximum sediment concentration at front end is about 0.745 kg/m^3.
3. Under the action of "Wipha" typhoon, maximum deposition depth in the intake open channel is 0.11 m. So after the implementation of the project, it didn't emerge obvious sudden deposition.

4. The mathematical model developed in this paper can be applied to predict deposition in other muddy coast port, but in order to ensure forecast accuracy of the model, basic data should be collected such as wave data, tidal current data and the sediment concentration data.

REFERENCES

Cancino L. & Neves R. 1994. 3D-numerical modelling of cohesive suspended sediment in the Western Scheldt estuary (The Netherlands). Netherlands Journal of Aquatic Ecology, 28.

Huang J.B. & Zhao X. & Li Z.Y. 2008. Numerical simulation for typhoon wave at north bay of the South Sea. Port & Waterway Engineering, (1): 7–10.

Luo Z.S. 2004. Sediment transport under the coexisting action of waves and currents and the prediction of sudden sedimentation in navigation channel. Journal of Sediment Research, (6): 1–9.

Ris R.C., Holthuijsen L.H., Booij N. 1994. A spectral model for waves in the near shore zone. Coastal Engineering, (1): 68–78.

Yu F.J. & Zhang Z.H. 2002. Implementation and application of nested numerical storm surge forecast model in t he East China Sea. Acta Oceanologica Sinica, 24 (4): 23–33.

Zhang N. & Yang H. & Yan B. & Zhao H.B. 2011. Three-dimensional numerical simulation of cohesive sediment transport due to a typhoon. Proceeding of the sixth international conference on Asian and Pacific coasts (APAC 2011).

Zhang N. & Yang H. & Yan B. & Zhao H.B. 2012. The application of SWAN/MOHID-based three-dimensional numerical model in 300,000 ton deep-water channel of Lianyungang Port. Port & Waterway Engineering, (9): 149–154.

Modeling and Computation in Engineering III – Zhang & Xie (Eds)
© 2014 Taylor & Francis Group, London, ISBN 978-1-138-02680-3

Hydrodynamic simulation of XiangShan harbor and adjacent waters based on the high resolution numerical model

Yu-Xiang Chen
Key Laboratory of Coastal Disaster and Defence, Ministry of Education, Hohai University, Nanjing, China
Key Laboratory of the Pearl River Estuarine Dynamics and Associated Process Regulation, Ministry of Water Resources, Guangzhou, China

Qi-Yan Ji & Qing Xu
Key Laboratory of Coastal Disaster and Defence, Ministry of Education, Hohai University, Nanjing, China

Hua-Zhi Zou
Key Laboratory of the Pearl River Estuarine Dynamics and Associated Process Regulation, Ministry of Water Resources, Guangzhou, China

Min Zhang
South China Sea Marine Prediction Center, State Oceanic Administration, Guangzhou, China

ABSTRACT: In this paper a High Resolution Hydrodynamic Simulation Model of XiangShan harbor and Adjacent Waters was build based on FVCOM. The results show: tidal type in XiangShan harbor belongs to the regular semidiurnal tide. The maximum amplitude of M_2, S_2, K_1 and O_1 constituents is 1.70 m, 0.70 m, 0.30 m and 0.24 m respectively. There is a tidal phase lag about six hours between the velocity of Fodu channel and Niubi mountain channel. The tidal current out the entrance of XiangShan harbor is rotary current and mainly shows clockwise. In the sea area between Gangpan mountain and the entrance of XiangShan harbor, the tidal current is reversing current, on the surface, the current ellipse is counterclockwise in the sea near the north shore and clockwise near the south shore, but the rotation is opposite on the bottom. The average tidal prism of XiangShan harbor is 12.21×10^8 m^3, and the water exchange rate is 44.22% in 30 days.

1 INTRODUCTION

XiangShan Harbor is located in the northern coast of ZheJiang province, 56 nautical miles away from BeiLun and Zhoushan and 179 nautical miles away from ShangHai. XiangShan Harbor is adjacent to the ZhouShan sea area by Fodu channel, ShuangQiMen channel at the northeast, and is connected to the open seas through NiuBi mountain channel at the southeast (Fig. 1). It is a semi-closed bay accessing from northeast to southwest to the inland. The shoreline of XiangShan Harbor twists and turns with many islands (more than 60), submarine topography is complex, with port in port, such as the XiHu harbor, HuangDun port and Tie harbor, the tidal flat is developing in these branch harbors.

Domestic scholars have done some research about the tidal characteristics of XiangShan harbor. Chi Wanqing (2004) do hydrodynamic numerical simulation of the XiangShan harbor using POM model; Wu XiaoYan & Guan Weibin (2009) simulate three-dimensional tide current of XiangShan harbor on the basis of POM combined with the moving boundary treatment technology which can better handle overbank problem; Miao QingSheng et al. 2010 establish three-dimensional tide current model in the XiangShan harbor based on ECOMSED (Estuarine Coastal Ocean Model and Sediment Transport Model). Due to the adopting of the

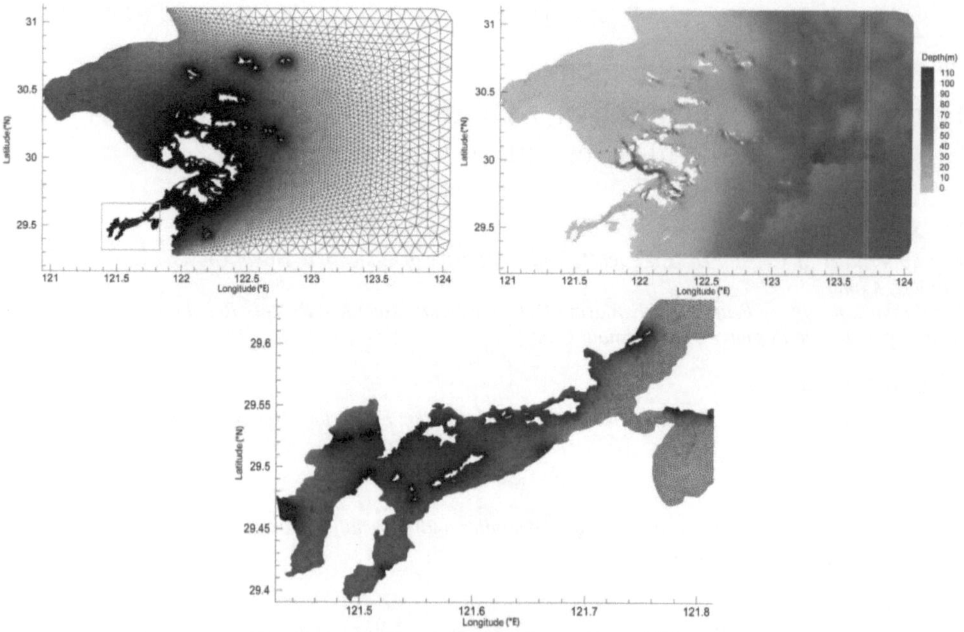

Figure 1. Mesh and depth in XiangShan harbor and around Zhoushan.

orthogonal curvilinear grid or rectangular grid, all these have obvious shortage of depicting shoreline for computing area and terrain features of channels and islands. FVCOM (Finite Volume Coastal Ocean Model) is favored by more and more scholars in recent years as it can better fit complex coast of islands and depict submarine topography characterization by adopting unstructured triangular mesh. Zhu Junzheng (2009) establish three-dimensional tide current model in the XiangShan harbor based on FVCOM; Cao Yin & Zhu Junzheng (2009) simulate the convection diffusion process ofwarm discharge water in XiangShan harbor by using FVCOM; Tu shanshan & Jie Jing (2012) establish three-dimensional hydrodynamic model in XiangShan harbor and do research about diffusing and drifting regularity of pollutants based on FVCOM; Due to selecting open boundaries on XiangShan harbor's two major waterways, the XiangShan harbor is treated as a single system, and the open boundary driver is obtained by interpolation of measured data, tidal results lack a good precision. This article place the open boundary far away from the harbor, ZhouShan islands and the HangZhou bay is included in the calculation; hydrodynamic model of XiangShan harbor is established with high precision. The verification results show that the model simulated values and measured values coincide well, the simulation accuracy is improved.

2 MODEL CONFIGURATION AND VERIFICATION

FVCOM is a nearshore marine numerical model developed by Chen Changsheng (2003) and his team in the university of Massachusetts. FVCOM have the following features: it use unstructured triangular mesh on horizontal direction and sigma coordinate transformation on vertical direction, use finite volume numerical method, solve control equations by triangle control volume flux integral, combine the advantages of fitting coastline of the finite element method with the advantages of high efficiency calculation of finite difference method; it is better able to ensure the conservation of quality and momentum for coastal and estuary area with complex topography and coastline. FVCOM also contains three dimensional wet/dry grid processing module which can handle overbank problem. So adoption of FVCOM model for the numerical simulation study on the XiangShan harbor in this paper is very suitable.

2.1 *Model configuration*

In order to better represent the tide of XiangShan harbor, this article include ZhouShan islands and the HangZhou bay into the numerical simulating area, the calculation area is set as 120°54′E–124°6′E, 29°18′N–31°6′N. In the model, the water depth of XiangShan harbor is measured by the second Marine institute of the state oceanic administration in 2012, the depth that out harbor come from the Navigational Chart published by the navigation guarantee department of the Chinese navy headquarters in 2006. The unstructured triangular mesh is used on horizontal direction, seven sigma level is set up on vertical direction. The paper do grid refinement near the complex coastline in XiangShan harbor and Zhoushan islands, as to better fit the topography. The calculation area include a total of 103,219 grid nodes, 198,019 grid triangles (cell), the minimum side of the grid rearch a length of 20 m (located in XiangShan harbor), the length of open boundary is about 10 km, the external integral step length of the model is 0.50 s, internal step length is 5 s. The Model is barotropic, with GOTM turbulence module opened. It is visible from the Figure 1, this mesh can better fit the complex coast of XiangShan harbor and ZhouShan islands.

2.2 *Boundary conditions*

The barotropic model in this paper does not consider thermohaline change, thermohaline are set as constant, 32 psu and 20°C. Model is set with zero initial conditions ($t = 0$, $U = V = \eta = 0$), water level and flow rate are set to zero; the water level drive of the 77 nodes on the Open boundary is obtained by the Tide Model Driver (TMD) software using eight major astronomical constituent (M_2, S_2, N_2, K_2, K_1, O_1, P_1, Q_1) harmonic constants for forecasting. TMD is a Matlab software package of tide model in high-latitude ocean, and can predict the tidal level and current. This paper simulate two time phase in summer and winter, integral time of numerical simulation is from July 20, 2011 to August 31, 2011 and February 1, 2012 to March 31, 2012.

2.3 *Model verification*

2.3.1 *Tidal level verification*

Figure 2 is a distribution diagram of water level and currents observation points. Due to space limitations, this article selects the hourly data from 0:00 on February 12 to 23:00

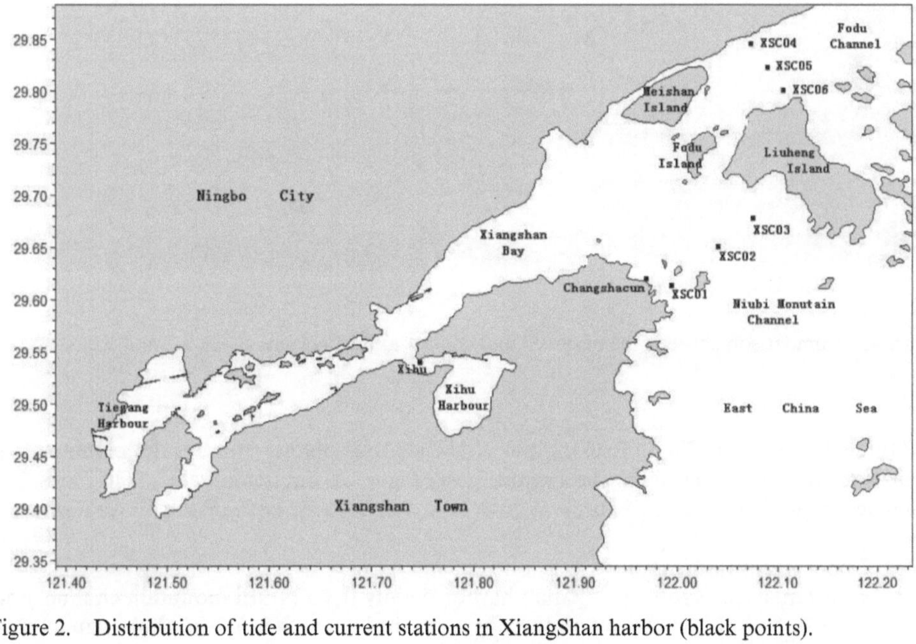

Figure 2. Distribution of tide and current stations in XiangShan harbor (black points).

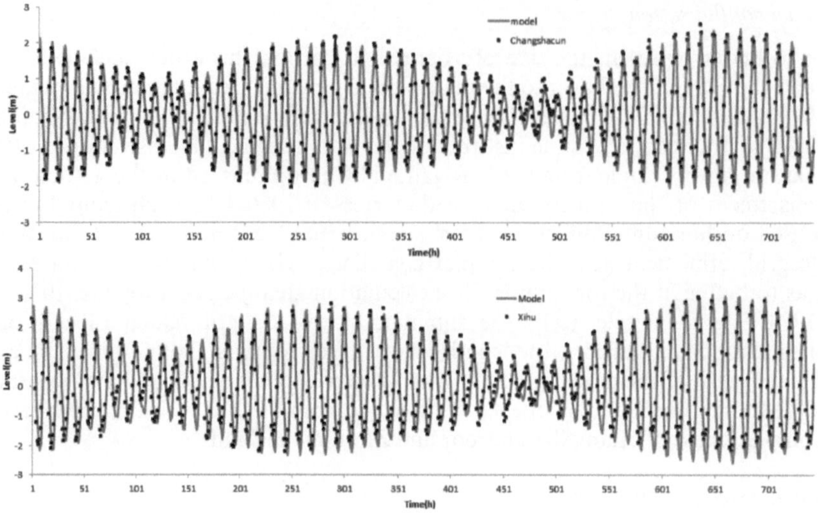

Figure 3. Comparison between the observed and computed tidal elevation at Changshacun and Xihu stations.

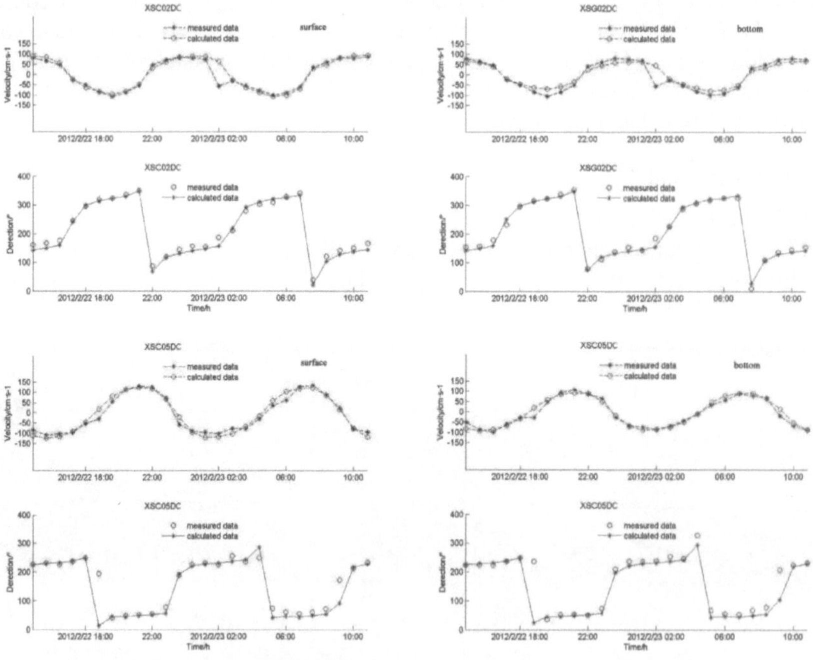

Figure 4. Comparison between the observed and computed tidal current at XSC02 and XSC05.

on March 13, 2012 of ChangShaCun and XiHu station to verify the results of numerical simulation. It can be seen from the Figure 3, the curve of simulated values and measured values accord well. The average absolute error is within 10% for both of the two stations.

2.3.2 *Tidal current verification*

The tidal currents access into XiangShan harbor mainly from NiuBi mountain channel and FoDu channel. So this article selects the observation data of XSC02 in Niubi cannel and

XSC05 in Fodu channel for verification of tidal flow. The time of measured data is from 15:00 on February 22, 2012 to 16:00 on February 23, 2012 in astronomical tide period. It can be seen from Figure 4 that the velocity and direction of tidal flow of simulated and measured values coincide well, the error of the simulation values from the surface to the bottom is in allowed range (15%).

Over all, the model about XiangShan harbor in this paper give precise result of tidal level and currents which can coincide well with measured data. The model established is credible for the research of tide characters in XiangShan harbor.

3 RESULTS ANALYSIS

3.1 *Tidal level analysis*

The tidal discriminate number F in XiangShan harbor is between 0.32 to 0.41, tidal nature belongs to regular semidiurnal tide. Harmonic analysis of the simulation results show that (Fig. 5), the four main tidal tidal constituent M_2, S_2, K_1, O_1 in XiangShan harbor are mainly introduced from the coast, amplitude is gradually increasing from the sea to the port, the amplitude of M_2 constituent can differ about 0.40 m between port and top of harbor; the maximum amplitude of the four main constituents M_2, S_2, K_1, O_1 in XiangShan harbor is 1.70 m, 0.70 m, 0.30 m and 0.24 m respectively, semidiurnal constituent is more dominant in the XiangShan harbor.

3.2 *Tidal currents analysis*

Figure 6 shows the average vertical tide currents at urgently rising and falling moments in XiangShan harbor during Spring. In the rising tide time, the water cross into XiangShan harbor through NiuBi mountain channel and FoDu channel. A branch is separated from the mainstream while arrive at middle area of XiangShan harbor and go into XiHu harbor. The mainstream is continually split into two branchs by BaiShiShan island, the two branchs go into HuangDun harbor and Tie harbor respectively. While it is ebb tide time, the water of each ports flow out of the entrance and go into the sea through the Niubi mountain channel and FoDu channel. The maximum velocity of tide currents occur in the southern area of GangPan mountain, it's 1.44 m/s and 1.29 m/s at urgently rising and falling moments respectively.

Figure 7 shows the average vertical tide currents at flood and ebb slack time. At flood slack time, one branch of the water come from NiuBi mountain channel access into

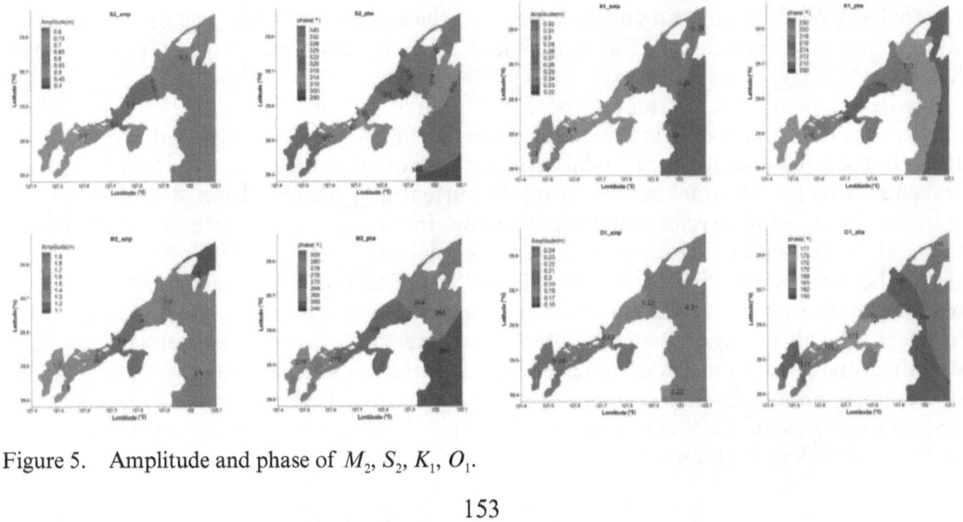

Figure 5. Amplitude and phase of M_2, S_2, K_1, O_1.

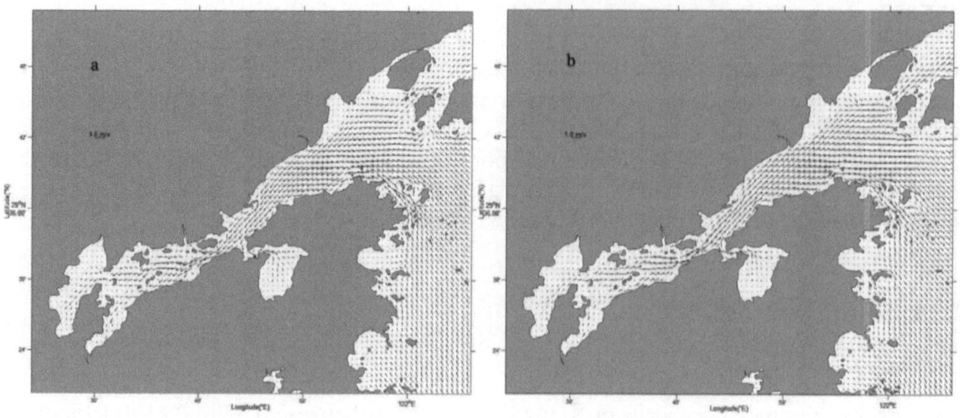

Figure 6. Tidal currents at urgently rising (a) and falling (b) moment.

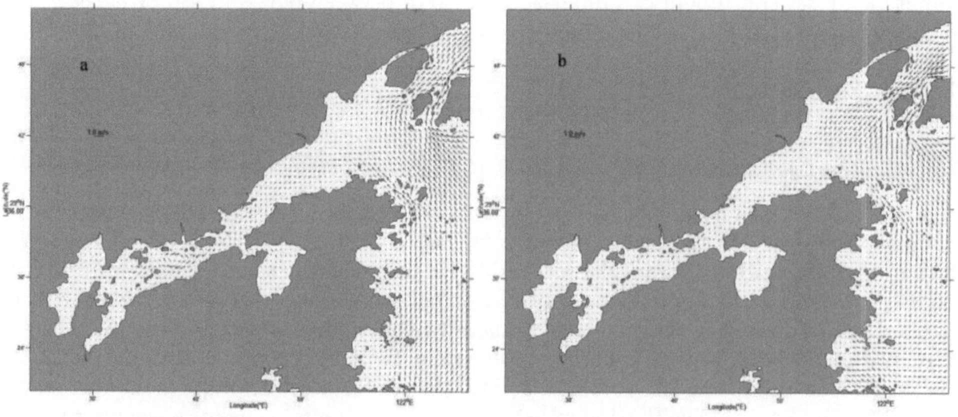

Figure 7. Distribution of tidal current in flood slack (a) and ebb slack (b).

XiangShan harbor and another branch flow out from FoDu channel. It is almost opposite at ebb slack time. This phenomenon is consistent with the above velocity verification curve. In Figure 4, there is a tidal phase lag about six hours between the velocity of FoDu channel and NiuBi mountain channel. It is also the same in summer. The phase lag has a significant impact in sediment transport, water exchange and water ecology in XiangShan harbor.

Figure 8 shows the tidal current ellipse distribution of M_2 and S_2 constituent in XiangShan harbor on surface and underlying layers (black circle: clockwise; gray circle: counterclockwise). It can be seen, the rotating rate of tidal current ellipse out the entrance of XiangShan harbor is relatively bigger, the tidal current is rotary current and mainly shows clockwise. Either surface or underlying layer have the phenomenon. In the sea area between GangPan mountain and the entrance of XiangShan harbor, the tidal current ellipse rotating rate is smaller, the tidal current is reversing current. On the surface layer, it is counterclockwise in sea area near the north shore and clockwise near the south shore, but it is opposite on the underlying layer. The phenomenon may be related to the tidal lag between two waterways, and this will affect the mixing process of water body in XiangShan harbor. In the western sea area of Gangpan mountains, tidal current ellipse rotating rate is also small, most of the flows are reversing currents, tidal current ellipse has staggered distribution, these are consistent on the surface and underlying layers.

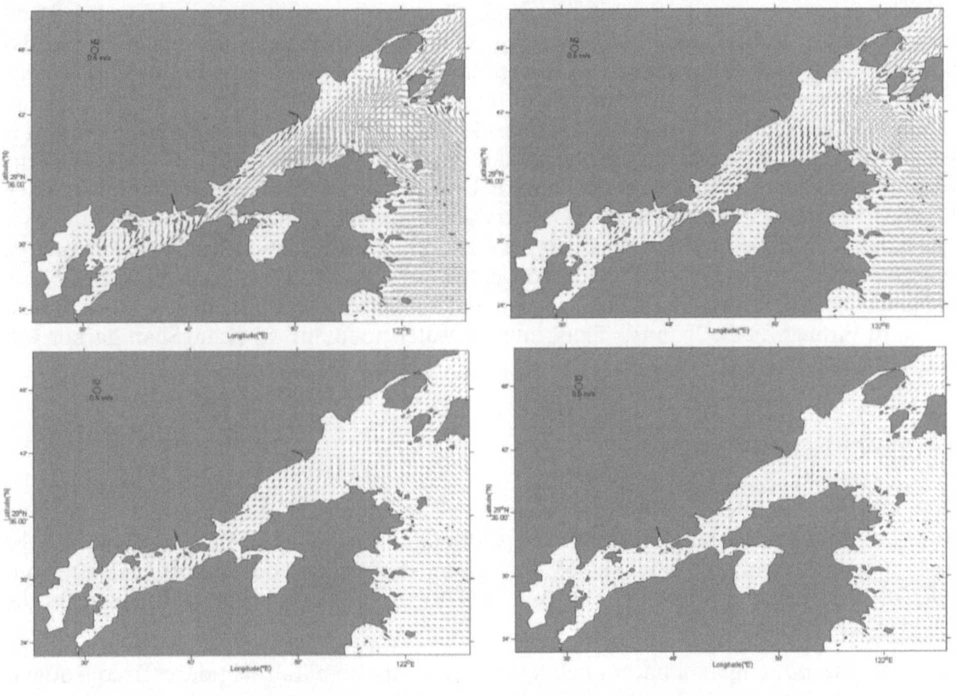

Figure 8. Distribution of current ellipse of M_2 and S_2 on the surface and underlying layers.

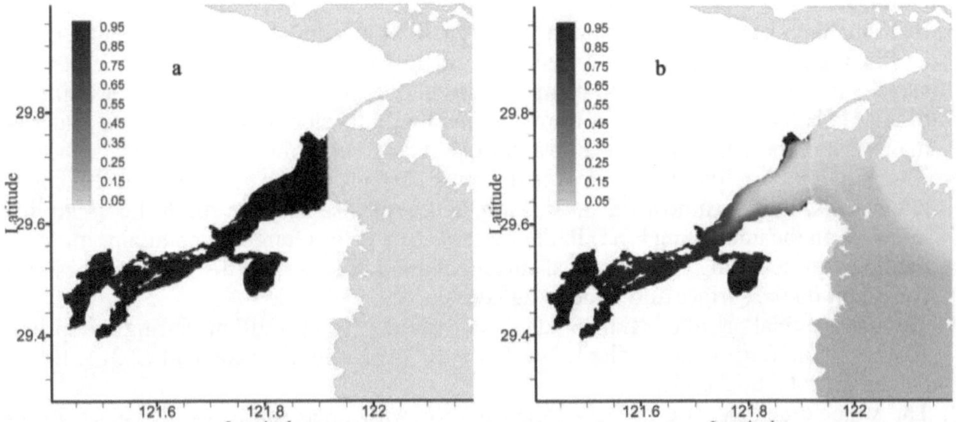

Figure 9. Distribution of Dye, (a) Begin, (b) After 30 days.

3.3 *Tidal prism and water exchange capacity*

The water quantity a bay can hold from low tide to high tide bay is called tidal prism. Tidal prism is one of the important indicators for hydrodynamic characterization of semi-closed harbor and its changes directly affect the tidal characteristics, water exchange with open sea, self-purification ability and environment capacity of the harbor, and may also break the erosion and deposition balance of harbor that will change the harbor's beach morphology. This paper adopts the calculation method mentioned by Chen Hongxia et al. 2009, the average tidal prism of XiangShan harbor is about 16.40×10^8 m^3 during the spring, 12.87×10^8 m^3 during the medium tide, 7.36×10^8 during the neap and 12.21×10^8 m^3 during the total tidal cycle.

Water exchange capacity is limited in XiangShan harbor because of its half closed and narrow nature, and its longitudinal change is obvious. This paper do numerical simulation about water exchange process in XiangShan harbor by using Dye module in FVCOM model, the result is shown in Figure 9(b). Initial concentration boundary line is shown in Figure 9(a), concentration in the port is set as 1 (black), the concentration outside the port is set as zero. Under the effect of tide, we simulate the water exchange for 30 days and define the water exchange rate as (1-Average content of Unit volume) × 100% (Zeng Xianming 2011). Figure 9(b) is the concentration distribution after 30 days. Calculated according to the concentration distribution, water exchange rate is 44.22% after 30 days. It can be seen from the concentration distribution after the exchange, the water go into Damu ocean through Niubi mountain channel from XiangShan harbor and spread to the south, but will not spread eastward through Fodu channel. Thus, internal water pollutants in XiangShan harbor will mainly affect south coast outside.

4 CONCLUSIONS

In this paper a high resolution hydrodynamic simulation model of XiangShan harbor and its adjacent waters was build based on FVCOM numerical model. In this model, we use the latest measured terrain data in XiangShan harbor. Eight major tidal constituents are considered. The simulated result and observed data are in good agreement. This paper has the following conclusions.

1. Tidal type in XiangShan harbor belongs to the regular semidiurnal tide, each constituent amplitude is gradually increasing from the sea into the port; the maximum amplitude of M_2, S_2, K_1 and O_1 constituent is 1.70 m, 0.70 m, 0.30 m and 0.24 m respectively. The amplitude of M_2 constituent at the top of bay is about 0.40 m larger than that at the entrance.
2. In XiangShan harbor, the maximum velocity is 1.44 m/s and 1.29 m/s at urgently rising and falling moments respectively. There is a tidal phase lag about six hours between the velocity of Fodu channel and Niubi mountain channel. The tidal current out the entrance of XiangShan harbor is rotary current and mainly shows clockwise, surface and underlying layer are the same. In the sea area between Gangpan mountain and the entrance of XiangShan harbor, the tidal current is reversing current. On the surface layer, it is counterclockwise in the sea near the north shore and clockwise near the south shore, but it is opposite on the underlying layer. In the western sea area of Gangpan mountains, most of the flows are reversing currents, tidal current ellipse has staggered distribution, these are consistent on the surface and underlying layers.
3. The average tidal prism of XiangShan harbor is about 16.40×10^8 m^3 during the spring, 12.87×10^8 m^3 during the medium tide, 7.36×10^8 m^3 during the neap and 12.21×10^8 m^3 during the total tidal cycle.
4. The water exchange rate of XiangShan harbor with open sea is 44.22% after 30 d's simulation. The water mainly go into Damu ocean through Niubi mountain channel from XiangShan harbor and spread to the south. Thus, internal water pollutants in XiangShan harbor will mainly affect south coast outside.

ACKNOWLEDGEMENT

This research was jointly supported by Ministry of Water Resources nonprofit sector project (201301073), open funded project of Pearl river water resources commission of the pearl river water conservancy science research institute ([2013]KJ02) and The state oceanic administration of Marine public welfare project (201105009). In the process of research for this paper, Key Laboratory of the Pearl River Estuarine Dynamics and Associated Process Regulation, Ministry of Water Resources provide great support. Thank you!

REFERENCES

Cao Ying & Zhu Junzheng. 2009. Numerical simulation of 3D cooling water based on FVCOM. *Chinese Journal of Hydrodynamics* 24(4):432–439. (Chinese).

Chen Changsheng et al. 2003. An Unstructured Grid, Finite-Volume, Three-Dimensional, Primitive Equations Ocean Model: Application to Coastal Ocean and Estuaries. *Journal of Atmospheric and Oceanic Technology* 20:159–186.

Chen Hongxia et al. 2009. Comparison Among Different Methods for Tidal Prism Calculation—Neap Tide of Jiaozhou Bay in Autumn 2006 as an Expounded Example. *Advances in Marine Science* 27(1):11–15. (Chinese).

Chi Wanqing. 2004. The Study of SusPended Sediment TransPortation and Distribution in Xiangshan Harbour. Qingdao: Ocean University of China. (Chinese).

Jie Jing et al. 2012. Numerical simulation of pollutant transport in Xiangshan bay. *Journal of Waterway and Harbor* 33(5):429–435. (Chinese).

Miao Qingsheng et al. 2010. Numerical Simulation and In-situ Measurement for heat discharge from XiangShangang power plant into sea. *Coastal Engineering* 29(4):1–11. (Chinese).

Tu Shanshan. 2012. Outfalls optimization settings of Xiangshan Bay. Dalian: Dalian University of Technology. (Chinese).

Wu Xiaoyan & Guan Weibin. 2009. 3D variable-boundary numerical modeling of tidal current in the Xiangshangang Bay. *Journal Of Marine Sciences* 27(2):23–31. (Chinese).

Zhu Junzheng. 2009. Numerical simulation of characteristic of 3-D tidal flow in Xiangshan bay. *Journal of Hydroelectric Engineering* 28(3):145–151. (Chinese).

Zeng Xiangming et al. 2011. Cumulative influence of long term reclamation on hydrodynamics in the Xiangshangang Bay. *Journal of Marine Sciences* 29(1):73–83. (Chinese).

Modeling and Computation in Engineering III – Zhang & Xie (Eds)
 ISBN 978-1-138-02680-3

Disinfection effectiveness of slightly acidic electrolyzed water in laying house

Y.T. Zang, C.H. Lu, Y.P. Tan, Y.N. Hu & S. Bing

Institute of Veterinary Medicine, Jiangsu Academy of Agricultural Sciences, Key Laboratory of Animal Diseases Diagnostic and Immunology, Ministry of Agriculture, National Center for Engineering Research of Veterinary Bio-products, Nanjing, China

X.S. Li, B.M. Li & W.C. Zheng

Key Laboratory of Structure and Environment in Agricultural Engineering, Ministry of Agriculture, China Agricultural University, Beijing, China

ABSTRACT: The efficacy of newly developed Slightly Acidic Electrolyzed Water (SAEW, pH 6.25–6.53) was investigated as a convenient and safe alternative sanitizer and it was compared to other sanitizers for inactivating microorganisms adhered to the facility and aerosolized in the air of the layer house. Also, the relationships between different ventilation level and concentrations of air microorganisms in the layer house were discussed. The SAEW, Potassium Persulfate Compound solution (PPC) and tap water were sprayed within the whole house. The ventilation level were divided into three degrees by opening or closing two doors (front and rear). SAEW with an Available Chlorine Concentration (ACC) of 200 mg l^{-1} can inhibit isolated microbes completely, the usage of SAEW (200 mg l^{-1}) resulted in a significant ($P < 0.05$) reduction in microbes on the surface of trough or cages (86.0%) and in the air (50.3%). Additionally, there were significant ($P < 0.05$) difference between the ventilation level.

1 INTRODUCTION

The poultry industry makes a significant contribution to the china economy. But the profitability of this industry was severely affected by the pathogenic microorganisms with huge economic consequences and chanced human infections (Graham et al. 2008). Disinfection as a considerable disease-preventing measure that is commonly used is important to prevent the introduction and spread of both endemic and epidemic infections (Sundheimet al. 1998; Rodríguez Ferri et al. 2010). Disinfection also has other objectives such as reducing or eliminating antibiotic use, improving productivity and achieving a higher health status of the chicken.

Slightly Acidic Electrolysed Water (SAEW) is one of the potential alternatives microbial decontamination that has been proven to exhibit strong bactericidal activity for inactivating many pathogens (Honda, 2003), SAEW is generated by electrolysis of dilute salt solution in a single chamber, and has a pH of 5.0–6.5, contains a high concentration of hypochlorous acid (HOCl, approximately 95%) and the hypochlorite ion (ClO, approximately 5%) (Hricova et al. 2008; Cao et al. 2009). Several studies have demonstrated that SAEW could be used as an environmental friendly sanitization method. Also, it has the advantage of reducing corrosion of surfaces and minimization of the potential for damage to human health (Abadias et al. 2008; Guentzel et al. 2008; Koide et al. 2009; Rahman et al. 2010). Several studies have demonstrated that SAEW could be used as a disinfectant with foods (Hricova et al. 2008; Huang et al. 2008), hospitals (Vorobjeva et al. 2004) and agricultural (Guentzel et al. 2010; Zhang et al. 2011) processes. As mentioned above, SAEW has been extensively studied. Hao et al. (2013) reported that SAEW can significantly reduce the microbes on the equipment or facility surfaces in the laying house. However, its application efficacy of reducing air microorganism in laying house has not been reported to the best of authors' knowledge.

The objective of this study was to evaluate the effectiveness of SAEW for inactivating the microorganisms in the air and on the surfaces of facilities in the laying house, the bactericidal efficiency of SAEW, tap water and commercial disinfectants. Potassium persulfate compound, which is routinely used on the farm, was compared simultaneously. Moreover, the relationships between different ventilation level and concentrations of air microorganisms in the layer house were discussed.

2 MATERIALS AND METHODS

2.1 *Experimental house*

Experiments were conducted in two different size neighboured layer houses with different dimensions of 20 m × 7 m × 3 m (length × width × height) and 17 m × 6 m × 3 m (length × width × height) located in the suburban district of Beijing, northern China. The study was carried out in the winter of 2013. These layer houses all containing approximately 1000 breeding hens (91 d old) confined in 2 rows of 2-tier stair-step cages and 3 paths. The house was ventilated with an automatic negative-pressure ventilation system with 5 fans located in the end wall (not use) and 2 doors each located in the front and rear. Feed was provided by the workers, water was supplied by drinking systems nipples.

2.2 *Preparation of disinfectants*

Slightly acidic electrolysed water was produced using a nonmembrane generator (Zhouji., Ltd., Beijing, China) to electrolysing NaCl (1 g/l). SAEW with a pH of 6.25–6.53, an oxidation–reduction potential (ORP) of 887–924 mV, and an Available Chlorine Concentration (ACC) of 200 mg/l was produced by the SAEW generator.

The physicochemical properties of SAEW were measured before use. The pH and ORP values were measured using a dual scale pH/ORP metre (CON60, TransWiggens, Singapore) with a pH electrode (PE02; range 0.00–14.00) and an ORP electrode (ORP06; range −999 to +999 mV). The ACC was determined using a digital chlorine test system (RC-2Z, Kasahara Chemical Instruments Corp., Saitama, Japan). The detection range was 0–320 mg/l. Potassium Persulfate Compound (PPC, kunyuan., Biotechnology Co., Ltd., Hunan) were purchased from commercial suppliers.

2.3 *Spraying treatment to the whole house*

In order to reduce the humidity of the layer house, the experiments were performed in two house and all treatments were repeated 3 times at 2 PM once every 2 days. Each different disinfectants (SAEW, PPC) and water were sprayed by hand in the two neighbouring different Layer house with High voltage electric cleaning device (Jingjiang nine long Motor Co., Model Y112M-4, nozzle radius 1 mm, pressure 0–14 Kpa). The larger house was sprayed at 120 ml/m^3, and the other house was sprayed at 100 ml/m^3, each disinfectants and water was sprayed for 20 min. A treatment–control study was performed in each site and sampling occurred in experimental areas (treatment) and neighboured untreated areas (control) in a house. Areas that were treated with the two disinfectants on each site were closed but not overlapping. The Colony-Forming Units (CFU) of microorganisms on the surfaces or in the air were measured before and after Spraying with the two disinfectants, respectively. And triplicate samples were collected at each area of each site.

Moisten sterile swabs with Buffered Peptone Water (BPW, Beijing Lanyi Bio-tech Co., Ltd., Beijing, China) were used to collect the surface samples (5 × 5 cm^2) one area per site on the trough or cages. The sterilized cotton swabs, which had been wiped back and forth for twenty times on the surfaces, were immediately transferred into 5 ml buffer solution (BPW; pH 7.0) tubes for microbiological analyses. Triplicate tests were carried out. Swab samples taken into BPW were returned to the laboratory under ambient conditions on the day of collection and processed. The total viable organisms in the buffer were quantified by plating appropriate serial dilutions onto Plate Count Agar (PCA, Beijing Land Bridge Technology Co., Ltd., Beijing, China) plates, and the CFU were counted. The percent reduction in microbes was calculated relative to the control using the following formula (Guentzel et al. 2008, Hao et al. 2013):

$$PR = 100(P_C - P_T)P_C \tag{1}$$

where PR is percent reduction (%); P_C is the survival populations of microbes in the control; P_T is the survival populations of microbes in the treatment.

Air sedimentation method was used to evaluate the concentrations of airborne microorganisms before and after disinfection, the air samples were collected in four different locations in laying house. The four sampling sites for each house are shown in Figure 1. The sampling time was one min. The PCA was used to count the total visible microorganisms. All samples were returned to the laboratory under ambient conditions on the day of collection and cultured at 37°C for 2 days.

The spraying disinfection efficiency on the air was measured with the percent of reduction in the total microbes and calculated is as follows:

$$PR = 100(P_B - P_A)/P_B \tag{2}$$

where PR is percent reduction (%); P_B is the survival populations of microbes in the air before treatment; P_A is the survival populations of microbes in the air after treatment.

2.4 *Ventilation level classification*

The ventilation level were divided into three degrees as good, ordinary and poor by opening two doors (front and rear), opening one door (front or rear), and closing two doors (front and rear). The concentrations of airborne microorganisms was evaluated by above mentioned method.

2.5 *Statistical analyses*

Means of bacterial populations from each treatment were calculated from three replications for each experiment. Data was expressed as the means ± standard errors. The results were analyzed using SPSS statistical package (SPSS Inc.,) and Significant differences ($P < 0.05$) between mean values were determined using Duncan's multiple range tests.

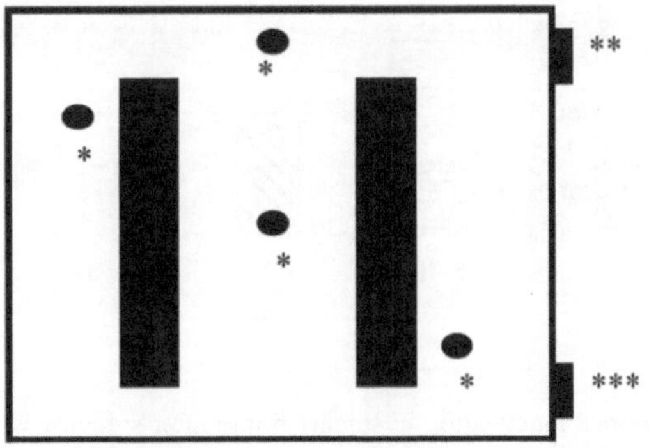

Figure 1. Diagram of the four sampling sites.
*) sampling sites **) front door ***) rear door.

3 RESULTS

3.1 *Microbial inactivation efficiency of SAEW in the air with spraying treatment*

The effect of spraying SAEW, PPC, and tap water to reduce contamination in the air inside the larger laying house is presented in Figure 2. The spraying time was 20 min. SAEW had the greatest disinfection, followed by PPC solution, but compared with spraying tap water, which had the highest microbial population in the air of the layer house. The airborne microbes decreased by 50.3%, 26.6%, and 14.3% after 30 min of exposure to SAEW, PPC solution and tap water. There was no significant difference ($P < 0.05$) microbial reduction between SAEW and PPC solution, also no significant difference ($P < 0.05$) between PPC solution and tap water. However, there was significant difference ($P < 0.05$) microbial reduction in the air between SAEW and tap water.

The effect of spraying SAEW, PPC, and tap water to reduce contamination in the air inside the smaller laying house is presented in Figure 3. The spraying time was 20 min. SAEW had the greatest disinfection, followed by PPC solution, but compared with spraying tap water, which had the highest microbial population in the air of the layer house. The airborne microbes decreased by 51.2%, 26.7%, and 16.8% after 30 min of exposure to SAEW, PPC solution and tap water. There was no significant difference ($P < 0.05$) microbial reduction between SAEW and PPC solution, also no significant difference ($P < 0.05$) between PPC solution and tap water. However, there was significant difference ($P < 0.05$) microbial reduction in the air between SAEW and tap water.

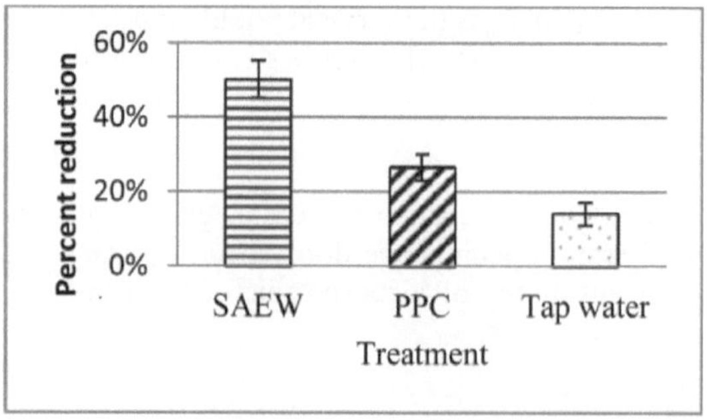

Figure 2. Airborne microbial inactivation in larger house after spraying with Slightly Acidic Electrolyzed Water (SAEW), Potassium Persulfate Compound solution (PPC) and tap water. Bars labeled with different letters indicate significant difference at $P < 0.05$. Vertical bars represent means of three replications ±SE.

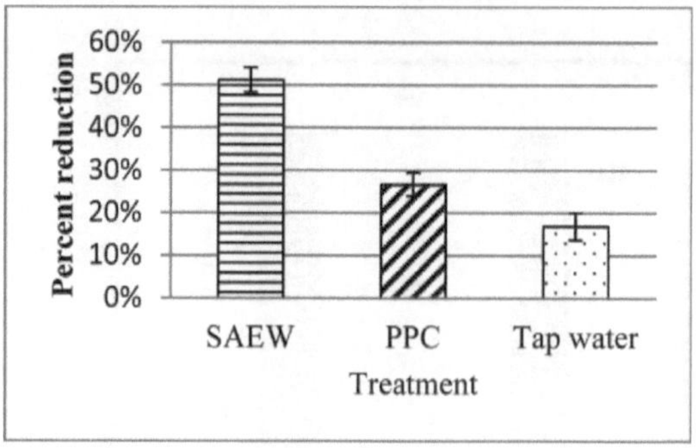

Figure 3. Airborne microbial inactivation in smaller house after spraying with Slightly Acidic Electrolyzed Water (SAEW), Potassium Persulfate Compound solution (PPC) and tap water. Bars labeled with different letters indicate significant difference at $P < 0.05$. Vertical bars represent means of three replications ±SE.

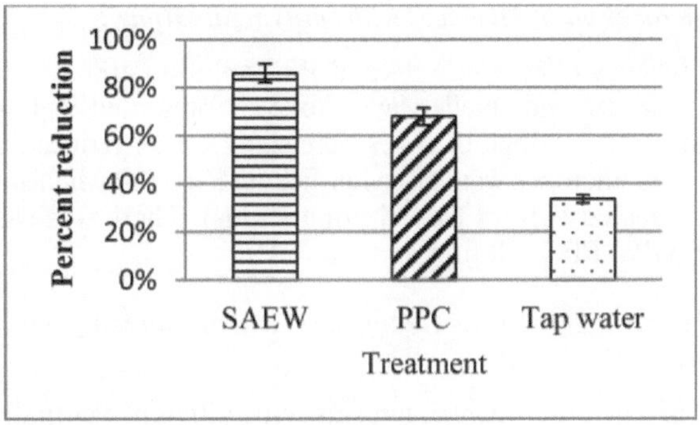

Figure 4. Microbial inactivation on surface of trough or cages in larger house after spraying with Slightly Acidic Electrolyzed Water (SAEW), Potassium Persulfate Compound solution (PPC) and tap water. Bars labeled with different letters indicate significant difference at $P < 0.05$. Vertical bars represent means of three replications ±SE.

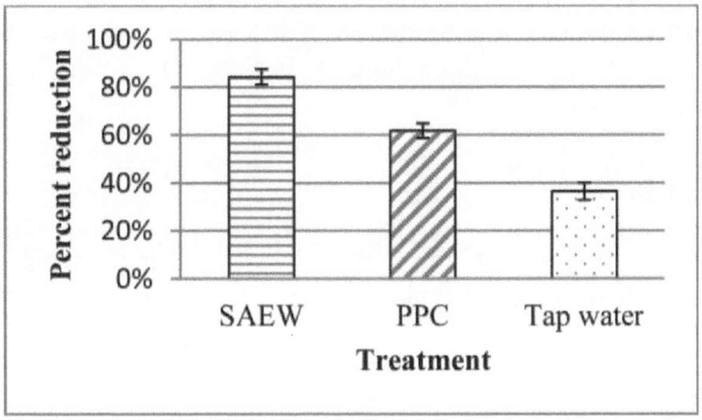

Figure 5. Microbial inactivation on surface of trough or cages in smaller house after spraying with Slightly Acidic Electrolyzed Water (SAEW), Potassium Persulfate Compound solution (PPC) and tap water. Bars labeled with different letters indicate significant difference at $P < 0.05$. Vertical bars represent means of three replications ±SE.

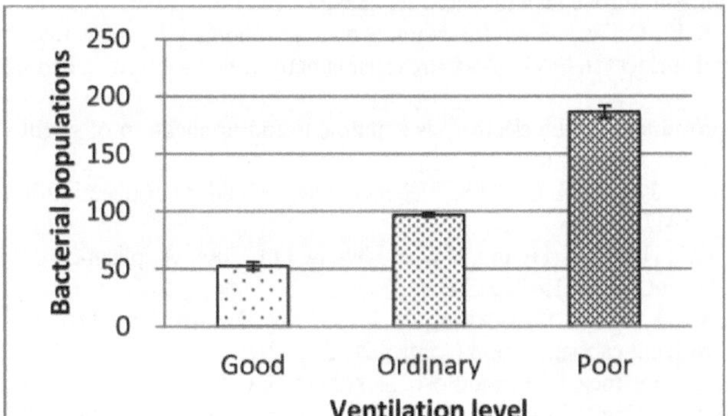

Figure 6. Airborne microorganisms with different ventilation level inside the layer house. Bars labeled with different letters indicate sinificant difference at $P < 0.05$. Vertical bars represent means of three replications ±SE.

3.2 *Reduction in microbes on the surfaces with spraying treatment*

The antimicrobial efficiency of spraying SAEW, PPC solution and tap water on the surface of trough or cages in larger and smaller layer house is presented in Figures 4 and 5. The microbial populations on the trough or cages were significantly reduced by SAEW spraying ($P < 0.05$). Reductions in microbes were also significant ($P < 0.05$) when spraying treated with PPC compared to untreated control (initial populations). There was significant difference ($P < 0.05$) between SAEW, PPC solution and tap water.

3.3 *Relationships between different ventilation level and air microorganisms in the layer house*

The concentrations of airborne microorganisms with different ventilation level inside the layer house is presented in Figure 6. The microbial populations between the good, ordinary and poor level were significant ($P < 0.05$) difference.

4 DISCUSSION

Slightly acidic electrolyzed water with a near-neutral pH value (5.0–6.5) is an effective way to reduce microorganisms in layer houses. It can significantly inactivate bacteria and fungi on facility surfaces or air ($P < 0.05$). The reduction of the airborne microorganisms concentrations with the increasing ventilation level reflects that the ventilation level should be better increased under the premise of warmth.

ACKNOWLEDGEMENTS

This work was supported by Chinese Universities Scientific Fund, Special Fund for Country Public profession (agriculture) special topic research (200903055).

REFERENCES

Abadias, M., Usall, J., Oliveira, M., Alegre, I. & Vinas, I. 2008. Efficacy of neutral electrolyzed water (NEW) for reducing microbial contamination on minimally processed vegetables. Int J Food Microbiol 123, 151–158.

Cao, W., Zhu, Z.W., Shi, Zh.X., Wang, Ch. Y. & Li, B.M. 2009. Efficiency of slightly acidic electrolyzed water for inactivation of Salmonella enteritidisand its contaminated shell eggs. Int J Food Microbiol 130, 88–93.

Graham, J.P., Leibler, J.H., Price, L.C., Otte, J.M., Pfeiffer, D.U., Tiensin, T. & Silbergeld, E.K. 2008. The animal–human interface and infectious disease in industrial food animal production: rethinking biosecurity and biocontainment. Public Health Rep. 123, 282–299.

Guentzel, J.L., Lam, K.L., Callan, M.A., Emmons, S.A. & Dunham, V.L. 2008. Reduction of bacteria on spinach, lettuce, and surfaces in food service areas using neutral electrolyzed oxidizing water. Food Microbiol 25, 36–41.

Honda, Y. 2003. Improvement of the electrolysis equipment and application of slightly acidic electrolyzed water for dairy farming. J. JSAM 65, 27–29.

Hricova, D., Stephan, R. & Zweifel, C. 2008. Electrolyzedwater and its application in the food industry. J Food Prot 71, 1934–1947.

Huang, Y.R., Hung, Y.C., Hsu, S.Y., Huang, Y.W. & Hwang, D.F. 2008. Application of electrolyzed water in the food industry. Food Control 19, 329–345.

Koide, S., Takeda, J., Shi, J., Shono, H. & Atungulu, G.G. 2009. Disinfection efficacy of slightly acidic electrolyzed water on fresh cut cabbage. Food Control 20, 294–297.

Rodríguez Ferri, E.F., S. Martínez, R. Frandoloso, S. Yubero, & C.B. Gutiérrez Martín. 2010. Comparative efficacy of several disinfectants in suspension and carrier tests against Haemophilus parasuisserovars 1 and 5. Res. Vet. Sci. 88:385–389.

Rahman, S.M.E., Ding, T. & Oh, D.H. 2010. Inactivation effect of newly developed low concentration electrolyzed water and other sanitizers against microorganisms on spinach. Food Control 21, 1383–1387.

Sundheim, G., Langsrud, S., Heir, E. & Holck, A.L. 1998. Bacterial resistance to disinfectants containing quaternary ammonium compounds. Int Biodeter Biodegr 41, 235–239.

Modeling and Computation in Engineering III – Zhang & Xie (Eds)
© 2014 Taylor & Francis Group, London, ISBN 978-1-138-02680-3

The evaluation of animal health risk management and resource allocation in economic

C.H. Lu, Y.N. Hu, Y.P. Tan, D.W. Yu, S. Bing & Y.T. Zang
Institute of Veterinary Medicine, Jiangsu Academy of Agricultural Sciences, Key Laboratory of Animal Diseases Diagnostic and Immunology, Ministry of Agriculture, National Center for Engineering Research of Veterinary Bio-products, Nanjing, China

ABSTRACT: With the current situation of the number of outbreaks of food-safety events, it has become essential for governments around the globe to increase awareness and preparedness for identifying and containing those agents. In order to protect the supply and public health safety of animal products effectively. This article introduces the basic concepts and methods of animal health risk analysis and animal health risk management decision. Additionally, the rub of the evaluation of animal health risk management decision was pointed out, and the view was proposed that different levels economic impacts and the risk of the animal disease introduction should be combined. Meanwhile, countermeasure was proposed taking shortage of current risk management mechanism on china into account.

1 INTRODUCTION

In recent years, the sustainable and healthy development of livestock industry has been severely damaged by continuous outbreaks of series food-safety case, vast concern have been raised to governments and the food industry. The best way to reduce incidences of food-borne diseases is to secure safe food supply. Although Hazard Analysis Critical Control Point (HACCP) system has been implemented in many food processing establishments, but, most outbreaks of food-borne illnesses still occurred. This situation indicates that hazards might still exist in the food supply systems. The World Health Organisation (WHO) has recommended animal health risk analysis as a important means to prevent the transmission of disease between infected and susceptible animals, eliminate reservoirs for disease spread and increase animal food safety standards. Risk management is the corresponding measures based on the results of risk assessment measures to reduce the risk of epidemic disease and its potential consequences to guarantee the healthy development of animal husbandry and related industries. As the most important part of risk analysis, it should be combined with the risk assessment process, however, China do not have the periodically evaluation program which was independent and reasonable applicated by scientists before the implementation of the risk management, the earn warning risk of livestock product quality and safety control can not be achieved because of the response and the remedy of the risk can not be timely, which is the cause of China's animal health risk management mechanism, can not be structured, fundamental and practical. We must strengthen the encouragement and guide scientific research departments have enough attention and research on animal health risk analysis.

2 THE BASIC STEPS OF ANIMAL HEALTH RISK MANAGEMENT DECISION

In the animal health management field, animal and human life and health risk has not been interrupted. The main body of animal health risk management is animal health administration departments. Including animal health administrative departments, supervision and law enforcement departments and the technical support departments. Animal health risk analysis is essentially a whole management decision process of the competent authorities at all levels in the implementation of risk management on animal health. For the effective supply of livestock products and public health and safety, to steadily improve Chinese animal health safety level. The basic steps for management decision of animal health risk and methods is establishing a scientific and reasonable mechanism. Animal health management decision process usually consists of five steps:

1. The definition of problems and opportunities;
2. The distinction between the process of interaction;
3. The collection of relevant information and the analysis of the process of interaction;
4. The decision making and implementation;
5. The evaluation of the final output.

Some risk decision index often need to be put forward aiming at the difficulty of management decision of animal health risk assessment, the index can be got by domestic and foreign literature, written questionnaires or on-the-spot investigation, the management of animal health risk can be finished by analysis of decision behavior: In consider of the three aspects of the incoming assessment, assessment and consequence assessment; The heoretical and empirical research of basic scientific animal health risk analysis and management of the formation, the development and application of information database and the risk early warning system. At the same time of the rational allocation of resources and animal health economics evaluation, according to the animal health management mechanism and the formation of livestock and poultry disease control investment cost—effect analysis, the animal health management data model of resources rational allocation was proposed to provide the basis for the establishment of Chinese effective animal health risk management and risk early warning mechanism.

3 THE ECONOMIC EVALUATION OF MANAGEMENT RESOURCES ALLOCATION SCHEME

3.1 *Internal research content*

1. Policy and capital investment, epidemic prevention and health system construction, livestock products, international trade and animal health risk management related information.
2. The starting from the microscopic point of view, analysis of resource investment, management and utilization mechanism of related information; analysis of the cause of livestock and poultry diseases, the veterinary service and drug which was considered as the special type of investment can be instead of improving management, making full use of other resources such as land, these inputs' economic efficiency can be increased by reducing disease, the combine of good veterinary input and other inputs will produce the best economic efficiency, the core problem is the countermeasure evaluation of the animal disease control effect.
3. In the intensive cultivation of large enterprises, the disease prevention and control measures should be considered in financial evaluation of economic losses caused by livestock and poultry disease.

3.2 *External research*

The external environment includes the development level of social and economic, the development level of science and technology, development of animal disease prevention and control measures. Before making the animal disease prevention and control measures, the cost and scheme of the prevention and control implementing effect need to be weighed. The lost of disease and the cost of the prevention measures should be comprehensive estimated.

Through the research on the above two parts, animal health resource reasonable configuration management and data model were proposed, and the national animal resources health risk management sharing platform was established.

3.3 *Methods of the evaluation*

The input and the output' technical and economic efficiency was determined based on the collection of certain information that was owned through the resource management mechanism investigation of China animal health institutions, departments, units; and health care services, animal health resource utilization, distribution of animal survival health status were analysised by the economic evaluation method.

Econometrics was analyzed through linear analysis and dynamic analysis methods, and risk decision theory was used as a guide to determin the influence elements and build data models.

4 THE ANIMAL HEALTH RISK MANAGEMENT RESOURCES ALLOCATION RECOMMENDATIONS

4.1 *Data acquisition problem*

In the project implementation process, data collection is a practical problem. But in the animal husbandry developed countries, such as Germany which data collection construction include three levels: enterprise self monitoring inspection, the neutral external agencies monitoring inspection and the government inspection monitoring system of the animal product safety. A lot of important basic data for risk assessment was provided by many accurate and reliable monitoring results. Therefore, the existing and potential food safety risk can be assessed and managed in timely through the Federal Bureau of consumer protection and food safety (BVL) monitoring coordination and European rapid alert system. However, animals and animal products safety monitoring and early warning system in China is not perfect, the lack of basic data has become the fundamental issues for implementation of the risk analysis, the existing risk assessment in china was occurred passive. Risk analysis as the animal and its products internationally recognized safety management means has become a priorities for Chinese animals and animal products safety work, some certain results have been achieved and improved by Chinese animals and animal products safety work. But comparing with developed country risk analysis field, the Chinese animals and animal product safety risk analysis is still a long way to go. We should speed up the construction pace of job separation, integration, system security, data research and technical support, active prevention and the evaluation of gradual realization of animals and animal products safety risk.

4.2 *The strengthen of economic evaluation of the rational allocation of animal disease risk assessment and risk management of resources*

Due to the problem of veterinary management system in our country, inspection and examination is not unified, exogenous animal epidemics and the risk assessment of spread should be joint with the Ministry of Agriculture veterinary department and the national quality inspection administration of animal and plant quarantine department. The possibility of the biological and economic consequence assessment presented in Figure 1 should be strengthened in animal disease prevention.

The animal health management measures can be proposed through the evaluation of the risk factors of the animal health and animal products safety in the production process, so as to continuously improve the ability of scientific prevention and control level of major animal epidemic. Animal health event can be early warned in advance through animal health risk analysis, and be prevented timely and effectively, and the effects of the sick animals itself and the tackle measures on the environment such as disinfection, culling, disposal can be reduced.

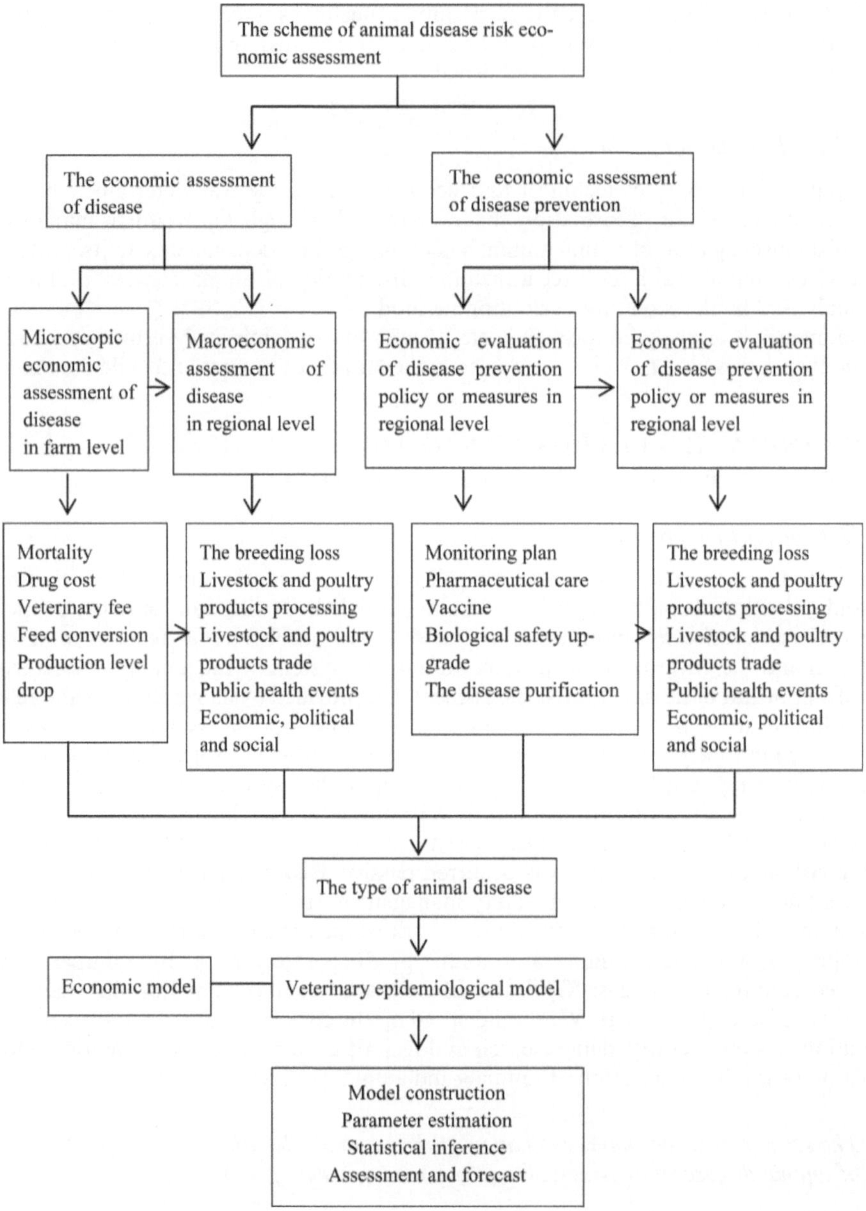

Figure 1. The scheme of animal disease risk economic assessment.

ACKNOWLEDGEMENTS

This work was supported by Country Public profession (agriculture) special topic research (200903055).

REFERENCES

Ad. ijkhuizen, R. s. Morris. 2006. The Animal Health Economics—principles and applications [M]. New Zealand: Post Graduate Foundation in Veterinary Science.

Geng R.L., Wang X.S., Huang H.W., 2007. Large-scale pig farms major epidemic diseases and comprehensive prevention and control strategy [J]. Journal of China animal husbandry and veterinary 12, 97–98.
He W.J., 2005. Risk management [M]. Beijing: China financial economic publishing house.
Jiang N.H., Xin X., Yin J., 2003. Animal products supply demand in China and trade behavior research [M]. Beijing: China agriculture press.
Li C. 2006., The food safety monitoring and warning system [M]. Beijing: chemical industry press.
Lu C.H., Wang C.J., Wu Z.M., 2006. Animal health economics and practice [M]. Beijing: China agricultural science and technology press.
Lu C.H., Wang C.J., He K.W., 2011. Animal and its products for animal health risk analysis [M]. China's agricultural science and technology press.
Wei H.Y., 2011. The development trend of 2011 swine disease and preventive measures [J]. Modern livestock and poultry breeding industry, 2: 60–61.
Yang H.C. 2007. China's large-scale pig disease epidemic characteristics and control countermeasures [J]. Modern livestock and poultry breeding industry, 2, 46–47.
Zhao Z.J., Liu X.M., 2005. Eoli O157: H7 infection outbreak assessment model caused by flies spread [J]. Journal of disease control, 9: 23–28.
Zhang Y.H. 2007. Food safety guarantee mechanism research [M]. Beijing: China water conservancy and hydropower press.
Zang Y.T., Tan Y.P., Hu Y.N., 2012. The construction of risk analysis model in Large scale pig farm [J]. Journal of agriculture in jiangsu, 28 (2): 365–369.

Modeling and Computation in Engineering III – Zhang & Xie (Eds)
© 2014 Taylor & Francis Group, London, ISBN 978-1-138-02680-3

Remote sensing dynamic monitoring of drought based on TVDI in dongting lake basin

Jun Liang, Shengmei Yang, Xianquan Han & Fangfang Zhou
Changjiang River Scientific Research Institute, Wuhan, Hubei, China

Li Zhou
Zhanghe Project Management Office, Jinmen, Hubei, China

Bo Hu
Institute of Geodesy and Geophysics, Chinese Academy of Sciences, Wuhan, Hubei, China

ABSTRACT: Normalized Difference Vegetation Index (NDVI) and Land Surface Temperature (LST) are both important index of drought monitoring. MODIS data is used in this paper, and Temperature Vegetation Dryness Index (TVDI) is obtained by establishing NDVI-Ts character space which inverts surface temperature by using NDVI and split window method, and it does well in spatial distribution of land surface drought degree, so as to realize the drought monitoring. It is dynamic monitored that the time and space characteristics of drought from different periods in a year in Dongting lake Hunan Province. The results show that the scope of the drought occurred is coinciding with the occurrence of the drought by TVDI; the main area of heavy drought in the distribution is basically consistent. We also used the related research results to test and verify drought monitor results, and it turned out to be credible and was able to provide a powerful support to decision-making departments.

1 INTRODUCTION

Drought is a natural hazard and complex phenomenon. Agricultural drought refers to a period with declining soil moisture content and consequent crop failure (Mishra and Singh, 2010). Drought is considered to be one of the most costly hazards (Wilhite, 2000) because it can lead to reduced water supply and consequently have substantial effects on agriculture and socioeconomic activities (Riebsame et al., 1991). The severity of drought depends on its duration, intensity, spatial extent, and local socioeconomic conditions. The detrimental impacts on crop production affect large areas (Lobell and Field, 2007).

Thus, monitoring droughts in the region is of primary importance for water planning and management to reduce possible impacts on crop production. Remote sensing has been recognized as a useful tool for large-scale area monitoring of droughts in terms of moisture content in the top few centimeter of soil (NASA, 1999). Several satellite-based indices have been developed and used to effectively detect and monitor droughts. The Normalized Difference Vegetation Index (NDVI) (Rouse et al., 1974) has been the most widely used for evaluating drought conditions (Henricksen and Durkin, 1986; Jain et al., 2009; Ji and Peters, 2003; Karnieli et al., 2010).

In this study, MODIS NDVI and LST data were used to investigate agricultural drought monitoring in Dongting Lake, Hunan Province. MODIS data have been widely used for regional drought monitoring (Li et al., 2008; Mallick et al., 2009; Patel et al., 2008; Rhee et al., 2010; Wan et al., 2004). Because NDVI provides little information about soil water content and LST is relatively related to water stress, the combination of LST and NDVI can provide better information on vegetation and moisture conditions at the surface. A simplified method, the temperature vegetation dryness index (TVDI) based on an NDVI–LST combination, has been suggested for assessment of surface moisture status (Sandholt et al., 2002; Liu Zhenboa et al., 2012).

2 STUDY AREA

Hunan is a province of China which located in the middle reaches of the Yangtze River and south of Lake Dongting between 109°–114° east longitude and 20°–30° north latitude. Lake Dongting (Fig. 1) is the largest lake in the province and the second largest freshwater lake of China. Due to the reclamation of land for agriculture, Lake Dongting has been subdivided into many smaller lakes, though there is now a trend to reverse some of the reclamation, which had damaged wetland habitats surrounding the lake. The east, south and west sides of the province are surrounded by mountains and hills which occupy more than 80% of the area and the plain comprises less than 20% of the whole province. The center and northern parts are somewhat low and a U-shaped basin, open in the north and with Lake Dongting as its center.

Hunan's climate is subtropical, with mild winters and plenty of precipitation. January temperatures average 3 to 8°C while July temperatures average around 27 to 30°C. Average annual precipitation is 1200 to 1700 mm. Hunan's traditional crop is rice. The Lake Dongting area is an important center of ramie production, and Hunan is also an important center of tea cultivation.

3 REMOTE SENSING DATA

Two kinds of MODIS data set is provided by Geospatial Data Cloud, Computer Network Information Center, Chinese Academy of Sciences: 1-day composite land surface temperature (LST, Fig. 2) MODLT1D product and the 1-day the normalized different vegetation index (NDVI, Fig. 3) MODND1D data. The two datasets (NDVI and LST), respectively each period, were acquired during the periods from 5 March 2005 to 10 Ocotober 2005. All raw pixel values of both MODIS data sets are expressed as gray scale digital numbers (DN)

Figure 1. Dongting lake in hunan province.

LT_050305 LT_050424 LT_050717 LT_050910 LT_051010

Figure 2. LST (land surface temperature) products in 2005.

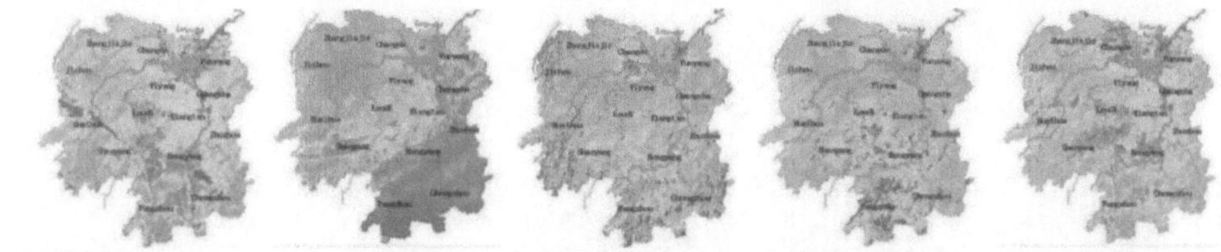

ND_050305 ND_050424 ND_050717 ND_050910 ND_051010

Figure 3. NDVI (normalized difference vegetation index) products in 2005.

and need to be converted into real LST and NDVI by applying scale factors and correction formulas with the HDF Explorer MODIS data processing software. Then, both kinds of images were geometrically rectified to the longitude–latitude coordinate projection system using a polynomial model available through MRT, and then re-sampled with the aid of the nearest neighbor method. One of the limitations of MODIS data was cloud contamination.

4 METHODS OF EXTRACTING DROUGHT INFORMATION

Remote sensing images were estimated from evaluating values of the Temperature Vegetation Dryness Index (TVDI) which combines information from the Normalized Difference Vegetation Index and Land Surface Temperature (Sandholt et al., 2002). A number of studies describing soil moisture as assessed by the combination of remotely sensed surface temperature (Ts) and vegetation index (e.g., the normalized difference vegetation index, NDVI) have been done (Moran et al., 1994; Bhuiyan et al., 2006; Goward et al., 2002). Sandholt et al. (2002) found that there were many isolines in Ts/NDVI feature space, so they developed a water stress index, i.e., Temperature Vegetation Dryness Index (TVDI) using simplified Ts/NDVI feature space (Fig. 4). In the Ts/NDVI feature space, the wet edge (Ts-min, maximum evapotranspiration and thereby unlimited water access) was processed as a straight line parallel with NDVI axes, and the dry edge (Ts-max, limited water availability) was linearly correlated with NDVI (Fig. 4).

TVDI value ranging from 0 to 1 for each pixel of image was calculated based on its position in Ts/NDVI feature space. TVDI values 1 at the "dry edge" correspond to limited water availability and 0 at the "wet edge" correspond to maximum evapotranspiration and thereby unlimited water access. The higher the TVDI value the lower the soil moisture content is. The formula of TVDI calculation used here is expressed as

$$\mathrm{TVDI} = \frac{\mathrm{Ts} - \mathrm{Ts}_{min}}{\mathrm{Ts}_{max} - \mathrm{Ts}_{min}} \tag{1}$$

where LST is the observed surface temperature at a given pixel; NDVI is the observed normalized difference vegetation index; and a and b are the intercept and slope of the dry edge (the upper straight line in the triangle) calculated from the NDVI–LST space regression with small intervals of NDVI (LST_{max} = a + bNDVI), where LST_{max} is the maximum surface temperature observation for a given NDVI. The lower horizontal line of the triangle represents the wet edge (LST_{min}). LST_{min} was calculated by averaging a group of points in the lower limits of the scatterplots (Wan et al., 2004). The TVDI values range from 0 to 1: TVDI = 1 at the dry edge, indicating no evaporation from the soil or limited moisture supply; and TVDI = 0 at the wet edge, indicating maximum evaporation from the soil or unlimited moisture supply. The TVDI is categorized into five classes describing drought conditions (Fig. 3): wetness (0–0.2); normal (0.2–0.4); slight drought (0.4–0.6); moderate drought (0.6–0.8); and severe drought (0.8–1) (Fig. 5).

According to the classification results of images in 2005, the percentage of arid areas were classified and counted in order to understanding the relationship between TVDI and the degree of drought in Dongting Lake Hunan Province (Fig. 6). TVDI class 1 to 5 were

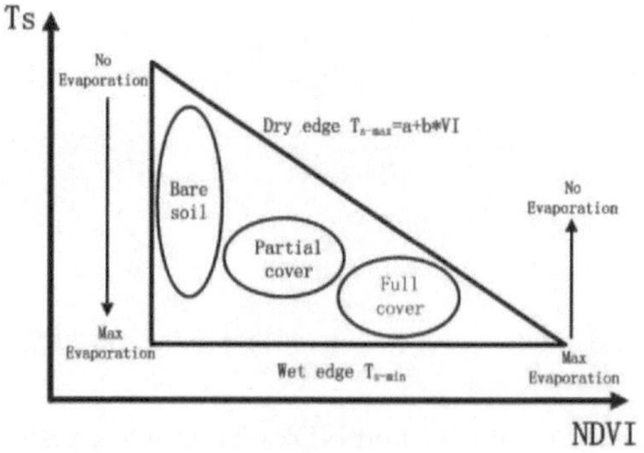

Figure 4. Simplified Ts/NDVI space (Sandholt et al., 2002).

TVDI_050305 TVDI_050424 TVDI_050717 TVDI_050910 TVDI_051010

Figure 5. TVDI (temperature vegetation dryness index) in 2005.

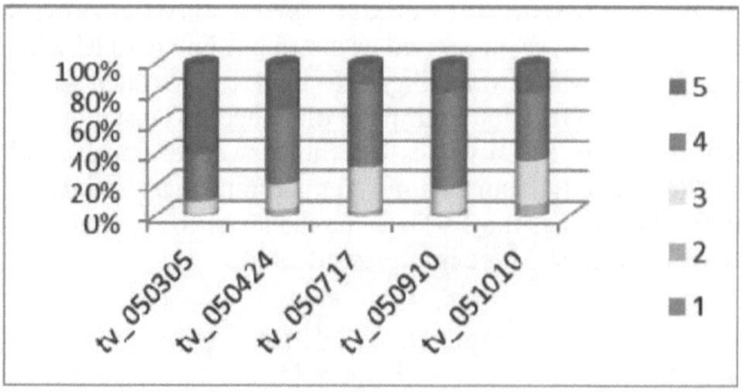

Figure 6. The percentage of arid areas on TVDI.

represented wetness, normal, slight drought, moderate drought and severe drought. It is easy to segment statistical for the variation tendency of drought from Figure 6.

5 CONCLUSIONS AND DISCUSSIONS

This study explored the use of MODIS NDVI and LST data for drought monitoring in Dongting Lake in Hunan Province and analyzed the degree of drought with time series data of NDVI, LST and TVDI, which make it possible to drought monitor real-time over vast area with low cost and simple operation.

It is demonstrated that the advantage of using MODIS NDVI and LST data for regional drought monitoring, a method that can be extended to other global locations for long-term monitoring of droughts over large-scale regions.

The main results were

1. The combination of LST and NDVI will enhance the ability to derive drought information without the utilization of time and resource intensive field work, which is especially important in the many areas of research that span more than small plots, but rather regional and global studies.
2. In Figure 6, it can be seen that the droughts began in the early dry season (March,) and likely returned to normal conditions in April with the onset of the rainy season. A larger area of moderate drought was observed in 2005 is July. The largest area affected by severe droughts was September during the 2005.
3. Findings from this study were consistent with the observations reported that the region had experienced droughts in 2005. Information on historical droughts in Dongting Lake obtained from this study could be important for agricultural drought risk management, water management, and irrigation scheduling for crop production.

ACKNOWLEDGEMENTS

This project was supported by a National Natural Science Foundation (grant no. 51209019, 41301434, 41204012) of China. Additional financial support was received from Basic Scientific Research Operating Expenses of Central-Level Public Academies and Institutes (Grant No. CKSF2012040/GC, CKSF2013022/GC, CKSF2014036/GC). The data set is provided by Geospatial Data Cloud, Computer Network Information Center, Chinese Academy of Sciences. (http://www.gscloud.cn).

REFERENCES

Adamson, P., 2005. Drought Study: Analysis, Forecasting, Planning and Management. *Mekong River Commission, Vientiane, Laos.*

Adamson, P., Bird, J., 2010. The Mekong: a drought-prone tropical environment International Journal of Water Resources Development 26, 579–594.

ADB, 2009. The Economics of Climate Change in Southeast Asia: A Regional Review. *Asian Development Bank, Manila, Philippines.*

Carlson, T.N., 2005. Review: an overview of the "triangle method" for estimating surface evapotranspiration and soil moisture from satellite imagery. *Sensors* 7, 1612–1629.

Chen, C.F., Son, N.T., Chang, L.Y., Chen, C.C., 2011. Monitoring of soil moisture variability in relation to rice cropping systems in the Vietnamese Mekong Delta using MODIS data. *Applied Geography* 31, 463–475.

Downing, T.E., Baker, K., 2000. Drought discourse and vulnerability. *In: Wilhite, D.A. (Ed.), Drought: A Global Assessment. Routledge, London, UK,* pp. 213–230.

Gao, B.C., 1996. NDWI—a normalized difference water index for remote sensing of vegetation liquid water from space. *Remote Sensing of Environment* 58, 257–266.

Goetz, S.J., 1997. Multisensor analysis of NDVI, surface temperature and biophysical variables at a mixed grassland site. *International Journal of Remote Sensing* 18, 71–94.

Gupta, A., Liew, S.C., 2007. The Mekong from satellite imagery: a quick look at a large river. *Geomorphology* 85, 259–274.

Hatfield, J.L., 1983. The utilization of thermal infrared radiation measurements from grain sorghum crops as a method of assessing their irrigation requirements. *Irrigation Science* 3, 259–268.

Henricksen, B.L., Durkin, J.W., 1986. Growing period and drought early warning in Africa using satellite data. *International Journal of Remote Sensing* 7, 1583–1608.

Hundertmark, W., 2008. Building drought management capacity in the Mekong River basin. *Irrigation and Drainage* 57, 279–287.

Liu Zhenbo, Shi Xuezheng, Eric Warner, 2008. Relationship between oriental migratory locust plague and soil moisture extracted from MODIS data. *International Journal of Applied Earth Observation and Geoinformation* 10 (2008), 84–91.

Son, N.T., Chena, C.F., 2012. Monitoring agricultural drought in the Lower Mekong Basin using MODIS NDVI and land surface temperature data. *International Journal of Applied Earth Observation and Geoinformation* 18(2012), 417–427.

Modeling and Computation in Engineering III – Zhang & Xie (Eds)
© 2014 Taylor & Francis Group, London, ISBN 978-1-138-02680-3

Numerical analysis on effects of deep foundation excavation to adjacent buildings

B.S. Xu, Q. Yan, W.T. Ding & Z.Y. Zeng
School of Civil and Hydraulic Engineering, Shandong University, Jinan, Shandong, China

W.D. Tang
Anhui Institute of Building Research and Design, Hefei, Anhui, China

ABSTRACT: In this paper, the New Traffic Building deep foundation excavation is studied. The effects of deep foundation excavation on neighboring buildings are analyzed using large finite difference software FLAC3D. The results show that the digging of the New Traffic Building foundation pit makes little effect on the settlement of surrounding buildings. The settlement and displacement of buildings increase with the increasing of excavation depth, and the settlement is mainly affected by construction conditions of the first, second and third steps.

1 INTRODUCTION

With the fast development of urban construction, lots of deep foundation pit projects have sprung up and they are usually surrounded by existing buildings and utilities. Therefore these deep excavations are subject to more severe restrictions from the surroundings. Studying the deformation of foundation pit during excavation and its effect on neighboring ground surface settlement presents important references for design and construction of deep foundation pit.

The ground surface settlement deformation during foundation pit excavation is induced by coupling of several factors. It is difficult for current computational theory to realize a complete and correct explanation for this. The numerical simulation methods, however, has been used extensively in engineering, which considers the nature of the soil, soil creep, soil-wall contact and other factors in calculation.

In this paper, the New Traffic Building deep foundation excavation is studied. The effects of deep foundation excavation and supporting on neighboring buildings are simulated and analyzed using large finite difference software FLAC3D.

2 PROJECT PROFILE

The New Traffic Building is located in the southeast intersection of Hefei railway line 1 and line 2. This building is for exhibition, office, commerce, meetings and dispatching. And it is serviced in the Communications Department and Railway control center. This project occupies 7366 m^2 of land (collectively, 11.05 acres), with a total construction area of 93,200 m^2. Its building includes high-rise towers (office), podium (including the control center) and basement located at the periphery of the towers and podium. All of the construction parts are placed on the same raft foundation chassis. The design working life of this project is 100 years, and its security level is the first class. This foundation pit is supported by two-in-one underground continuous wall, and constructed using horizontal structure inverse construction method. The maximum excavation depth of this foundation pit is 26.4 m. Adjacent buildings surrounding the project include Phoenix Bridge No. 10, Hefei Federation of Trade Unions dormitory Block B No. 1 and No. 2.

The terrain of the foundation pit is quite flat. It is dominated by channel and floodplain. And there is bedrock of 25 meters deep underneath the foundation pit. The sedimentary soil layers between ground and roof of bedrock mainly include mixed clay, silt and sand sediments. There is some soft soil in local regions, which mainly include beige and black silty clay. The layers with the maximum exploration depth of 60.00 m, according to sedimentary years and genetic type, can be divided into 3 sets of strata, respectively called artificial stacked layer, the quaternary holocene diluvial layer and the cretaceous bedrock layer.

3 NUMERICAL ANALYSIS

3.1 *Numerical model*

The 3D numerical model of the New Traffic Building foundation pit is created by FLAC3D. According to the project design conditions, the computational domain is defined as follows: The upper boundary of the model is ground surface; the bottom boundary is three times of excavation depth below the bottom of foundation pit; other horizontal boundaries extend to five times of excavation depth outward from pit edges. As shown in Figure 1, this foundation pit model is 360 m long, 340 m wide and 100 m high, including 33,407 nodes and 29,580 elements.

The bottom of the model is fixed. The upper boundary of calculation model is the ground, which is free. And unidirectional constraints are applied on the lateral boundaries of model. The constitutive model of soil is the classical Mohr-Coulomb ideal elastic-plastic constitutive model. The mechanical parameters of soil are shown in Table 1.

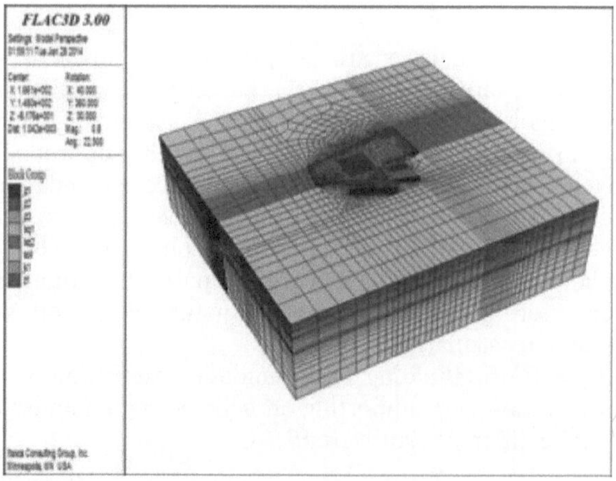

Figure 1. 3D numerical model.

Table 1. Mechanical parameters of soil.

Soil	Density γ(kN/m^3)	Cohesion c (kPa)	Frictional angle φ(°)	Elasticity modulus E (MPa)	Poisson ratio ν
Miscellaneous fill	17.5	0	8	3	0.3
Silty clay	19.7	30	15	10	0.25
Silt	20.3	25	16	11	0.3
Silty-fine sand	20.5	0	25	13	0.3
Highly weathered argillaceous sandstone	21.0	35	30	35	0.3
Moderately weathered argillaceous sandstone	22.0	35	30	80	0.3

According to on-site construction program, the New Traffic Building foundation pit is constructed using reverse construction method. There are five steps in total. Build the floor after completing the excavation of the first step, and finish the temporary support construction of main floor. Repeat the above steps until the excavation of five-storey basement is finished. And then dismantle the temporary support and build up floors from the bottom up. Numerical simulation use a total of 12 working conditions, including an initial equilibrium step, five excavation steps and six steps of support demolition and floors building, to simulate the actual construction conditions.

3.2 *The simulation of underground continuous wall*

Block element is used to simulate enclosure structure composed of underground continuous wall. The calculation model of enclosure structure is shown in Figure 2. The enclosure structure of the New Traffic Building foundation pit is partly coincident with the enclosure structure of rail transport line 1, 2. The overlapped part is 1.2 meters wide (underground continuous wall 1, blue zone), while the others are 1.0 meters wide (underground continuous wall 2, red zone).

During the simulation, interfaces are applied to simulate the interaction between enclosure structure and surrounding soil, and Mohr-Coulomb elastic-plastic constitutive model is used. The equivalent strength parameters of enclosure structure can be acquired through the method of equivalent stiffness. The calculation of equivalent stiffness refers to Equation 1.

$$EI = E_1 I_1 + E_2 I_2 \tag{1}$$

where E is the equivalent elasticity modulus of the beam element; E_1 and E_2 are elasticity modulus of concrete and rebar, respectively; I is the equivalent inertia moment of beam element; I_1 and I_2 are equivalent inertia moment of concrete and rebar, respectively. Parameters of equivalent beam element are shown in Table 2.

3.3 *Analysis of calculation results*

The final settlement of the New Traffic Building is shown in Figure 3. This picture is a top plan view of 3D model and shows the ground surface settlement at elevation ±0.00 meters. Affected by the excavation, the soil around foundation generates settlements of different degrees.

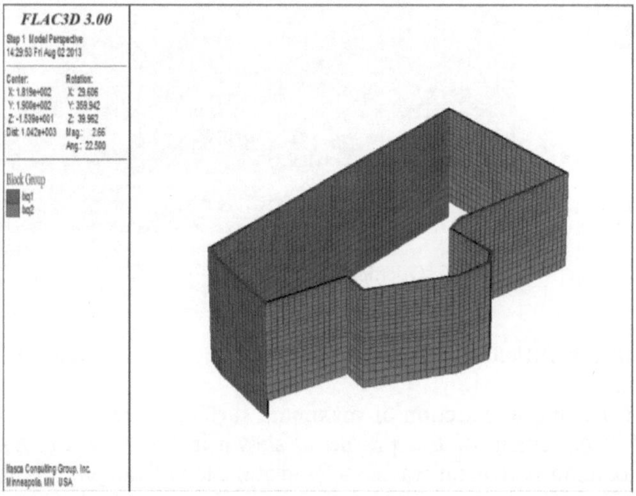

Figure 2. Calculation model of enclosure structure.

Table 2. Mechanical parameters of enclosure structure.

Enclosure structure	Elasticity modulus E (GPa)	Poisson ratio ν	Density γ(kg/m^3)
Underground continuous wall 1	42.6	0.3	2000
Underground continuous wall 2	44.5	0.3	2000

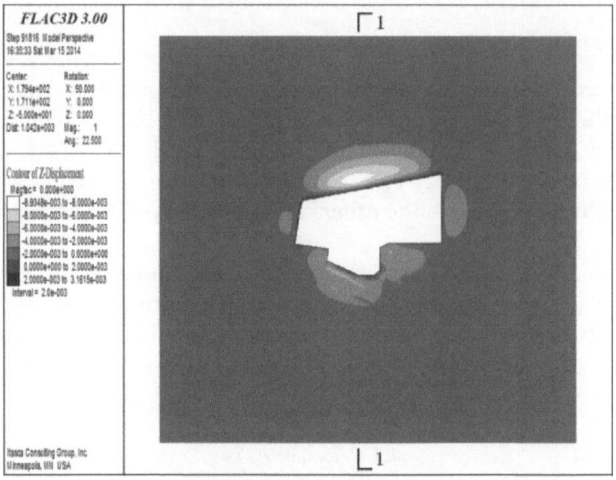

Figure 3. Contour map of ground settlement.

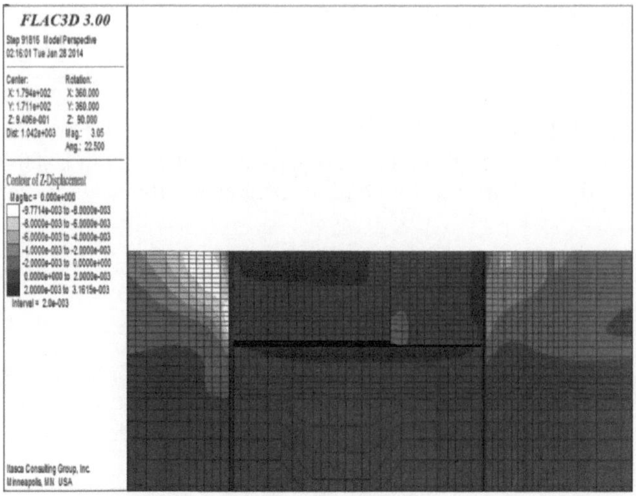

Figure 4. Contour map of 1-1 profile settlement.

The maximum surface settlement appears along foundation pit and reaches a displacement value of 8.93 mm.

A profile 1-1 is taken at the section of maximum surface settlement happens. So we can gain the vertical displacement of this profile, as shown in Figure 4. From this picture, we know that the maximum settlement reaches 9.77 m at the left of the pit. It is about 5–6 meters from the edge of pit, and the contour of settlement displacement looks like a funnel. Due to

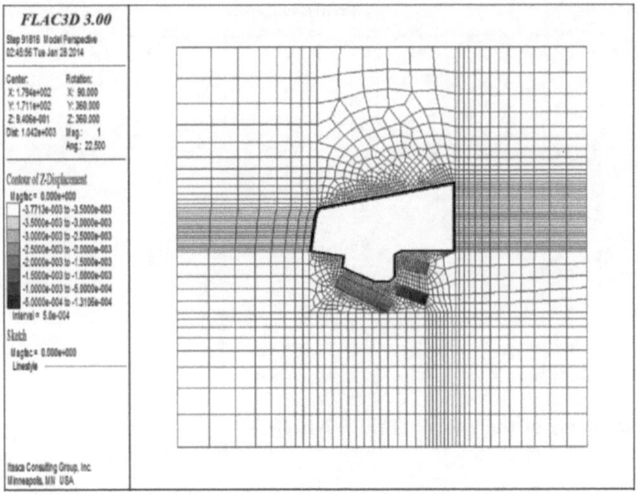

Figure 5. Contour map of surrounding buildings settlement.

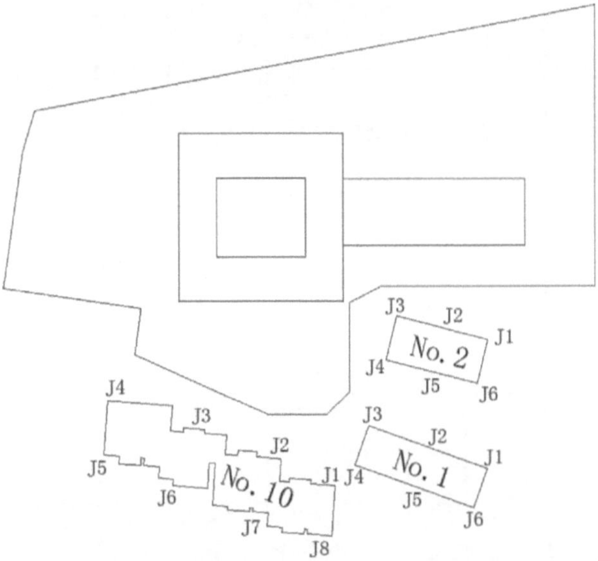

Figure 6. Arrangement of monitoring points.

the influence of building foundations, however, there are not subsidence funnels at the right side of foundation pit.

Figure 5 is a ground surface settlement contour of buildings. There is an obvious settlement deformation near the outer edge of foundation pit. The maximum vertical displacement values of this three buildings are, from left to right, respectively 3.77 mm, 4.30 mm and 7.12 mm.

To study the settlement effect of deep foundation excavation to buildings, displacement monitoring points are arranged within the coverage of buildings. The arrangement of monitoring points is shown in Figure 6. Settlement curves of monitoring points are shown in Figures 7–9, respectively.

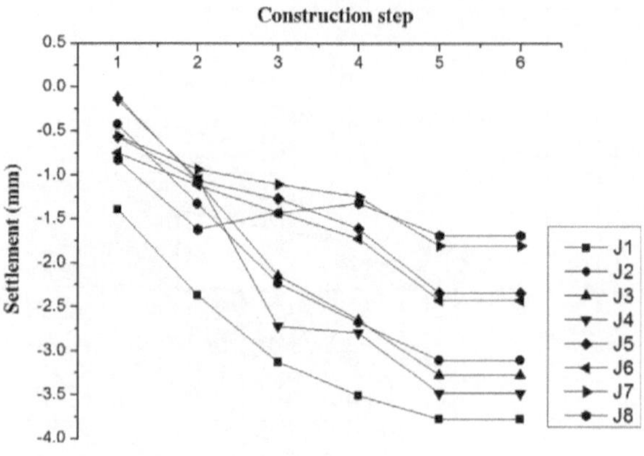

Figure 7. Settlement curve of the Phoenix Bridge No. 10.

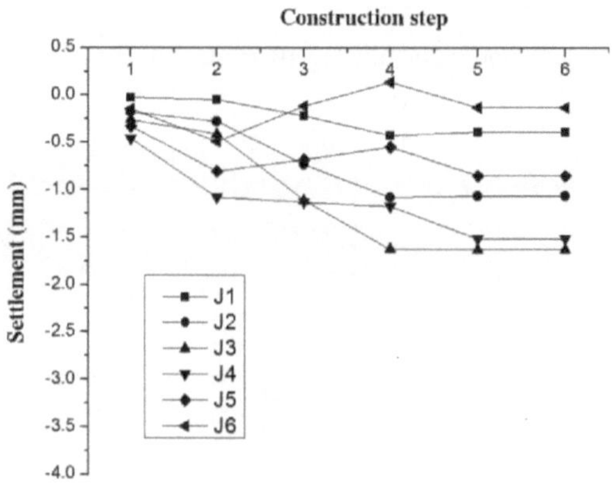

Figure 8. Settlement curve of the Federation of Trade Unions dormitory Block B No. 1.

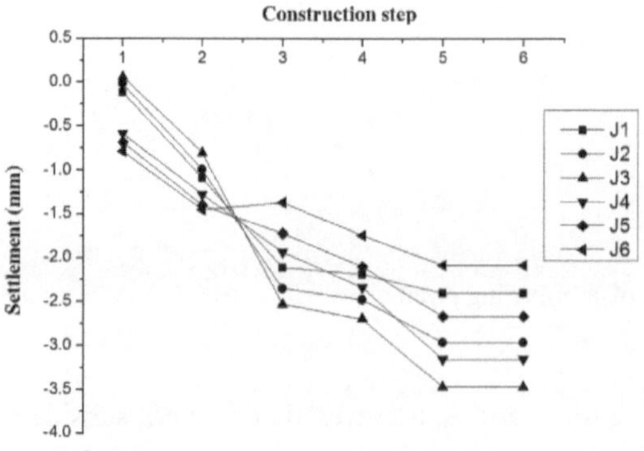

Figure 9. Settlement curve of the Federation of Trade Unions dormitory Block B No. 2.

4 CONCLUSIONS

From the simulation results, the digging of the New Traffic Building foundation pit makes little effect on the settlement of surrounding buildings. The main conclusions are as follows:

1. The site of maximum ground surface displacement locates in the north of the main foundation pit, the maximum value reaching 8.93 mm. The maximum vertical displacement of

Phoenix Bridge No. 10, Hefei Federation of Trade Unions dormitory Block B No. 1 and No. 2 are respectively 3.77 mm, 4.30 mm and 7.12 mm. The maximum settlement points of these buildings all locate adjacent to the foundation pit.

2. The displacement of buildings increase with the increasing of excavation depth, and the settlement is mainly affected by construction conditions of the first, second and third excavations.

REFERENCES

Chen, Y.M. & Xu, D.P. 2008. *The base and engineering examples of FLAC/FLAC3D*. Beijing: China Water Power Press (in Chinese).

Huang, M.S. et al. 2012. A Review of Recent Advances in the Underground Engineering and Deep Excavations in Soft Soils. *China Civil Engineering Journal* 45(6): 146–161 (in Chinese).

Jiang, X.L. et al. 2007. Construction monitoring and numerical simulation for a deep excavation in Tianjin. *China Civil Engineering Journal* 40(2): 79–103 (in Chinese).

Ma, F. 2012. Numerical Simulation on Effect of Soil Deformation Caused by Excavating Deep Foundation Pit of Subway Station. *Journal of Liaoning Technical University* 31(3): 295–299 (in Chinese).

Yao, A.J. et al. 2012. Numerical Analysis of the Influence of Isolation Piles in Metro Tunnel Construction of Adjacent Buildings. *Procedia Earth and Planetary Science* 5(0): 150–154.

Zhang, X.W. et al. 2013. Numerical Simulation of the Stability of Surrounding Rock in the Construction Process of Tunnel Engineering at Gucheng Hydropower Station. *Safety and Environmental Engineering* 20(3): 121–125 (in Chinese).

Zhang, Z.G. et al. 2013. Evaluation of deformation response for adjacent tunnels due to soil unloading in excavation engineering. *Tunnelling and Underground Space Technology* 38(0): 244–253.

Modeling and Computation in Engineering III – Zhang & Xie (Eds)
© 2014 Taylor & Francis Group, London, ISBN 978-1-138-02680-3

Method of MTF and SNR comprehensive simulation for Time-Delayed-Integration Charge Coupled Device on camera of mapping satellite

Xiaoming Gao, Xinming Tang & Qingxing Yue
Satellite Surveying and Mapping Application Center, NASG, Beijing, China

Tengteng Li
Shandong University of Science and Technology, Qingdao, China

ABSTRACT: The mechanism of MTF (Modulation Transfer Function) and SNR (Signal Noise Ration) degradation in the TDI CCD (Time-Delayed-Integration Charge Coupled Device) optical camera imaging process was studied in this paper. The authors discussed the descriptive means for the signal and simulation method of MTF and SNR degradation in every stage by using the mathematic models and the signal transmission process. The MTF and SNR simulation flow had been established. Meanwhile, the simulation test was conducted for TDI CCD camera of MTF and SNR by using this method.

1 INTRODUCTION

The space-borne optical camera with TDI CCD (Time-Delay-Integration Charge Coupled Device) using time delay method to get multiple exposures on the same target has the distinct characteristics of wide field and high resolution. The SNR (Signal Noise Ratio) raise with the increase of the integration number, but the range of increasing varies with the proportion of all kinds of noise (Y. Hao & B. Yang, 1999). The resolution of TDI CCD camera is one of the indicators that designers and users are most concerned about. As a measure of geometrical resolution of the camera, ground resolution is the minimum size of ground targets that the camera can distinguish. The camera factors such as optics system, detectors and electronic circuit usually cause decrease of image resolution. In addition to the camera itself, the factor such as the ground target contrast, atmospheric conditions, platform movement and processing of on-board and ground data also can cause the loss of image resolution in different extent. Under the condition of actual observation, the images can't reach the theoretical resolution for contrast decline and noise disturbing. (Z. Wan et al. 2008).

Modulation Transfer Function (MTF) and Signal Noise Ratio (SNR) are the two quantitative indicators to analyze the decrease of the imaging signal contrast and radiation performance. Resolution of the image can be considered as outcome of combining action of ground targets, MTF and SNR. For TDI CCD camera, the MTF and SNR have the relationship of mutual restriction. With the increases of integral number, SNR will raise, but the MTF decreases with the increase of the integral number, which is caused by the attitude motion and speed mismatch. As a result, the overall image quality is not proportional to the integral number. The paper analyzed the descriptive means for the signal and simulation method of MTF and SNR degradation in every stage by using the mathematic models and established comprehensive simulation method of MTF and SNR with the main line of the signal transmission process.

2 MTF MODEL

2.1 *The definition of MTF*

An imaging series can be effective as a spatial frequency filters, its imaging features and quality evaluation can be described by the ratios of frequency between the objective and the image, and the frequency comparison feature is Optical Transfer Function (OTF). The amplitude of OTF is Modulation Transfer Function (MTF) and the phase part is called Phase Transfer Function (PTF). PTF makes a shift for the image point of relatively ideal position, which does not affect the imaging resolution. In fact, MTF is mainly used to evaluate imaging quality (S. Chen, 2003).

MTF is a function of spatial frequency and it's generally a curve that decreases with the increase of spatial frequency. Generally speaking, the high frequency part reflects the transmission of the objects details, and intermediate frequency part and low frequency part reflect transmission of the level of the objects and the objects outline respectively.

2.2 *Factors of MTF for TDI CCD camera*

The MTF of Linear array push-broom sensor imaging system is the product of static MTF and dynamic MTF. It includes atmospheric MTF_{atm}, dynamic imaging MTF_{mot}, optical system MTF_{opt}, detector MTF_{det}, electronic circuitry MTF_{ele} and data processing MTF_{del}, which the combined effects results of MTF_{opt}, MTF_{det} and MTF_{ele} can be expressed as a unified camera system MTF_{cam}. Assuming that the modulation of input scene is M_{obj}, the modulation of the output image is M_{img}, it can be expressed as Equation 1 below:

$$M_{img} = M_{obj} \bullet MTF_{atm} \bullet MTF_{mot} \bullet MTF_{opt} \bullet MTF_{det} \bullet MTF_{ele} \bullet MTF_{del} \quad (1)$$

3 SNR MODEL

3.1 *The definition of SNR*

In this article, the ground reflectance field is considered as the signal field. The most ideal imaging should be that there is a constant proportional relationship between the output DN value of each pixel and the average reflectance of the corresponding ground area in the imaging projection direction. Based on this 'standard', any factors causing the signal floating and attenuation can be treated as noise.

Generally speaking, the SNR can be defined as the ratio of output signal voltage and random noise Root Mean Square (RSM) voltage (S. Chen, 2003). It can be expressed as Equation 2 below:

$$SNR = V_s / V_n \quad \text{or} \quad SNR = 20 \lg V_s / V_n \quad (2)$$

where V_s = output signal voltage; V_n = random noise Root Mean Square (RSM) voltage.

3.2 *Factors of SNR for TDI CCD camera*

The main factors are four aspects which include atmospheric, optical system, detector and electronic circuitry.

The atmospheric effects on signal are mainly on the transmission attenuation for sun light, scattering and radiation for ground targets, attenuation for the signal of reflection or radiation from the ground and atmospheric back scattering. The last two factors increase the non-target signals, and compress the scope of the target signals substantially.

The influence of the optical system for signal radiation mainly includes the transmission and attenuation of the camera lens, the vignetting effects of the camera lens and stray radiation.

The detector effects on signal include Poisson random noise and Gaussian random noise. The pseudo random number generator is structured by the standard deviation to generate the charge distribution for simulating graupel noise. The Gaussian pseudo-random number generator is constructed to simulate the charge distribution of Gaussian random noise.

In the process of converting the number of the charge into digital number, there will produce a loss of signal and other additive noise.

4 SIGNAL TRANSMISSION MODEL

If the decline of neighboring pixel signal contrast that caused by various filter effects in the process of signal transmission is not considered, image plane illumination can be computed by the following Equation 3 (Z. Wan et al. 2008 & S. Chen, 2003):

$$E = Kn(\alpha)\cos^n \alpha / 4F^2 \int_{\lambda_1}^{\lambda_2} [E(\lambda)\rho(\lambda)\tau_\alpha(\lambda) + \pi L_u(\lambda)]\tau_0(\lambda)d\lambda + \int_{\lambda_1}^{\lambda_2} E_{\lambda f} d\lambda \tag{3}$$

where K_n: vignetting coefficient of the lens; F: the reciprocal of relative aperture; $E(\lambda)$: ground target spectrum irradiance of wavelength at λ; $\rho(\lambda)$: ground target reflectivity; $L_u(\lambda)$: atmospheric scattering radiance; $\tau_\alpha(\lambda)$: atmospheric transmittance; $\tau_0(\lambda)$: the transmittance of camera; $E_{\lambda f}$: Miscellaneous illumination; α: Semi-FOV of lens.

It can obtain ground receiving irradiance in different solar elevation angle through the look-up table which is prepared in atmospheric correction software.

If the CCD area is A and integration time is T, the light energy during the time is the following Equation 4:

$$Q = T \bullet A \tag{4}$$

According to quantum theory, photon number can calculate by Equation 5:

$$n = Q\lambda / hc = \lambda T \bullet A / hc$$

where λ: the average wavelength of the light; h: quantum efficiency; c: speed of light in vacuum.

If the quantum efficiency is η, the charge number $n_e = \eta n$; when integration number is M, $n_e = M\eta n$ (Y. Hao et al. 1999 & Farrier M.G. et al. 1980).

The charge can be transform to digital number by gain and analog to digital conversion as following Equation 5:

$$DN = 2^{GRD} / n_{ful} \bullet G \bullet n_e \tag{5}$$

where GRD: quantization levels; n_{ful}: number of saturated electrons; G: gain; n_e: charge.

5 MTF AND SNR COMPREHENSIVE SIMULATION

5.1 *The flow of simulation*

The strict noise simulation should be done sequentiality and gradually, because the physical meaning and the description method of the signal noise are different at the unequal stages of the signal transmission.

It can obtain the ground reflectivity field by the input images and retrieval parameters, according to the sun angle, azimuth angle and atmospheric conditions of the simulation of the image, through the look-up table to obtain ground receiving irradiance, then it can get the ground radiance or radiant exitance. Then there is the comprehensive stimulate of MTF and SNR of signal transmission process.

5.2 *The modeling of MTF and SNR comprehensive*

5.2.1 *Atmospheric transmission stage*

The input signal is the ground radiance or radiant exitance, the output is the pupil radiance. MTF show the proximity effect caused by atmospheric turbulence. SNR show the weakness caused by transmission attenuation and atmospheric backscattering.

$$L' = f_{IEFT}[f_{FFT}(L_0 \bullet \tau_{up}) \bullet MTF_{atm}] + L_u \tag{6}$$

where f_{FFT}: fast fourier transform; f_{IFFT}: inverse fast fourier transform; L_0: ground target radiance; τ_{up}: atmospheric transmittance; L_u: atmospheric backscattering radiance.

5.2.2 *Optical system stage*

The input signal is the pupil radiance and the output is the sum of the brightness and miscellaneous brightness that attenuated and filtered by optical system, so the mechanism of the MTF includes two aspects. Firstly, the filtering effects caused by the aperture diffraction of the optical system itself, geometric difference and defocus. Secondly, the filtering effects caused by platform inherent speed, the image motion speed, charge transfer speed mismatch and posture movement. The mechanism of SNR is the weakness caused by transmission attenuation and vignetting of lens.

$$L'' = f_{IFFT}\{f_{FFT}\{[f_{IEEF}(f_{EEF}(L') \bullet MTF_{opt} \bullet K(\alpha)) + L_f] \bullet MTF_{mot}\}\} \tag{7}$$

where $K(\alpha)$: correlation coefficient of FOV; L_f: stray light radiance.

5.2.3 *Detector stage*

The input signal is focal plane irradiance and the output is the sum of the light generating charge and detector noise charge (Farrier M.G. et al. 1980). MTF show the geometry size of detector, charge transfer efficiency filtering effect caused by charge diffusion. SNR show the typical noise of TDI CCD under the given conditions.

$$N_e = f_{IEET}[f_{EET}(N_0) \bullet MTF_{det}] + N_{nos} \tag{8}$$

where N_0: number of charge-generating of irradiance; N_{nos}: charge number of noise; N_e: output charge.

5.2.4 *Electronic circuits stage*

The input charge is the sum of signal and noise, output is the digital number include noise. MTF show the filtering effect of electronic circuits. SNR show analog to digital conversion (B. Tian et al. 2009 & Feng Y. 2003).

$$DN = f_{IFFT}[f_{FFT}(DN_0) \bullet MTF_{ele}] + DN_{nos} \tag{9}$$

where DN_0: digital signal of N_e; DN_{nos}: digital signal of electronic circuits stage noise; DN: the sum of digital number.

5.2.5 *Camera system stage*

The input is the entrance pupil radiance, the output is the digital number of analog image.

$$DN = f_{IFFT}[f_{FFT}(DN_0) \bullet MTF_{cam}] + DN_{cam} \tag{10}$$

where DN_0: the digital number of radiation response parameters of camera system; DN_{cam}: the noise caused by inconsistent of camera system response.

5.3 *The experiment of simulation*

Simulation design pathway is 500 km, TDI CCD integral number to 24, CCD size of 7 μm, input image resolution of 0.5 m, analog image resolution of 2.1 m on the ground. It gives overall processing for MTF and SNR of the camera system in the simulation, the influence of atmospheric MTF is negligible in the simulation experiment.

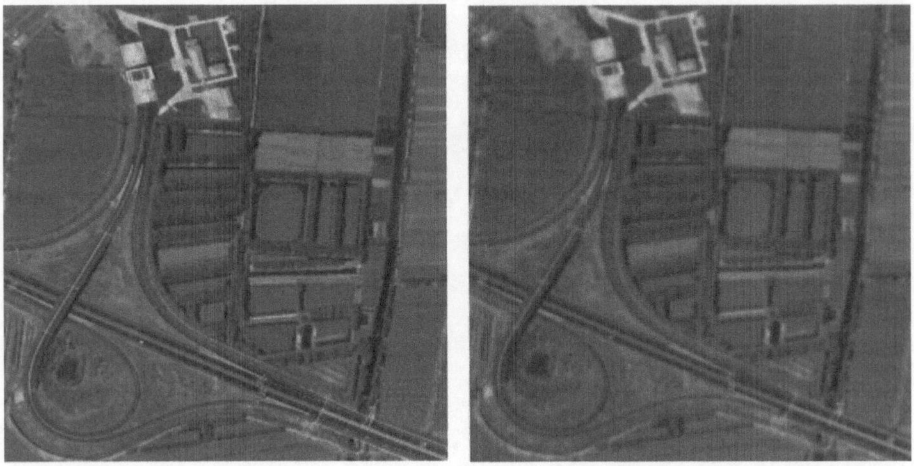

Figure 1. Comparison between images before (left) and after (right) MTF-SNR degradation.

Factors considered for Simulation include: firstly, signal overall compression caused by the atmospheric transmittance and optical system transmittance; secondly, signal overall uplift caused by the atmosphere backscatter, the stray light and other factors; thirdly, random noise of the camera system.

6 THE CONCLUSION

MTF and SNR are the two important factors for image resolution. This paper had established the MTF and SNR comprehensive simulation flow on the basis of mathematic models and the signal transmission process. This work is very important and gives a way for quantitative analysis of image quality by MTF and SNR.

ACKNOWLEDGEMENTS

This study was funded by Science and Technology Programme of National Administration of Surveying, Mapping and Geoinformation named 'Optimization and demonstration of technical indicators of optical surveying and mapping satellite', National Science and Technology Support Programme (No. 2011BAB01) and National Defense Science and Technology Programme 'Research on key technologies of data processing, application and in-orbit testing for ZY3'.

REFERENCES

Bret-Dibat T., Albouys V., Berthon J., et al. 1999. Tests of a high resolution three mirrors anastigmatic telescope. SPIE 3870:126–137.

Chen S. 2003. Space camera design and experiment. Beijing: Space Press.

Farrier M.G., Dyck R.H. 1980. A large area TDI image sensor for low light level imaging. IEEE Transactions on Electron Devices 27(8):1688–1693.

Feng Y. 2003. Model of a TDI line scan camera and its electronics. Proceedings of the 29th Annual Conference of the IEEE Industrial Electronics Society. Roanoke, VA, USA: IEEE:2215–2220.

Hao Y., B. Yang. 1999. Characteristics and development tendency of TDI CCD remote sensor with long focal length. Spacecraft Recovery and Remote Sensing 20(1):13–19.

Tan B., C. Jing, X. Zhang, et al. 2009. Imagings imulation of electrooptical system for space target. Computer Simulation 26(5):240–243.

Wan Z., J. Ren, X. Li, et al. 2008. Analysis of signal-to-noise ratio for remote sensing TDI CCD camera based on radioactive transfer model. Infrared and Laser Engineering 37(3):497–500.

Modeling and Computation in Engineering III – Zhang & Xie (Eds)

Stability reliability analysis of high earth and rockfill dam slope

Yang Zhou & Degao Zou
Faculty of Infrastructure Engineering, Dalian Technology of University, Dalian, China

Xuejing Deng
Department of Engineering Mechanics, China University of Petroleum (East China), Dongying, China

Shiqiang Zhao
Faculty of Infrastructure Engineering, Dalian Technology of University, Dalian, China

ABSTRACT: Combining dynamic finite element method with reliability theory based on probability statistical analysis, the extreme response of the high earth and rockfill dam is calculated under earthquake. In this paper, the probability density of the minimum safety factor as one of these responses is fitted by kernel density estimation method. Based on reliability theory, the method of high earth and rockfill dam slop safety probability is proposed. The results show that the method applied in earth and rockfill dam slope stability reliability is feasible.

1 INTRODUCTION

Dynamic response and seismic stability problems of high earth and rockfill dam under earthquake are intensively researched in the field of soil dynamic. Safety factor based on deterministic method is extensively adopted as an evaluation index in China. To get an appropriate safety factor, the pseudo-static method is widely used (Xu *et al.*, 2012; Liu *et al.*, 2001; Deng *et al.*, 2010). The seismic loading is equivalent to a static load applied on the dam and based on limit equilibrium method the stability safety factor is given. This method is simple and widely used, but dynamic characteristics such as time of duration cannot be taken into account in practical application, and meanwhile safety factor calculated by pseudo-static method cannot correctly evaluate dynamic stability of high earth and rockfill dam (Luan *et al.*, 2007). Dynamic finite element stability method is necessary to be used in stability analysis of high earth and rockfill dam slope.

Seismic stress at every time during earthquake is used in dynamic finite element stability method to get the safety factor. The nonlinear stress-strain behavior of the materials can be better considered (Xu *et al.*, 2012). Despite its advantage to pseudo-static method, however, it still exist drawbacks in traditional way as it tends to choose one deterministic seismic wave as seismic input. The deterministic safety factor is given (Wu *et al.*, 2007). In fact, due to indeterminacy of seismic strength, distance and propagation medium, the seismic wave cannot be a deterministic one.

Reliability theory and analysis method is needed to consider uncertainties in practical engineering. On this basis, structural failure probability is used as an evaluation index. Most study on current reliability methods place particular emphasis on uncertainty of material parameters other than load, especially seismic load (Shu, 2012). However, for high earth and rockfill dam, effect of randomness of ground motion on dynamic stability is much greater than variability of material parameters. Although first-passage time probability theory is used to study randomness of ground motion in early period, the process is very complex and not conducive to realize. In practice, when a large number of random seismic waves in accord with the same ground motion model are introduced as ground motion inputs, the results of nonlinear calculations show that some extreme response of the dam meets certain probability distribution. Combined with reliability theory, structural failure probability and reliability index can be given. This approach is simple, easy to operate, and very good considering the randomness of ground motion (Deng, 2014).

Dynamic finite element stability method is combined with reliability analysis method based on probability statistics. Through a large number of nonlinear calculations with random seismic inputs, the probability density of the minimum safety factor as one of these responses is fitted by kernel density estimation method. Based on reliability theory, the method of high earth and rockfill dam slop safety probability is proposed.

2 THE THEORY OF DAM SLOPE STABILITY RELIABILITY

The limit state of dam slop is the evaluation criterion whether it is safe or not and could be described by the performance function. Requirement of dam slop designing based on the limit state is given by

$$g(X_1, X_2, \dots X_n) \geq 0 \tag{1}$$

where $g(\cdot)$ is performance function. $X_i (i = 1, 2, \dots n)$ is the random variables which refer to various effects, material properties, geometry parameters of the structure and so on. With only response and resistance of the structure as basic variables, requirement of dam slop designing based on the limit state is also written as

$$Z = R - S \geq 0 \tag{2}$$

where R refers to structural resistance and S refers to structural responses.

If R and S fit the normal distribution, the performance function Z will fit the normal distribution as well. Let μ_R, μ_S and σ_R, σ_S be the mean value and standard deviation of R and S, based on probability statistics, the mean value of Z can be given by $\mu_Z = \mu_R - \mu_S$, the standard deviation of Z can be given by

$$\sigma_Z = \sqrt{\sigma_R^2 + \sigma_S^2}$$

And the probability density function of Z can be expressed as

$$f_Z(z) = \frac{1}{\sqrt{2\pi}\sigma_z} \exp\left[-\frac{(z - \mu_z)^2}{2\sigma_z^2} \right] \tag{3}$$

The reliability of the dam slop is represented by p_s which means probability of the performance function $Z \geq 0$, the failure probability of the dam slop is represented by p_f which means probability of the performance function $Z < 0$ and the equation of p_f is given by

$$p_f = p(Z < 0) = \int_{-\infty}^{0} f_Z(z) dz = \int_{-\infty}^{0} \frac{1}{\sqrt{2\pi}\sigma_z} \exp\left[-\frac{(z - \mu_z)^2}{2\sigma_z^2} \right] dz \tag{4}$$

$$p_s + p_f = 1 \tag{5}$$

Standardizing the normal distribution variables, the equation (4) will be transformed into

$$p_f = \frac{1}{\sqrt{2\pi}} \int_{-\infty}^{-\frac{\mu_z}{\sigma_z}} \exp\left(-\frac{y^2}{2} \right) dy = \Phi\left(-\frac{\mu_z}{\sigma_z} \right) \tag{6}$$

Table 1. Corresponding relation between failure probability and reliability index.

Failure probability	10^{-1}	10^{-2}	10^{-3}	10^{-4}	10^{-5}	10^{-6}	10^{-7}
Reliability index	1.28	2.33	3.09	3.72	4.27	4.75	4.20

With

$$\beta = \frac{\mu_z}{\sigma_z} = \frac{\mu_R - \mu_S}{\sqrt{\sigma_R^2 + \sigma_S^2}}$$

p_f can be expressed as

$$p_f = \Phi(-\beta) = 1 - \Phi(\beta) = 1 - p_s$$

where β is reliability index, the corresponding relation between failure probability and reliability index is listed in Table 1 (Liu, 2008).

For the actual dam slope, when the distribution of R and S is non-normal distribution, it can be converted into a normal distribution and calculated its failure probability and reliability index. Some standard for structural reliability design points that the goal reliability index is usually between 1 and 2 for composite effect under seismic loads (Liu, 2008).

3 NUMERICAL EXAMPLES

In the paper, analysis program of GEODYNA and FEMSTABLE are adopted for two dimensional (2-d) nonlinear finite element numerical calculation. According to the ground motion model suggested in the paper (Deng, 2014), 200 seismic waves with epicenter distance of 20 km and also 200 seismic waves of 50 km are made to compute and compare the probability distribution under the same magnitude ($M = 8.0$). For convenience of description, the case of $M = 8.0$, $R = 20$ km is named as Case I while the case of $M = 8.0$, $R = 50$ km is named as Case II.

3.1 *Numerical model*

The model is a concrete face rockfill dam with the height of 200 m, the upstream slope gradient and the downstream are 1:1.4 and 1:1.6 respectively. The equivalent linear viscoelastic model is adopted in dynamic calculation. The model formulas and parameters are given in Tables 2 and 3 (Kong *et al.*, 2013). The 2-d finite element mesh is shown in Figure 1.

3.2 *Random seismic wave samples analysis*

Two typical random seismic waves are picked from the two cases respectively and their acceleration time histories are in Figure 2. Figure 3 shows the probability statistics of peak acceleration of the 200 seismic waves with two different epicentral distances and it can be seen that peak acceleration is not deterministic but meets certain probability distribution.

3.3 *Stability results analysis*

Dynamic finite element stability method is used to calculate the safety factor at the downstream slope. The probability distribution of the minimum safety factors under two cases is given in Figure 4 and the parameters of the distribution are given in Table 4. It is shown that the probability distribution of the minimum safety factors meets the normal distribution and the mean value μ under Case I is less than that under Case II. It means that with shorter

Table 2. Parameters of static model.

Material	ρ/(g/cm^3)	φ_0/(°)	$\Delta\varphi$/(°)	K	n	R_f	K_b	m
Rockfill	2.15	54	11	1027	0.2	0.77	410	0.13

Table 3. Parameters of dynamic model.

Material	K	n	υ
Rockfill	2339	0.5	0.4

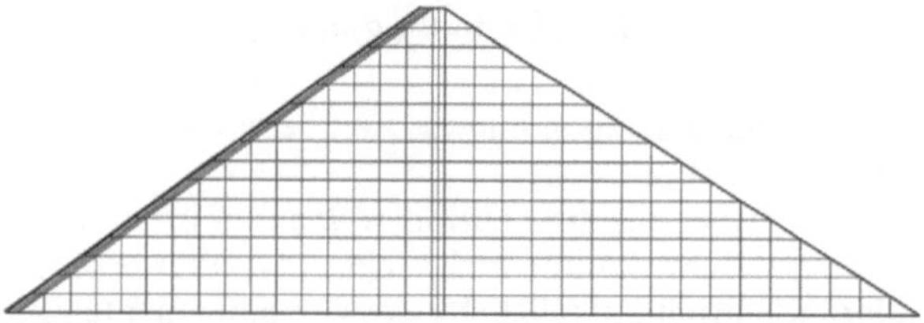

Figure 1. 2-d finite element mesh.

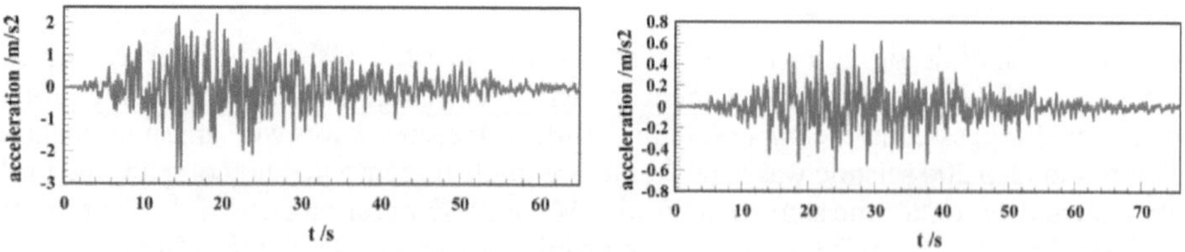

(a) Case I (M=8, R=20km) (b) Case I (M=8, R=50km)

Figure 2. Typical seismic waves acceleration time histories.

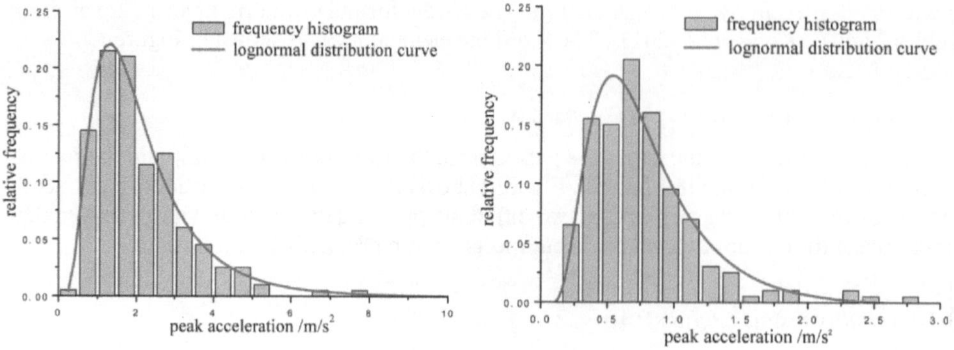

(a) Case I (M=8, R=20km) (b) Case I (M=8, R=50km)

Figure 3. Distribution of synthetic seismic waves peak acceleration.

Table 4. Parameters of the normal distribution.

Cases	Parameters
I ($M = 8$, $R = 20$ km)	$\mu = 0.9739$, $\sigma = 0.3189$
II ($M = 8$, $R = 50$ km)	$\mu = 1.4709$, $\sigma = 0.2145$

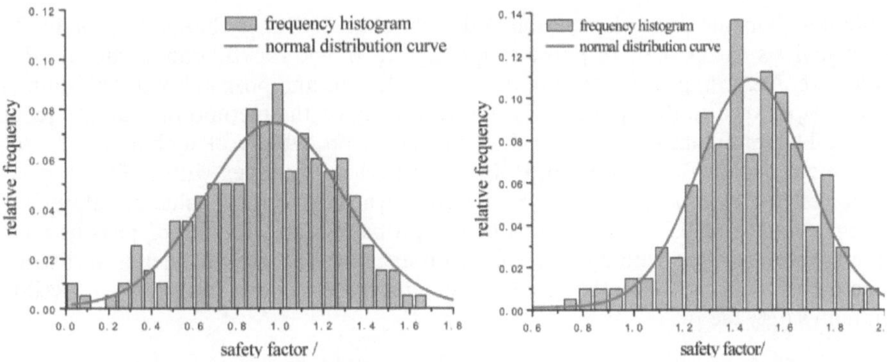

(a) Case I (M=8, R=20km) (b) Case II (M=8, R=50km)

Figure 4. Distribution of the minimum safety factor.

Table 5. The failure probability and reliability index of the dam slop.

	Critical value of safety factor			
Cases	0.7	0.8	0.9	1.0
I ($M = 8$, $R = 20$)	$p_f = 19.52\%$ $\beta = 0.859$	$p_f = 29.28\%$ $\beta = 0.545$	$p_f = 40.84\%$ $\beta = 0.232$	$p_f = 53.26\%$ $\beta = -0.082$
II ($M = 8$, $R = 50$)	$p_f = 0.016\%$ $\beta = 3.593$	$p_f = 0.089\%$ $\beta = 3.127$	$p_f = 0.39\%$ $\beta = 2.661$	$p_f = 1.41\%$ $\beta = 2.195$

epicentral distance, the dam tends to be safer. Furthermore, the distribution of the minimum safety factors tends to be more discrete which leading to a great probability to damage due to a strong randomness. It is necessary to make the study of quantifying the impact on the minimum safety factor made by the ground motion with different epicentral distances (Hou *et al.*, 2012).

3.4 *Reliability results analysis*

The probability density function of the resistance $R(x_1,x_2,\dots x_n)$ and the response $S(x_1,x_2,\dots x_n)$ is used to calculate the failure probability of the dam slop. The minimum safety factor is used as the response in the performance function and its probability density function is fitted by kernel density estimation method (Deng, 2014) which is already given in previous section. At some moment under seismic load, if there is an instantaneous overload moment at which the safety factor is less than 1.0 calculated by dynamic finite element method stability, it does not mean that the dam will completely be instable and out of work. So the resistance $R(x_1,x_2,\dots x_n)$ corresponding to the minimum safety factor is supposed as a set of values 0.7, 0.8, 0.9, 1.0. The reliability results are given out in Table 5.

It is shown that under the same earthquake, with the increase of critical value of safety factor, the failure probability is increasing, in contrast, reliability index is decreasing. Meanwhile, the failure probability with epicentral distance of 20 km is far larger than the case with epicentral distance of 50 km.

4 CONCLUSION

Combining dynamic finite element method with reliability theory based on probability statistical analysis, the extreme response of the high earth and rockfill dam is calculated under earthquake. According to a large number of calculations and probability distribution analysis, this paper verifies the applicability and feasibility of the method on stability reliability analysis. The minimum safety factor and slop safety probability of high earth and rockfill dam are calculated. Both failure probability and reliability index with different epicentral distance of 20 km and 50 km ($M = 8$) are given when the critical value of safety factor is respectively assigned 0.7, 0.8, 0.9 and 1.0. The probability distribution of peak acceleration response indicates that with smaller epicentral distance, the peak acceleration distribution of the dam top A_{max} tends to be more discrete and have a wider value range as well as the minimum safety factor.

ACKNOWLEDGEMENTS

This research was supported by Earthquake Industry Special Funds for Scientific Research Projects (No. 201208013), National Natural Science Foundation of China (No. 51121005, 51138001, 51279025). The supports are gratefully acknowledged.

REFERENCES

Deng, X.J & Kong, X.J. & Zou, D.G. 2010. Pseudo-static analysis for evaluation of earthquake stability of landfills. *Chinese Journal of Geotechnical Engineering* 32(8):1301–1308.

Deng, X.J. 2014. Extreme Earthquake Response Distribution Method of Geotechnical Slope Stability reliability.

Hou, H.Y. & Guo, D.C. & Liu, H.S. 2012. Effects of Input Ground Motion Characters on Safety Factor of Rock Slope. *Journal of Hydroelectric Engineering* 38(10):17–20.

Kong, X.J. & Zhang, Y. & Zou, D.G. 2013. Study on the stress distribution characteristics of face-slab of high concrete-face earth and rockfill dam. *Journal of Hydraulic* Engineering 44(6):631–639.

Liu, L.P. & Lei, Z.Y. & Zhou, F.C. 2001. The evaluation of seismic slope stability analysis methods. *Journal of Chongqing Jiaotong University* 20(3):83–88.

Liu, Y.F. 2008. Reliability Calculation and Application Research of Soil Slop Based on Matlab, Changsha, China: Central South University.

Luan, M.T.Z. Li & Fan, Q.L. 2007. Analysis and evaluation of pseudo-static aseismic stability and seism-induced sliding movement of earth-rock dams. *Rock and Soil Mechanics* 28(2):224–230.

Shu, S.X. 2012. Analysis on Random Reliability of Soil Slope based on Finite Element Method, Wuhan, China: Huazhong University of Science & Technology.

Wu, Z.G & Han, G.C. and Lin, G. 1992. *Generality of random soil dynamics*. Dalian, China: Dalian Technology of University Publisher.

Xu, B. & Zou, D.G. & Kong, X.J. 2012. Seismic stability of slopes of high rockfill dams. *Chinese Journal of* Geotechnical *Engineering* 34(1):139–144.

Modeling and Computation in Engineering III – Zhang & Xie (Eds)
 ISBN 978-1-138-02680-3

Research on relationship between different soil granularity fractal characteristics of reservoir water conservation district and soil physic-chemical characteristics

Xi-Bing Shen
Qinzhou University, Qinzhou, China

ABSTRACT: According to the fractal theory, this thesis attempts to study the soil structure and its physic-chemical properties of the five different forests (Chestnut forest, glauca forest, Christmas tree plantation, mixed forest, Chinese fir forest) in Dongfeng reservoir of Yuxi, so as to establish a correlation formula between soil volume percentage content and its fractal dimension of those forests, as well as the regression model of fractal dimension of soil volume percentage content and corresponding index of soil's physic-chemical properties. The result shows that fractal dimension of different soil granularity is between 2.358 to 2.442, that significant linear correlation exists between soil volume percentage content, whose particle size is less than 0.02 mm, and fractal dimension. Also the result tells us that soil volume fractal dimension is significantly positively correlated with density, total porosity, total nitrogen, available phosphorus, available potassium, total phosphorus, total potassium and hydrolysable nitrogen respectively. Thus it is so clear that the application of fractal theory in studying the forest soil's physic-chemical properties provides a new method for woodland evaluation.

1 INTRODUCTION

Soil, a very irregular geometry, is a porous medium bonded by solid component and pores with different size and shape in a certain form (Lin. 1996). The soil solid component's size, quantity (Sheng. 2011), shape and its combination mode determines the character and structure of soil (Wang. 2012), thereby affecting the physical properties of soil. The research also indicates that soil is a system with fractal feature (Turcotte D.L. 1986), which can characterize soil particle size (Rieu M. 1990), the degree of uniformity of tenure (Falconer K.J. 1989), the soil permeability and fertility (Bartoli F. 1991; Rieu M. 1990; Young I.M. 1991; Tyler S.W. 1990). Many researches on soil aggregate structure through the use of fractal model by Li Baoguo (Li Baoguo, 1994), Wu Chengzhen (Wu. 1999), Liu Jinfu (Liu. 2002), He Dongjin (He. 2001), etc. have achieved good results. Based on the previous research, this paper is aimed at studying the fractal dimension of soil structure of different forests in Yuxi Dongfeng Reservoir Headwater using fractal theory, and the establishment of a relational model between different forest soil dimension and soil physic-chemical properties, so as to provide the scientific basis for choosing water conservation forest in water district and evaluating forest soil.

Located in the upstream section of Qujiang River in Sand Village (a village in northeast Hongta district of Yuxi City, Yunnan, China), Dongfeng Reservoir, whose geographic position is 102°33′~102°43′E, 24°12′~24°28′N, is a first tributary belonging to the Nanpan River of Xijiang River system of Pearl River basin. Being subtropical plateau monsoon climate, the Dongfeng Reservoir district's annual average temperature and annual here is 67%, and its vegetation types mainly are *Pinus yunnanensis faranch, Cyclobalanopsis glauca, Alnus nepalensis, Cunninghamialanceolata, etc.*

2 RESEARCH METHOD

2.1 *Sample collection method*

June 2013, five representative stands, named Chestnut forest (A1), Glauc forest (A2), Christmas tree plantation (A3), mixed forest (A4), Chinese fir forest (A4), were chosen to do the research. Three 20 m × 20 m sample plots, settled in each stands by GPS, were chosen three sections on the upside, middle part and lower part of the plots, whose soil were stratified sampled vertically according to 0~20 cm, 20~40 cm, 40~60 cm with a cutting ring. Each three soil sample collection were repeated near the cutting point, and the soil samples were put into a prepared aluminum box, and bulk samples were taken for spare at the same time.

2.2 *Measure method*

In this paper, three measure methods are mentioned and applied, which are ways of determining soil bulk and density according to Zhang Wanru, determinating soil chemical elements by conventional methods, and determining the percentage content of soil particle size by a MS2000 particle size analyzer.

2.3 *Fractal dimension model of soil structure*

The data can be obtained from soil particle that is got by using laser particle analyzer. According to the American soil texture classification system, the soil practical diameter can be divided into seven levels. The fractal dimension (*D*) of soil particle volume can be calculated according to the following formula:

$$D = 3 - \frac{\lg(V_i / V_o)}{\lg(d_i / d_{max})}$$

In this formula, *D* is the fractal dimension; D_i, the average diameter between D_{i-1} and D_{i+1}; D_{max}, the largest average diameter. Meanwhile, V_i is the accumulated volume percentage whose diameter is less than D_i, and V_o stands for the total volume percentage of soil samples. In the specific application, the first step is to calculate the $\lg(V_i/V_o)$ and $\lg(D_i/D_{max})$ of different D_i of the soil samples, then the slope(*K*) can be calculated by linear regression analyzing the data mentioned above, and the last step is to get the fractal dimension, which is $D = 3 - K$.

2.4 *The correlation model of soil properties with fractal dimension*

To establish a correlated model between soil physic–chemical properties (*Y*) and fractal dimension (*D*) by using a linear and nonlinear regression.

$$Y = a + bD \tag{1}$$
$$Y = aD^b \tag{2}$$
$$Y = \ln(a + bx) \tag{3}$$

In the formula, *Y* stands for the index content of forest soil properties, *D* for the fractal dimension, and *a* and *b* are parameters.

3 RESULT AND ANALYSIS

3.1 *The relationship between fractal dimension and soil granularity*

Soil is a porous medium with fractal features. The fractal dimension of different soil granularities can be got through the regression model relationship of soil structure's fractal dimension (Table 1). In Table 1, it is clear that the fractal dimension of different forest soil grades

Table 1. The percentage content of different diameter and fractal dimension of soil granularity.

Different types forests	The percentage content of different diameter							Fractal dimension
	<0.02 mm	0.02~ 0.002 mm	0.05~ 0.02 mm	0.1~ 0.05 mm	0.25~ 0.1 mm	0.5~ 0.25 mm	1~ 0.5 mm	
Chestnut forest	21.24	42.5	15.15	7.96	9.21	3.94	0.09	2.442
Oak forest	11.82	44.59	15.91	14.34	10.13	2.63	0.58	2.358
The Christmas tree plantation	14.55	32.35	17.59	18.03	12.86	4.21	0.41	2.388
Mixed forests	13.34	45.55	12.5	11.57	13.58	3.36	0.08	2.375
Chinese fir forest	13.51	43.73	11.63	7.99	14.91	7.91	0.32	2.377

Table 2. The status of soil pore components in different silvicultural density.

Different types forests	Deep/cm	Bulk density/ (g/cm³)	Porosity/%		
			Capillary porosity/%	Non-capillary porosity/%	Total porosity/%
Chestnut forest	0~20	1.40	33.74	13.43	47.17
Oak forest	0~20	1.33	35.31	14.37	49.68
The Christmas tree plantation	0~20	1.37	34.80	14.26	49.06
Mixed forests	0~20	1.36	38.26	10.42	48.68
Chinese fir forest	0~20	1.38	37.74	10.29	48.03

Table 3. The soil chemical characteristics in different silvicultural density.

Different types forests	Organic matter/ (g/kg)	Total N/ (mg/kg)	Total P/ (mg/kg)	Total K/ (mg/kg)	Hydrolytic N/(mg/kg)	Available P/(mg/kg)	Available K/(mg/kg)
Chestnut forest	1.67	3.09×10^3	530	6.95×10^3	213	32.8	78.5
Oak forest	1.46	2.05×10^3	455	5.74×10^3	88	19.1	46.0
The Christmas tree plantation	1.62	2.18×10^3	495	5.95×10^3	146	27.9	60.0
Mixed forests	1.49	1.70×10^3	478	5.76×10^3	107	15.9	51.5
Chinese fir forest	1.51	1.65×10^3	480	5.80×10^3	101	13.3	61.0

is between 2.358~2.442. Correlation analysis between fractal dimensions and volume percentage content of the following seven different grain diameter shows that the relationship between fractal dimension and soil volume percentage content, whose particle diameter is less than 0.02 mm, is $D = 2.259 + 0.009X$, $R^2 = 0.9927$, which is a significant linear correlation ($P<0.01$). It also tells us that the fractal dimension can objectively characterize the structural traits of soil and provide a new indicator for soil physical properties.

3.2 *Relationship between fractal dimension of different forests' soil structure and soil physic-chemical properties*

Fractal dimension of soil structure can be used to characterize soil fertility that can be showed by various indexes like soil physic-chemical properties. The fractal dimension of soil granularity of different forests can be ranked in an order of A1 (2.442) > A3 (2.388) > A5 (2.377) > A4 (2.375) > A2 (2.358). By calculating the measurement results of different forests soil physic-chemical properties (Tables 2 and 3) and the corresponding soil volume fractal

Table 4. The relation model between fractal dimension and soil physical and chemical properties of different silvicultural density.

Item (y)	Model	Correlation coefficient
Bulk density	$y = -0.29 + 0.69D$	0.564
	$y = 0.48D^{1.202}$	0.643
	$y = \ln(-2.58 + 2.72D)$	0.566
Capillary porosity	$y = 1.231 + 0.365D$	0.148
	$y = 308.43D^{-2.47}$	0.139
	$y = \ln(3E17 - E17D)$	0.215
Non-capillary porosity	$y = -0.139 + 0.111D$	0.293
	$y = 1.491D^{2.436}$	0.300
	$y = \ln(4.5E6 - 2E6D)$	0.309
Total porosity	$y = 1.09 - 0.25D$	0.617
	$y = 146.17D^{-1.267}$	0.619
	$y = \ln(5E22 - 2E22D)$	0.564
Organic matter	$y = -4.50 + 2.533D$	0.744
	$y = 0.057D^{3.799}$	0.736
	$y = \ln(-25.4 + 12.6D)$	0.761
Total N	$y = -32.93 + 14.67D$	0.666
	$y = (2.2E - 6)D^{15.791}$	0.711
	$y = \ln(-0.27 + 0.12D)$	0.499
Total P	$y = -1.53 + 0.85D$	0.947
	$y = 14.47D^{4.04}$	0.939
	$y = \ln(-4.68 + 1.97D)$	0.826
Total K	$y = -31.4 + 15.7D$	0.935
	$y = 0.031D^{6.07}$	0.948
	$y = \ln(-1.57 + 0.67D)$	0.784
Hydrolytic N	$y = -35.7 + 15.5D$	0.945
	$y = (8.9E - 8)D^{24.2}$	0.925
	$y = \ln(-0.15 + 0.65D)$	0.895
Available P	$y = -4.70 + 2.06D$	0.515
	$y = (5.2E - 7)D^{20.1}$	0.521
	$y = \ln(-5.03 + 2.63D)$	0.514
Available K	$y = -8.24 + 3.70D$	0.895
	$y = (4.5E - 4)D^{13.5}$	0.873
	$y = \ln(-15.3 + 7.15D)$	0.907

dimension, a related model between various indexes like soil physic-chemical properties and fractal dimension of soil volume is gained (Table 4). From Table 4, it is clear that soil volume fractal dimension is significantly positively correlated with density, total porosity, total nitrogen, available phosphorus, available potassium, total phosphorus, total potassium and hydrolysable nitrogen respectively. Based on the relationship of fractal dimension and soil physic-chemical properties, this paper tells us that fractal dimension can show soil physic-chemical properties in a certain degree, especially the soil fertility, which provides a new indicter for studying the soil fertility status.

4 DISCUSSION AND CONCLUSION

Domestic scholars have done researches on granularity fractal dimension with different utilization patterns of soil. For example, Huai Tai (Huai. 2008) and Li Jin-feng (Li. 2007) have studied the relationship between the volume fractal dimension of soil particle size distribution and the particle content, as well as the relationship between volume fractal dimension and soil organic matter and pH; Ju Li not only studied the relation between volume fractal dimension and volume percentage content, but also made correlation analysis of fractal dimension and organic matter, total nitrogen, and total phosphorus (Ju. 2011); Liu Xia made research on soil structure of different forest stands, and studied the relationship between soil structural fractal dimension and soil total porosity, capillary porosity and no capillary porosity (Liu. 2005). Based on the researches above, this paper attempts to study the soil structural fractal dimension of different forests in in Dongfeng reservoir of Yuxi by using fractal theory, and establish the relationship model of soil fractal dimension and soil physic-chemical properties, so as to make it clear that soil volume fractal dimension (2.358~2.442) is significantly positively correlated with density, total porosity, total nitrogen, available phosphorus, available potassium, total phosphorus, total potassium and hydrolysable nitrogen respectively, providing the scientific basis for choosing water conservation forest in water district and evaluating forest soil.

REFERENCES

Bartoli F., Philippy R., Doirisse M., et al. Structure and self–similarity in silty and sandy soils: the fractal approach [J] Soil Science, 1991, 42(2):167–185.

Falconer K.J. Fractal Geometry [M]. New York: John wily and Sons, 1989:89–159.

He Dongjin, Hong Wei, Wu Chengzhen, et al. Study on characteristic index of soil fertility in Chinese Fir plantation mixed with alniphyllum fortunei [J]. Journal of Mountain Science, 2001, 19(Suppl):98–102.

Huai Tai & Peng Jiangli & Wen Qing. The fractal characteristics of soil particle size distribution under different land utilization patterns [J]. Ecology and Rural Environment Journal, 2008, 24(1):41–44.

Ju Li & Wen An bang & Guo Jin. Soil particle fractal characteristics of different land utilization types in Three Gorges Reservoir [J]. Soil and Water Conservation Journal, 2011, 25(5):234–237.

Li Baoguo. Applicant and development of fractul theory in soil science [J]. Progress in Soil Science, 1994, 22(1): 1–10.

Li Jinfeng & Gong Yuanbo & Chen Linwu. Soil fractal characteristics of different land utilization types of Guangyuan [J]. Soil and Water Conservation Journal, 2007, 21(5):167–170.

Lin Kaimin, Yu Xintuo, He Zhiying, et al. Studies on biomass structure and soil fertilize in different density of Cunninghamia lanceolatast and [J]. Scientia Silvae Sinicae, 1996, 32(5):385–391.

Liu Jinfu, Hong Wei, Wu Chengzhen. Fractal features of soil clusters under some precious hardwood standsin the central subtropical region, China [J]. Acta Ecol Sin, 2002, (2):198–205.

Liu Xia & Wang Li & Zhang Guangcan. Soil structural characteristics of different forest types of stone mountain in central Shandong Province [J]. Soil and Water Conservation Journal, 2005, 19(6):49–52.

Rieu M, Sposito G. Fractal fragmentation, soil porosity and soil water properties: Applications [J]. Soil Science Society of America Journal, 1990, 55(5):1239–1244.

Rieu M, Sposito G. Fractal fragmentation, soil porosity and soil water properties: Theory [J] Soil Science Society of America Journal, 1990, 55(5):1231–1238.

Sheng Weitong. A long-term study on development and succession of undergrowth vegetation in Chinese Fir plantation with different density [J]. Forest research, 2001, 14(5): 463–471.

Tyler S.W, Wheatcraft S.W. Fractal scaling of soil particle–size distributions: analysis and limitations [J]. Soil Science Society of America Journal, 1990, 56(2):362–369.

Turcotte D.L. Fractal and fragmentation [J]. Journal of Geophysical research, 1986, 91(2): 1921–1926.

Wu Chengzhen, Hong Wei. Study on fractal features of soil aggregate structure under different managenment patterns [J]. Atcta PedolSin, 1999, 36(2):162–167.

Wang Yin, Yang Zhangqi, Zhang Zhenglin, et al. Effect of the density of Masson Pine on the chemical nature of the land of its artificial plantation [J]. Journal of Anhui Agricultural Science, 2010, 38(36):20934–20935.

Young I.M, Crawford J.W. The fractal structure of soil aggregates: measurement and interpretation [J]. Journal of Soil Science, 1991, 42:187–192.

Modeling and Computation in Engineering III – Zhang & Xie (Eds)
© 2014 Taylor & Francis Group, London, ISBN 978-1-138-02680-3

Water exchange ability of an offshore artificial island in a semi-open bay

Bing Yan
Tianjin Research Institute for Water Transport Engineering, M.O.T., China

ABSTRACT: It shows an increasing tendency in the construction of artificial island because of exploitation of ocean areas and resources in china. The water exchange is an important problem about environment, which should be considered in the layout of artificial islands. The mathematical model method is used to evaluate the water exchange ability of the offshore artificial island (Lotus Island) in a semi-open bay with the out sea in this paper. The results show that after the second artificial island is finished, tidal currents around and within island are still regular. The half-life time is 1.05 days. Lotus Island satisfies the PINAC's rule about the environment protection that the concentration of internal water is less than 37% after eight periods of tide.

1 INTRODUCTION

It shows an increasing tendency in the construction of artificial island because of exploitation of ocean areas and resources. There are two important problems about environment, which should be considered in the layout of artificial islands (Wang et al., 2008). On the one hand, how and to what extent do the artificial islands change the hydrodynamic environment around the projects? On the other hand, some artificial islands have the internal water systems, which connect to the out sea. Are internal water systems health and how to estimate and predict the health? It is often advantageous to model the water exchange as a separate system within a larger water body connected to the out sea. The water quality of the system depends crucially on the water exchange ability. The exchange ability is governed by the tidal exchanges between the system and the outer sea. The mathematical model method is used to evaluate the water exchange ability of the offshore artificial island in a semi-open bay with the out sea in this paper.

The artificial island will be built on the western coast of Jin Meng bay. Jin Meng bay is in north of Bohai sea (Fig. 1). The island named Louts Island because that the shape of it is like a lotus. It is composed of the main island and a breakwater. The main island has an internal water system, which is connected to outer sea with six channels for water exchange (Fig. 2). Another The artificial island, namely Hailuo island is building near the Lotus Island. The filed observations show that the tidal currents of Jin Meng bay flow basically along the coastline (Fig. 3).

2 NUMERICAL MODELS

2.1 *Tide current model*

The tidal current field at project sea area are studied by the plane two-dimensional numerical model which developed by Danish Hydraulics Research institute (MIKE21).

$$\frac{\partial \zeta}{\partial t} + \frac{\partial uh}{\partial x} + \frac{\partial vh}{\partial y} = hS \tag{1}$$

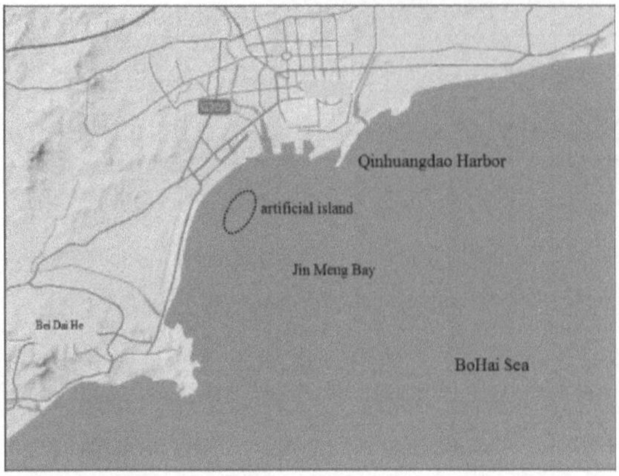

Figure 1. The position of the project.

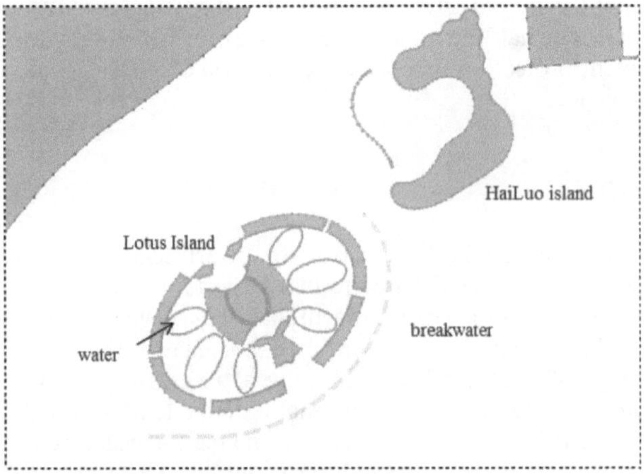

Figure 2. The layout of the Louts Island.

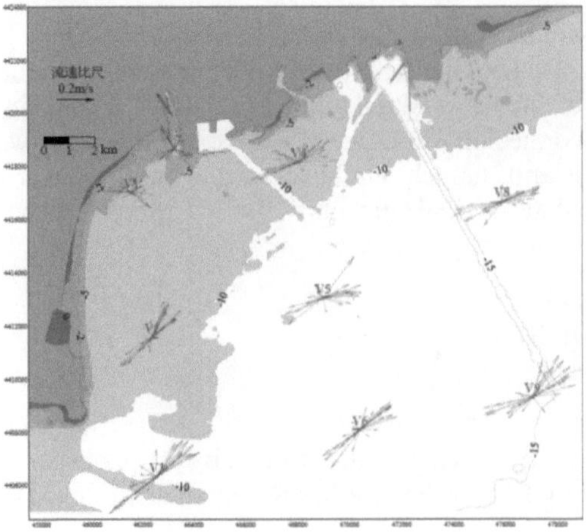

Figure 3. The tidal currents in Jin Meng bay.

$$\frac{\partial u}{\partial t}+\frac{\partial u^2}{\partial x}+\frac{\partial uv}{\partial y}=fv-g\frac{\partial \zeta}{\partial x}+\frac{\tau_{sx}-\tau_{bx}}{\rho h}+E_x\left(\frac{\partial^2 u}{\partial x^2}+\frac{\partial^2 u}{\partial y^2}\right)-\frac{1}{\rho}\left(\frac{\partial S_{xx}}{\partial x}+\frac{\partial S_{xy}}{\partial y}\right)+u_s S \quad (2)$$

$$\frac{\partial v}{\partial t}+\frac{\partial uv}{\partial x}+\frac{\partial v^2}{\partial y}=-fu-g\frac{\partial \zeta}{\partial y}+\frac{\tau_{sy}-\tau_{by}}{\rho h}+E_y\left(\frac{\partial^2 v}{\partial x^2}+\frac{\partial^2 v}{\partial y^2}\right)-\frac{1}{\rho}\left(\frac{\partial S_{yx}}{\partial x}+\frac{\partial S_{yy}}{\partial y}\right)+v_s S \quad (3)$$

where ζ is tide level, means the distance between water surface and datum plane; x and y are the Cartesian co-ordinates; h is water depth; g is gravitational acceleration; u and v are the velocity components in the x and y; t is time; f is Coriolis parameter; ρ is the density of

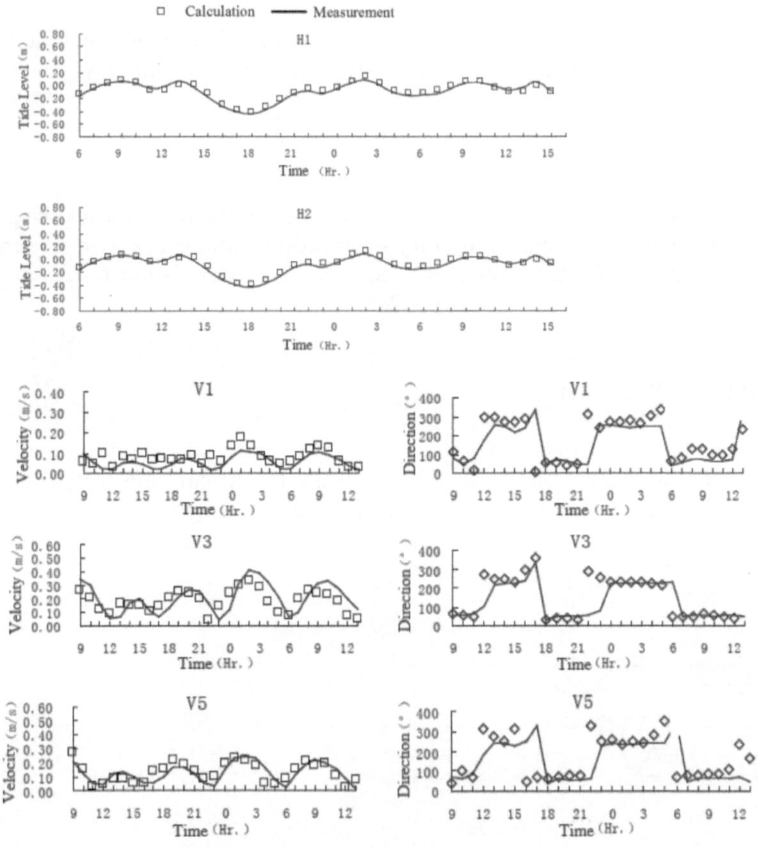

Figure 4. Comparison of computed and measured tide level and current.

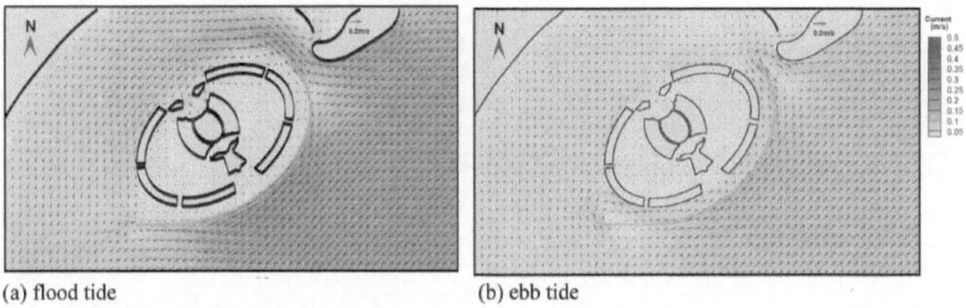

(a) flood tide (b) ebb tide

Figure 5. The currents after the Louts Island.

water; S is the magnitude of the discharge due to point sources and (u_s, v_s) is the velocity by which the water is discharged into the ambient water; S_{xx}, S_{xy}, S_{yx} and S_{yy} are components of the radiation stress tensor; (τ_{sx}, τ_{sy}) and (τ_{bx}, τ_{by}) are the x and y components of the surface wind and bottom stresses; E_x and E_y are the horizontal eddy viscosity in the x and y.

2.2 *Water exchange model l*

The conservation equation for a scalar quantity is given by

$$\frac{\partial(hc)}{\partial t}+\frac{\partial(uhc)}{\partial x}+\frac{\partial(vhc)}{\partial y}=\frac{\partial}{\partial x}\left(h\cdot D_x\cdot\frac{\partial c}{\partial x}\right)+\frac{\partial}{\partial y}\left(h\cdot D_y\cdot\frac{\partial c}{\partial y}\right)-F\cdot h\cdot c+S \tag{4}$$

where c is the concentration of the scalar quantity; D_x and D_y are the horizontal diffusion coefficient in the x and y; F is the linear decay rate of the scalar quantity; S is the magnitude of the discharge due to point sources.

2.3 *Verification*

The models have been calibrated with field observations and the results would be given here. Tide and current data from nine temporary observation stations are used to verified and validate the present model. By comparison, the numerical calculation results are in good agreement with the measured data (see Figs. 2–5).

3 WATER CHANGE ABILITY

Figure 5 shows the current fields after the Lotus Island during flood tide and ebb tide. The tidal currents around and within island still flow regularly. The current speed within island is less than 0.1 m/s, which is disadvantageous for the water exchanges.

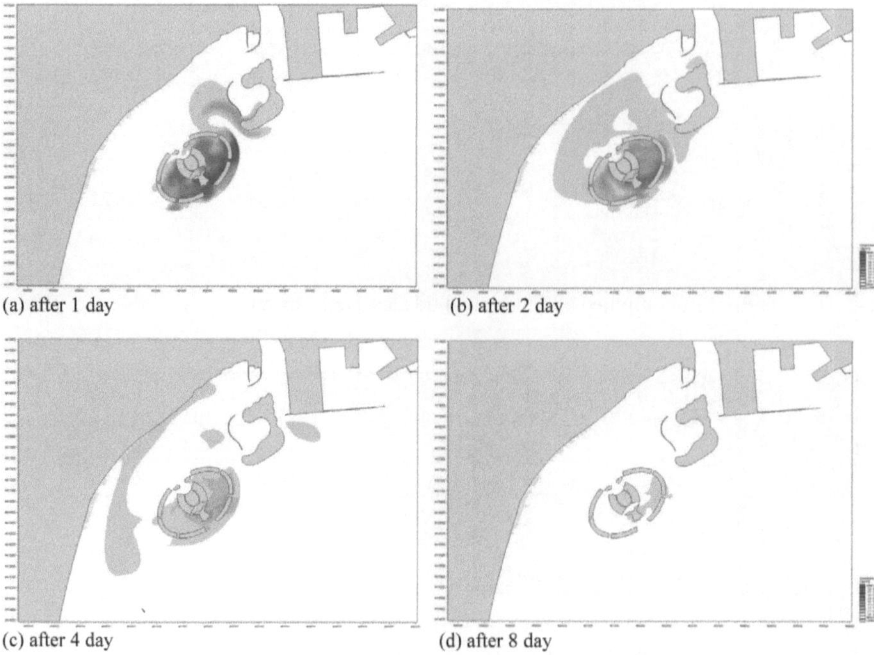

Figure 6. The concentrations after 1, 2, 4 and 8 days in action of currents.

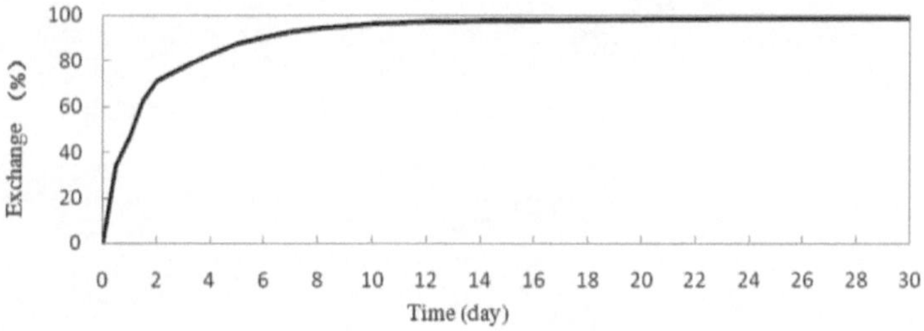

Figure 7. Rates of water exchange.

The initial concentration of the water inside Louts Island is set to 1.0 unit and one of water outside is 0 unit. The open boundary is zero gradient. Figure 6 shows the changes of the concentration after 1, 2, 4 and 8 days in the action of tidal currents. Figure 7 shows that the rates of water exchange. Based on the analyses of the internal concentration before and after the project, the following conclusions are reached:

1. Water exchange is more quickly on the mouth of the water channel than inner water of island.
2. The water of Lotus Island affect the Hailuo island. However the exchange of water inside Lotus Island with Hailuo island is not obvious.
3. The mean water exchange rate of the waters in Lotus Island is 38% of its initial value after 1 days, 65% after 2 days, 80% after 4 days and 93% after 8 days.
4. The half-life time of Lotus Island is 1.05 days.
5. The mean water exchange rate of Lotus Island reaches 80% after 4 days, which satisfies the PINAC's rule about the environment protection that the concentration of internal water is less than 37% after eight periods of tide.

4 CONCLUSION

1. The model is able to predict the tidal current around the artificial island reasonable. The simulated water level and current basically agreed with the measured data.
2. After the second artificial island is finished, tidal currents around and within island are still regular.
3. The half-life time of Lotus Island is 1.05 days.
4. Lotus Island satisfies the PINAC's rule about the environment protection that the concentration of internal water is less than 37% after eight periods of tide.

ACKNOWLEDGMENTS

The research was supported by the National Natural Science Foundation of China under grant no. 51209111.

REFERENCES

MIKE21 & MIKE 3 FLOW FM hydrodynamic and transport module Scientific Documentation, MIKE by DHI 2009.

Wang, H., Chen, P., et al. 2008. Advance in the research on water exchange in the sea area. *South China Fisheries Science*, 4(2): 75–80. (Chinese).

Yan, B. 2013. Report of water exchange of Lotus Island based on numerical simulation. *Tianjin Research Institute for Water Transport Engineering.* (Chinese).

Modeling and Computation in Engineering III – Zhang & Xie (Eds)
© 2014 Taylor & Francis Group, London, ISBN 978-1-138-02680-3

Inventory optimization and modeling techniques applied within manufacturing company

H. Fidlerová
Faculty of Materials Science and Technology, Slovak University of Technology, Bratislava, Slovakia

J. Prachař
European Polytechnical Institute, s.r.o, Czech Republic

M. Horváthová
Faculty of Materials Science and Technology, Slovak University of Technology, Bratislava, Slovakia

ABSTRACT: Inventory and storage management has represented an important aspect of logistics in manufacturing enterprises or companies for many years. Recently however it has become increasingly recognized as a cross function of logistics. The objective of this paper is to present an overview of the most common forms of inventory and storage modeling techniques. The paper addresses the optimization of inventory management and storage in a manufacturing company, through the combination of several methods and techniques to produce a synergic effect. Firstly ABC modeling techniques were adopted for analysis of the purchased components. Following the ABC analysis, the turnover for selected items from the A class was identified and the application of a Kanban system proposed using EDI. The paper is a part of KEGA project No. 037STU-4/2012 "Implementation of the subject "Corporate Social Responsibility Entrepreneurship" into the study programme Industrial management in the second degree at MTF STU Trnava".

1 INTRODUCTION

Inventory and warehouse management is considered to be an integral part of any company and therefore should not be regarded in the company only as a process in which stocks are stored until they will be further used. Effective inventory and storage management can also result in significant cost savings, through the reduction of inventories held. As a result, gradually a synchronized material flow is reached (Mlkva et al 2012).

Inventory and warehouse management is a process of continual optimization and the search for new opportunities to streamline these activities and eliminate any risks that could stop processes in the enterprise. According to Vidová (2009) here is a potential area for savings through the application of lean logistics techniques in order to reduce the inventory held by industrial enterprises.

The paper considers the optimization of inventory management in the manufacturing company which has been provided initially as a result of ABC analysis and follow owed by analysis of inventory turnover according to the term of the last issue to identify dead stocks.

2 OPTIMIZATION OF INVENTORY AND STORAGE MANAGEMENT IN A MANUFACTURING COMPANY

The production function of the manufacturing company is focused upon the following products: linear cage and cage thrust assemblies, needle cage radial cylindrical roller bearings, pulleys, welded needle cage and control systems. Within the framework of the inventory optimization in the enterprise, components purchased from external firm were analyzed, where the company had allocated a large amount of capital.

According to the Handbook of Logistics and Distribution Management (Rushton 1989), one means of tackling these problems is to be more aware of the range of products held. This can be achieved by using Pareto analysis (or ABC analysis).

2.1 *The application of ABC analysis modeling of inventory in the manufacturing company*

The ABC analysis provides a mechanism to identify items that will have a significant impact on overall inventory cost, while also providing a mechanism for identifying different categories of stock that will require different management and controls (Vollmann 2011).

According Rushton (1989) clearly this policy will differ according to product type, industry type, service level requirements, etc. The essential point is to be aware of the appropriate stock-holding costs and recover the costs accordingly.

All purchased components of the company were divided through the application of ABC analysis modeling into three groups. It was firstly established, which components purchased from external companies were the most funds invested, because it would be inefficient to pay the same attention to all material items. The ABC analysis modeling has been conducted on the 4011 material items of purchased components from external companies. The aim was to find out in which items the company holds most of the funds. From the analysis (Fig. 1) it is obvious that in the category A of material items is the most of the financial funds (75% of the total value of inventory), and their percentage of the total number of items is small (8%).

In the group A could fall also some material items, which account for a large amount of capital investment but on the other hand, the enterprise did not have to consume them at all for the last period. **So it was decided to continue the optimization of stock of the components purchased from external companies from category A with analysis of turnover and inventory analysis according to the date of the last issue.**

2.2 *Analysis of inventory turnover in the manufacturing company*

The aim of the inventory turnover analysis was to sort the category items and material from the ABC analysis according to their speed and identify high-turnover, moderate-turnover, low-turnover and non-turnover inventory in warehouses. Within the framework of inventory optimization is one of the most important aims of the company to identify stocks of non-turnover and in an appropriate manner to discard—either scrapping or selling them.

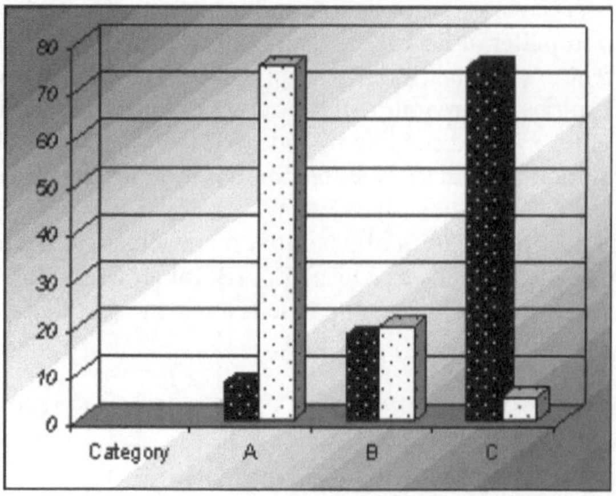

Figure 1. Distribution of ABC classes in the ABC analysis in the manufacturing company.

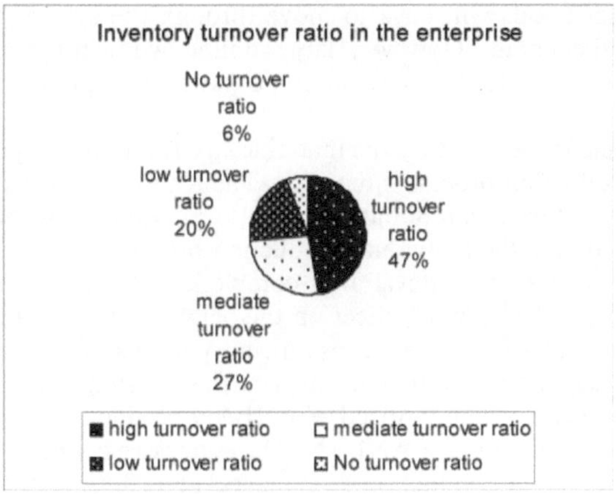

Figure 2. Inventory turnover ratio in the industry company.

Non-turnover stocks ties up the firm's capital, attaching storage area, affecting the clarity in stock and storage, increasing demands on handling (the movement of pallets, etc.).

For each material item from A category, the consumption has been detected over the past year.

An increasing stock turnover figure or one much larger than the average for an industry segment may indicate poor inventory management. The stock turnover was calculated as annual sales (consumption)/the average store held.

Application of inventory turnover analysis showed which material items from the category A (from the ABC analysis) had no turnover ratio over the last year; it was up to 6.2% of the value of inventory category A (Fig. 2).

From this point of view non-turnover items were marked, and on the basis of the analysis of the final consumption it could be determined whether these items were consumed at all in past years.

2.3 *Analysis of inventory according to the term of the last issue*

The principle of this analysis is in the identification of the date of the last issue of a specific material in the past. Without moving the items represent so-called "dead" stocks, which is subsequently to be disposed in an appropriate manner (sale or scrapping). The priority of the analysis was to find these "dead" materials without moving. Such items have a financial cost, and also the cost of the maintenance and the requirement for a storage area which affects the clarity of registering and storage and increases the demands on handling (the movement of pallets, etc.). For each item, where there is no last date of consumption, it is a component purchased from external companies, although the material was obtained, but never consumed.

3 APPLICATION OF KANBAN SYSTEM IN MANUFACTURING COMPANY

The word 'Kanban' (the signal) refers to a system of cards, which is used to organize the progress of materials through the manufacturing process. It may be easier to understand the system if squares are marked on the factory floor. The squares contain work-in-progress required by the next step in the manufacturing process. The squares become empty as the materials are used. The next batch of materials may only move into a square when the square is empty.

This approach is replicated as materials move progressively from one step to the next. Thus no build-up of goods occurs, and materials move through the system in an orderly fashion.

The problem is that goods will have to move through the system at the speed of the slowest element in the chain. However, large online work-in-progress stocks will be eliminated. This too contributes to the reduction of working capital being used by the system (Rushton 1989).

The Kanban principle is the fact that the materials, parts and developed production come at a time when the production process is required. The Kanban system not only satisfies the information function (informs you about their work, providing information of ordering), but also helps to coordinate the individual production processes (coordinates the supply of parts to the assembly line, their removal after completion of the assembly), minimizes the individual processes with the ultimate effect on the height of the stock. In the company, it appears to be the case that a purchaser sends an empty means of transport with a dispatch note, which is a label (Japanese "kanban"), which carries the standard order (Fig. 4).

Kanban in the enterprise is standardized. KANBAN production has a white base color, the color yellow has a collection of KANBAN. The enterprise connects the Kanban system with the Pull-principle, when production is managed according to the consumption with defined inventory.

In inventory control several basic levels of inventory should be monitored. These are:

- maximum stock—the highest level that is reached just as the arrival of new supplies at the warehouse
- minimum stock—represents the state of the stock, just before the arrival of new supplies in stock. It is given by the sum insured, strategic and technological supplies

Signal stocks—(ordering supplies, point of order) respects such inventory levels at which it is necessary to issue order to enable the delivery arrived at the warehouse later than the time when stocks reach a minimum (Sixta & Žižka).

The following process always pulls material from the previous operation when it is finished with the old order. This ensures that what is produced, is only what it required. It does not operate when nothing is required. If the warehouse is filled up with a maximum amount/level of production it shuts down the previous operation, or can adjust the level of production.

The discussed system has an overall impact on inventory levels and production in the company.

From a logistics point of view it is important to mention an example of the Kanban application on the segment 07-in the enterprise.

The situation before the application of Kanban, when inventories were not co-ordinated, in this way was that the company had more storage costs, and capital was held in goods, which were not required by the customer at that time. After the application of the principle of streamlining the production process there was a positive impact on the amount of inventory in stock.

Stocks and consumption are regulated in the enterprise and are visible in the "Supermarket-Shelf" style as a central element of the consumption management. A Supermarket is a device, in which a customer can actually get what they need, when they need it, as well as in the quantity required. A Supermarket has a discreet folder for each material and works on the principle of "First In-First Out" (first-in, first-out). Via the Kanban system there is a signal that should be sent to the previous process to request a production of a defined quantity.

4 SOLUTION TO OPTIMIZE THE STORAGE AND INVENTORY MANAGEMENT IN THE INDUSTRIAL ENTERPRISE

The prime objective of most warehouses is to facilitate the movement of goods through the supply chain to the end consumer.

In the context of inventory optimization it is recommended that the company focuses on the level of stocks already in the procurement of the basic material. From the above mentioned analysis it is obvious that in the enterprise is a relatively high value/level of stocks with low, or no turnover.

This means a potential for improving the optimization of the process and thus the level of the stock holding. The main challenge would be to implement such a system in undertaking measures to simplify and speed up the process of procurement of the basic material. It may be through the implementation of electronic communication via EDI between the company and its suppliers.

EDI has been defined as: *computer-to-computer exchange of structured data for automatic processing*. EDI is used by supply chain partners to exchange essential information necessary for the effective running of their businesses. These structural links are usually set up between organizations that have a long-term trading relationship. The main advantages of using EDI are information needs to be entered on to the computer system only once; speed of transactions; reduced cost and error rates (Rushton 1989).

Through electronic data interchange it is easier to process orders of base materials. The company should with their suppliers agreed levels of stock levels for a particular type of basic material (whether the maximum or even minimal). The supplier has an overview and information of the inventory in the enterprise of the purchaser and monitors movements within the framework of the agreed maximum or minimum limits. As soon as the material has reached the minimum level, through electronic data interchange is the signal to the vendor. The supplier would automatically deliver the required type of material in the agreed amount of inventory.

On the other hand, the company as a customer would then receive an exact overview of the type and quantities of a specific item a supplier sends.

On the basis of this information the production process can be planned. Therefore can be achieved better disposal of the material and the reduction of capital in inventory. In addition, by means of the use of electronic data interchange, the supplier of the material could send a message about the transport not only the company, but also to the forwarder.

During the analysis of the stock management in a chosen industry enterprise some possible causes of low-turnover, or non-turnover stocks were identified. The causes were for example: a change of input material for the product, or the customer cancelled the contract for certain products and then a purchased segment had no use.

There are two main options to deal with this material:

- To determine whether there is a possibility of consumption in other companies within the group;
- To find out whether it is possible to use this material as an alternative material for production of other products of the company or to modify segment.

These options should provide a better return on investment for these types of materials in the enterprise. If it is not possible to use either of these options, the firm should proceed to another alternative solution which on the one hand reduces the inventory supply, but on the other hand, must account for the loss. Firstly, alternatives were considered to address the supplier of the materials, whether it is possible to sell this material back. The price depends on several factors, for example the life of the material, storage period, the current market price of the commodity, etc. However, the company must consider the fact that they will not get back the full value originally paid for the material. On the other hand, this method can help in reducing the inventory and at the gets some free financial capital, which can also be used. A last solution, which is not financially favorable for the enterprise is to scrap the materials.

5 CONCLUSION

The objective of inventory management analysis in the manufacturing company was for the identification of items which hold the largest capital investment. The authors determined as it is the first stage is ABC analysis, it is necessary to understand stock control issues. Through ABC analysis it was concluded that the manufacturing company has a high level of stocks with no turnover. As a suitable alternative for inventory management optimization of the inventory level through the application of Kanban is also considered. The use of EDI the has also been identified within the organization as the process will create additional benefits.

REFERENCES

Ferenčáková, L. 2011. *Optimalizace skladování a řízení zásob v podniku INA Skalica*. Bachelor thesis. EPI Kunovice.

Fidlerová, H. & Prachař, J. & Sakál, P. 2014. Application of material requirements planning as method for enhancement of production logistics in industrial company. In: *Applied Mechanics and Materials*. ISSN 1660-9336. Vol. 474, pp. 49–54.

INA Skalica, spol. s.r.o. 2009. *Move-Akademie Supermarket Kanban V06*. Skalica.

Mĺkva, M. & Kučerová, M. & Večeřa, P. 2012. Application of tools and methods within "Lean manufacturing. In: *Carpathian Logistics Congress, Czech Republic*. Ostrava: TANGER s r.o, ISBN 978-80-87294-33-8.

Rushton Alan R. 1989. *The Handbook of Logistics and Distribution Management*. Kogan Page Limited, ISBN 978 0 7494 5714 3

Sixta, J. & Žižka, M. 2009. *Logistika: používané metody*. Brno: Computer Press, a.s. 237 p. ISBN 978-80-251-2563.

Vidová, H. 2009. Lean logistics—the importance and utilisation in Slovak industrial practice. In: *Research papers Faculty of Materials Science and Technology Slovak University of Technology in Trnava*. ISSN 1336-1589. no. 27, pp. 119–125.

Vollmann Thomas E. 2011. *Manufacturing planning and control systems for supply chain management*. ISBN-10: 0071750312.

Modeling and Computation in Engineering III – Zhang & Xie (Eds)
© 2014 Taylor & Francis Group, London, ISBN 978-1-138-02680-3

Overtopping risk analysis of Xiashan Reservoir

Ying Sun
School of Continuing Education of Beijing University of Technology, Beijing, China

Zunliang Zhu
Xiashan Reservoir Administration, Weifang, China

Limei Teng
SINOPEC Engineering Incorporation, Beijing, China

Haibo Xu
Xiashan Reservoir Administration, Weifang, China

ABSTRACT: Xiashan Reservoir is a large reservoir lies in Weifang City, Shandong Province. With the development of economy and society, the deficit in water resource has increasingly arisen in many cities, such as Weifang City. Considering thoroughly the uncertainties of flood, wind, storage volume and discharge capacity, the overtopping risk analysis for earth or rockfill dam against the simultaneous actions of flood and wind wave during flood period has been established in the paper and then its application to Xiashan Reservoir is described. By taking 10^{-6} as acceptable risk, or 99.999% as acceptable reliability, the result of risk analysis show that the earth dam of Xiashan Reservoir is very reliable against the design series of flood events accompanied with the runup generated by wind wave during flood period. Besides, in order to efficiently utilize the valuable water resources (including flood period) upstream of the reservoir, on the basis of risk analysis, we recommended that its designed limiting reservoir level before flood, say 37.4 m, may be safely raised to an new elevation of 38.0 m. The correspondent increment of water volume should be 7765×10^4 m^3. It would bring a great deal of direct benefits and social benefits.

1 INTRODUCTION

Along with the development of industry and agriculture as well as the growth of population, deficit in water resources has increasingly arisen in some region, particularly the North China, and it will restrict the further development of economy if the problem be not solved properly. On the other hand, the annual distribution of rainfall is very non-uniform, such as the rainfall is concentrated in July, August and September in China. Therefore, the problem of flood management is serious. There are some non-structural measures to control flood. In this study, we take the overtopping risk analysis as a non-structural measure to control flood. Considering thoroughly the uncertainties of flood, wind, storage volume and discharge capacity, risk model is developed for the earth dam to analyze its risk of overtopping against the design series of flood events accompanied with the runup generated by wind wave during flood period, and then applied to the earth dam of Xiashan Reservoir. By taking 10^{-6} as acceptable risk, or 99.999% as acceptable reliability, the evaluation of safety of Xiashan Reservoir against overtopping is made. Besides, in order to utilize efficiently the valuable water resources (including flood period) upstream of the reservoir, on the basis of risk analysis, we recommended that its original limiting reservoir level before flood, say 37.4 m elevation, may be safely raised to an new elevation of 38.0 m, the correspondent increment of water volume storage should be 7765×10^4 m^3, it would bring a great deal of direct benefits and social benefits.

2 STUDY METHODS

We use the 'Theory of overtopping risk analysis' as our study methods, which will be introduced as following.

2.1 *Uncertainties*

For earth dam the overtopping is strictly forbade unless it is previously permitted during designing. Once overtopping of earth dam occurs, the dam should be considered as beginning of dam break which may result in a catastrophe. For a reservoir already constructed such as in our study, the flood, wind, volume of reservoir and discharge capacity are all considered as random variables.

2.1.1 *Inflow flood*

Inflow flood, either design one or forecasted or measured one, is considered in the traditional reservoir design as a random process. For design flood, its uncertainties are initiated from the stochastic characteristics of the historical hydrological data (storm rainfall, peak flow of flood, etc.), of typical flood hydrograph, of loss of rainfall, and of unit hydrograph. For the forecasted flood, its uncertainties are initiated from the stochastic characteristics of the rainfall distribution in both space and time, and of the forecasted model used and parameter values taken. For the measured flood, its uncertainties are initiated from its stochastic characteristics of measuring in space and time, and of the measuring errors. For Xiashan Reservoir, its different series of flood follow P-III Type of distribution based on hypothesis testing.

2.1.2 *Wind*

We know that wind of any magnitude from any direction is a random event, so the wind setup, wave height and runup generated by wind are then random too. As for the overtopping risk of the earth dam, the wind toward the dam during flood period is major concern and is defined as "effective wind for overtopping" by us. Strictly, speaking the statistics of wind series should take the data of wind corresponding each flood. Unfortunately, such data is difficult to obtain in present time. And therefore we take the series data of maximum wind during flood period, which will bring the results of risk analysis on the safe side. The effective wind series for Xiashan dam should be the series of maximum wind toward the dam during flood period. Based on hypothesis testing, it follow I Type of extreme distribution.

2.1.3 *Reservoir area and storage volume*

As we known, the contour lines plotted by different surveyors for a given reservoir topography may be different. Under the same contours, the calculated reservoir area and storage volume at a given water level may be different due to variety of computation philosophy and the instruments used. Moreover, the sediment transport during and after flood may change the underwater topography and thereby the reservoir area and storage volume unless the underwater contours be surveyed after each flood immediately. So, the reservoir area and storage volume are considered as random variables. A Normal Distribution is assumed for function F(z) and V(z) based on hypothesis testing.

2.1.4 *Discharge capacities*

The discharge capacities of Xiashan Reservoir are consist of right-bank discharging tunnel, left-bank spillway tunnel and emergency spillway. Their uncertainties arise from many sources, such as the simplification of 3-D flow to 1-D flow model, measurement errors and roughness estimate etc. By hypothesis testing, a normal distribution is assumed for discharge function Q(z).

2.2 *Risk model*

The risk model of overtopping was developed and may be expressed as follows:

$$\overline{R} = P(Z(t) \geq Z_c) = P\left(Z_0 + H_{\max} + e + R_p \geq Z_c\right) \tag{1}$$

where $\overline{R}$ is risk against simultaneous actions of a series of flood and effective wind events; Z_0 is initial reservoir level; $H_{\max}$ is reservoir level increases due to flood; e is wind setup; R_p is wave runup on upstream slope of the dam; Z_c is predetermined critical elevation, such as the dam crest elevation or top elevation of parapet.

In the case of concurrence of flood event $[Q_{i-1}, Q_i]$ and wind event $[W_{j-1}, W_j]$, the risk P_{ij} is:

$$P_{ij} = P\left(Z_0 + H_{\max} + e_{ij} + R_{P_{ij}} \geq Z_c\right) \tag{2}$$

Then, the total risk may be expressed as follows:

$$\overline{R} = \sum_{i=1}^{\infty} f(Q_i)dQ_i \left[\sum_{j=1}^{\infty} f_w(W_j)dW_j P_{ij}\right] \tag{3}$$

In the above equations, the mean value of wind set-up may be calculated according related design code.

It was well known that wave heights due to wind follow Rayleigh distribution, so the runup also follow Rayleigh distribution. Because wind events are stochastic, the variable R_p are composite random variable. It may be shown that the probability of P_{ij} of occurrence of R_p when the wind velocity is within the interval $[W_{j-1}, W_j]$ is

$$P_{wi} = P\{R_p < x\} = \int_0^x f_{R_p}(R_p)dR_p \tag{4}$$

where $f_{Rp}(R_p)$ is the probability density function of Rayleigh distribution for the interval $[W_{j-1}, W_j]$. Therefore, the distribution function of R_p may be expressed as

$$F(x) = \sum_{i=1}^{\infty} f_w(W_i)P_{wi}dW_i = \sum_{i=1}^{\infty} f_w(W_i)dW_i \int_0^x f_{R_p}(R_p)dR_p \tag{5}$$

For single inflow, the key problem in overtopping risk is to determine $Z(t)$ in equation (1), which may be solved by reservoir routing. Now, the routing process is a random process and both storage volume $V(t)$ and reservoir water level $H(t)$, are Markov processes and independently increasing processes. Then, the random differential equation of reservoir routing may be expressed as follows:

$$d\overline{H}(t) = \frac{\overline{Q}(t) - \overline{q}(H,C)}{\overline{F}(H)}dt + \frac{dW_n(t)}{\overline{F}(H)} \tag{6}$$

$$\overline{H}(t_0) = \overline{H}_0 \tag{7}$$

where $\overline{H}$ is the mean value of reservoir water level, $\overline{H}(t_0)$ is initial condition of $\overline{H}_0$, $\overline{H}_0$ is water level before flood, $\overline{Q}(t)$ is flood process, $\overline{q}(H,C)$ is the mean volume of outflow discharge which is a function of reservoir water level H and hydraulic parameter C expressing the effect of the type and size as well as discharge coefficient of spillway and outlet works, $\overline{F}(H)$ is the mean value of reservoir and $W_n(t)$ is Wiener process.

Solution of above equations gives us the overtopping risk with predetermined critical mode.

2.3 *Critical mode*

Critical mode is the value selection of Z_c which in equation (1). General, there are two mode were analyzed here. The first taking the crest elevation of earth dam, or the top elevation of anti-seepage parts as the critical elevation Z_{c1} for flood and wind setup, The second, taking the top elevation of parapet as the critical elevation Z_{c2} for flood, wind setup and wave runup.

In Xiashan case, the crest elevation of the main dam and four auxiliary dams is 44.0 m, and the top elevation of parapet is 45.0 m. The elevation of Shuicheng auxiliary dam is only 41.2 m due to it hasn't been constructed at present. So we should take different elevations as the first and second critical elevation in different case.

3 OVERTOPPING RISK CRITERION

Based on the investigations on the data of dam overtopping and dam break around the world, it was concluded that the probability of overtopping for earth dam is 10^{-5} per year per dam. Once overtopping and dam break occur, their consequences are very serious and therefore we should take overtopping risk of the order of magnitude 10^{-6} as an acceptable risk. Before the publication of national standard, we use 10^{-6} as a risk criterion temporarily, thus the safety reliability of the dam against simultaneous actions of flood and wind series is 99.999%. It is believed that with such reliability, there exists sufficient sense of security for any decision makers.

4 XIASHAN RESERVOIR CHARACTERISTICS AND REGULATION SCHEME

4.1 *Reservoir characteristics*

Xiashan Reservoir lies in the lower reaches of Weihe River in Shandong Peninsula, Weifang City, Shandong Province, China. It controls a catchment area of 4210 km^2. Its design flood has a frequency of 0.1% and the extreme flood has a frequency of 0.01%. The maximum storage capacity of reservoir is 14.05×10^8 m^3.

Xiashan Reservoir has one main dam and five auxiliary dams which are rolled homogeneous earth dams. The maximum height of the main dam is 21.0 m, crest length is 2750 m and crest elevation is 44.0 m. Top elevation of the parapet on the main dam crest is 45.0 m. The maximum heights of Nanxin, Wulan, Zhenggong and Liujiagou auxiliary dam are 14.5 m, 13.5 m, 11.9 m and 8.5 m respectively. The crest lengths of them are 4900 m, 10050 m, 12380 m and 1270 m respectively, and their crest elevations are all 44.0 m. Among the five auxiliary dams, only Nanxin auxiliary dam has parapet on its dam crest, which top elevation is 45.0 m. The elevation of Shuicheng auxiliary dam is only 41.2 m due to it hasn't been constructed at present.

The maximum discharge of the spillway tunnels is 15853 m^3/s and for the five discharging tunnels are 42 m^3/s, 3.5 m^3/s, 29.5 m^3/s, 54 m^3/s, and 8.03 m^3/s respectively.

4.2 *Regulation scheme of Xiashan Reservoir*

Considering Shuicheng auxiliary dam hasn't been constructed yet, its ground elevation is only 41.2 m. The regulation of Xiashan Reservoir taking 36.8 m as limiting reservoir level before coming flood, or simply Limiting Level Before Flood (LLBF). The regulation of reservoir in 2012 was given as follows:

1. With the coming flood, the LLBF maintains 36.8 in flood season. If heavy rain occurred in July, the reservoir water level is allowed to 37.4 m after August 1.
2. When the flood frequency of coming flood is 5%, the reservoir water level is 38.2 m, the maximum discharge is controlled not over 3000 m^3/s so as to ensure the safety of the downstream.

3. When the flood frequency of coming flood is less than 5% and more than 2%, the reservoir water level is 39.25 m, the maximum discharge is controlled not over 3500 m^3/s so as to ensure the safety of the downstream.
4. When the flood frequency of coming flood is less than 2%, the reservoir water level is more than 39.25 m, the gates in discharging tunnel and spillway tunnel are all opened to discharge the flood, making the highest water level not exceed 41.20 m. meanwhile the reservoir management should be ready to do rescue word.
5. When the flood peak is over, the reservoir level should be lowered to the limiting level before flood, say 36.8 m in order to keep the dam safely against the coming flood with multiple peaks or the flood with frequency scarcely encountered.

5 RESULTS OF RISK ANALYSES

5.1 *Overtopping risk analyses for 2012 reservoir regulation*

As we know, Shuicheng auxiliary dam hasn't been constructed, So the ground elevation of Shuicheng auxiliary dam, say 41.2 m was taken as the critical elevation Z_c for reservoir regulation in 2012. The overtopping risk of 2012 reservoir regulation is listed in Table 1 in which $\bar{R}$ is the computed risk corresponding to the critical elevation Z_c.

If we take a risk of order of magnitude of 10^{-6} as an acceptable risk, it can be seen from Table 1 that the computed overtopping risk $\bar{R}$ against flood series (flood frequency is greater than 1%) and wind events during flood period are less than 10^{-6}, or the safety reliability more than 99.999%. If flood frequency is not more than 1%, the computed overtopping risk $\bar{R}$ will be more than acceptable risk 10^{-6}, or the safety reliability will be less than 99.999%.

5.2 *Overtopping risk analyses for designed reservoir regulation*

According to the preliminary design of the reservoir, only the main dam and Nanxin auxiliary dam have parapet, the remaining four auxiliary dam didn't have parapet. So overtopping risks of two case were computed.

The overtopping risk of dam without parapet under designed reservoir regulation is listed in Table 2, in which the crest elevation, say 44.0 m is taken as the critical elevation Z_c, $\bar{R}$ is the computed risk corresponding to the critical elevation Z_c, and Q_{max} is maximum discharge of reservoir during regulation.

The overtopping risk of dam with parapet under designed reservoir regulation is listed in Table 3, in which the crest elevation of the dam, say 44.0 m, and the top elevation of parapet, say 45.0 m were taken as the first and second critical elevation Z_{c1} and Z_{c2}, $\bar{R}_1$ and $\bar{R}_2$ are the computed risk corresponding to the critical elevation Z_{c1} and Z_{c2}, and Q_{max} is maximum discharge of reservoir during regulation.

If we take a risk of order of magnitude of 10^{-6} as an acceptable risk, according the designed reservoir regulation of reservoir, it can be seen from Table 2 that the computed overtopping risk $\bar{R}$ against different flood series and wind events during flood period are all less than

Table 1. Overtopping risk of reservoir regulation in 2012.

The upper limit of flood series (flood frequency)	LLBF (m)	Q_{max} (m^3/s)	$\bar{R}$
5%	36.8	3000	5.0096443×10^{-8}
2%	36.8	3500	1.0584093×10^{-7}
1%	36.8	9077	6.6743257×10^{-7}
0.1%	36.8	12202	1.4837884×10^{-4}
0.05%	36.8	13279	3.3314669×10^{-4}
0.01%	36.8	15579	5.7851017×10^{-4}

Table 2. Overtopping risk of dam without parapet under designed reservoir regulation.

The upper limit of flood series (flood frequency)	LLBF (m)	Q_{max} (m^3/s)	$\bar{R}$
5%	37.4	3000	$<1.0000000 \times 10^{-36}$
2%	37.4	3500	$1.3383493 \times 10^{-10}$
1%	37.4	10404	$6.7350231 \times 10^{-10}$
0.1%	37.4	12677	1.3565608×10^{-8}
0.05%	37.4	13629	5.7066977×10^{-8}
0.01%	37.4	15853	6.3763781×10^{-7}

Table 3. Overtopping risk of dam with parapet under designed reservoir regulation.

The upper limit of flood series (flood frequency)	LLBF (m)	Q_{max} (m^3/s)	$\bar{R}_1$	$\bar{R}_2$
5%	37.4	3000	$<1.0000000 \times 10^{-36}$	$<1.0000000 \times 10^{-36}$
2%	37.4	3500	$<1.0000000 \times 10^{-36}$	$<1.0000000 \times 10^{-36}$
1%	37.4	10404	$<1.0000000 \times 10^{-36}$	$<1.0000000 \times 10^{-36}$
0.1%	37.4	12677	$<1.0000000 \times 10^{-36}$	$1.1599505 \times 10^{-10}$
0.05%	37.4	13629	$<1.0000000 \times 10^{-36}$	$9.5348840 \times 10^{-10}$
0.01%	37.4	15853	$<1.0000000 \times 10^{-36}$	2.5402553×10^{-8}

10^{-6}, or the safety reliability more than 99.999% in the first case. From Table 3, It can be seen that the computed overtopping risk $\bar{R}_1$ and $\bar{R}_2$ against different flood series and wind events during flood period are all less than 10^{-6}, or the safety reliability more than 99.999% in the second case, but its risk is less than the risk in the first case.

6 CONCLUSIONS AND RECOMMENDATIONS

On the basis of risk analysis against overtopping for Xiashan Dam, when Shuicheng auxiliary dam will have been constructed as design, we can declare that the dam of Xiashan Reservoir is safe against overtopping under its designed regulation. Besides, in order to efficiently utilize the valuable water resources (including flood period) upstream of the reservoir, on the basis of overtopping risk analysis, some recommendations are given as follows:

6.1 *Constructing Shuicheng auxiliary dam as soon as possible*

From Table 1, we can see that if flood series which frequency is not more than 1%, the computed overtopping risk $\bar{R}$ will be more than 10^{-6} due to Shuicheng auxiliary dam hasn't been constructed, its ground elevation only 41.2 m, it severely limits the use of the reservoir.

6.2 *Built parapet on other four auxiliary dam*

By comparing Tables 2 and 3, we can see that the overtopping risk in Table 3 is more less than in Table 2. So we recommend that constructing parapet on the four auxiliary dam which haven't parapets in design. If all the dams of reservoir have parapets on them, the safety reliability of reservoir will be greatly improved.

6.3 *Raising the designed Limiting Level Before Flood (LLBF) of Xiashan Reservoir*

From Table 2, we can see that according to the designed limiting reservoir level before coming flood, say 37.4 m, the corresponding overtopping risk against different flood series and wind

Table 4. Overtopping risk of designed reservoir regulation under 38.0 m.

The upper limit of flood series (flood frequency)	LLBF (m)	Q_{max} (m^3/s)	$\bar{R}_1$	$\bar{R}_2$
5%	38.0	3000	$<1.0000000 \times 10^{-36}$	$<1.0000000 \times 10^{-36}$
2%	38.0	3500	$<1.0000000 \times 10^{-36}$	2.3865743×10^{-9}
1%	38.0	10559	$<1.0000000 \times 10^{-36}$	1.3738546×10^{-9}
0.1%	38.0	12994	$<1.0000000 \times 10^{-36}$	3.5142151×10^{-8}
0.05%	38.0	13909	$<1.0000000 \times 10^{-36}$	1.1023844×10^{-7}
0.01%	38.0	16085	$<1.0000000 \times 10^{-36}$	9.4097305×10^{-7}

events during flood period are all less than 10^{-6}. So on the basis of overtopping risk computation and analyses, we raise the designed Limiting Level Before Flood (LLBF) of Xiashan Reservoir to 38.0 m, and take the crest elevation, say 44.0 m as the critical elevation Z_{c1} and Z_{c2}, the computed overtopping risk is showed in Table 4.

It can see from Table 2 that although the limiting reservoir level before coming flood is raised to 38.0 m, the corresponding overtopping risk against flood series (which flood frequency is not less than 1%) and wind events during flood period are also far less than 10^{-6}. Besides, the correspondent increment of water volume storage should be 7765×10^4 m^3, it would bring a great deal of direct benefits and social benefits. For the flood series which frequency is less than 1%, based on security considerations, its limiting reservoir level before coming flood is not changed, say 37.4 m.

REFERENCES

Chen. Z.H., Li. Q.J. and Sun. Y. et al., Research report: Overtopping risk for earth dams and risk criterion, Postgraduates School, Beijing Administrator's Inst. Of Water Resources and Hydropower, Dec. 1996.

Li. Q.J. and Chen. Z.H., 2008, Theory and application of overtopping risk for earth-rock fill dams, China Water & Power Press.

Sun. Y., Wang. Y. and Li. Q.J., 2010, A study of flood risk management of Xueye Reservoir, Journal of Xi'an University of Technology. 111–115.

Sun. Y., Chen. Z.H., and Li. Q.J., 1999, Overtopping risk analysis as a non-structure measure, Proceedings of 99 International Symposium on Flood Control, Beijing, China, 89–97.

The Ministry of Water Resources of the People's Republic of China, 2002, Design code for rolled earth-rock fill dams.

Modeling and Computation in Engineering III – Zhang & Xie (Eds)
© 2014 Taylor & Francis Group, London, ISBN 978-1-138-02680-3

Urban hydraulic structures discharge capacity study with CFD

Sui Ping Zeng
School of Architecture, Tianjin University, Tianjin, P.R. China

Jian Tian
Town and Country Planning and Design Institute, CAUPD, Beijing, P.R. China

Jie Zeng
South Florida Water Management District, FL, USA

ABSTRACT: This paper presents a new approach of Computational Fluid Dynamics (CFD) applied to the unban hydraulics. The approach focuses on the investigation of flow behavior and stage-discharge relationship for flood control and/or drainage hydraulic control structures in urban areas. To obtain accurate stage–discharge relationship in unban hydraulic structures, such as spillway and culvert, is one of key elements of decision support and operation of flood control and drainage management in urban areas. Two successful CFD applications are displayed in the paper. One is the simulation for gate spillway, another focuses on investigation of flow passing through a gated-culvert. These two are typical flood control and drainage hydraulic control structure in urban areas. Traditionally, the flow behavior and discharge capacity are investigated by laboratory experiment or field measurement, which are expensive and time-consuming. These two case studies demonstrate that CFD-based numerical approach is cost-effective and can obtain accuracy flow data for urban hydraulic structure. This method is believed can be extended and applied to many other similar studies.

1 INTRODUCTION

Today, more than half of the world's population is living in towns or cities, and this tendency of urbanization is keeping on accelerating. The subsequent of urbanization has many environmental impacts, such as increasing unban energy consumption demand, urban pollutant dispersion problem and heat island effect. In addition, urbanization will cause land cover change which may significantly affect hydrologic variations. To adopt the hydrology change, a lot of new engineering measures, such as drainage and flood control system will be required. In addition, non-structural engineering measures such as real time monitoring and decision support and operation system is important. These real time hydro-data monitoring and decision support and operation system can make it possible to optimize the control and operation of canal network and hydraulic structures to avoid flood disasters.

However, to obtain reliable hydro-data for decision support and operation for a drainage and flood control system is a challenge task, especially for the discharge information through a control structure. Due to the difficulties of direct measurement of discharge, flows through most hydro-structures are computed using flow rating equations based on real time monitoring water stages and structure operational information as the input. In order to obtain accurate structure-specific flow computation equations, collecting field flow data or conducting physical experiment for validation, calibration and verification of the flow computation equations is essential. Data collection, however, can be expensive and time consuming task. Usually, it is difficult to cover all flow ranges; this will limit the accuracy of flow computation through a structure.

In recent year, the development CFD technique and advent of high-performance computers make it possible to investigate hydrodynamic performance with numerical simulation. Many successful CFD applications such as Savage & Johnson (2001), Zhao & Misra (2004), Gessler, D. (2005), Johnson & Savage (2006), Weber et al., (2007), Bhosekar, V. et al., (2012), and among others have been reported that CFD can be a very useful tool to solve hydraulic engineering practice problem.

In this paper, CFD simulation was used to study complex three-dimensional phenomena of flow through prototype spillway and culvert. By investigation of the discharge capacity of these control structures, and apply generated flow data and their corresponding water stages and operation data, a set of good flow equations for accurate flow computation can be obtained. The presented CFD applications relies on solving Reynolds-Averaged Navier-Stocks (RANS) equations with κ-ε turbulence closure model, and resolving free-surface with Volume of Fluid (VOF) method. Two successful applications have been given in the paper. One is the CFD applicant to gate spillway, another is investigation of culvert. These two are typical flood control and drainage hydraulic control structure in urban areas. These case studies demonstrate that CFD-based approach are accuracy and cost-effective, and it can be applied to stage-discharge study of hydraulic structures in urban areas.

2 NUMERICAL MODEL

In this study, the parallel, pressure-based, non-hydrostatic commercial solver ANSYS-FLUENT was employed for the simulations. In Fluent, the RANS solver employs implicit and a cell-centered, finite-volume scheme, in which the gradients of the variables are computed by means of the Green-Gauss' theorem. The diffusion terms are discretized using a second order accurate central scheme. The convective terms are discretized using the second order upwind scheme such that the overall scheme is second order accurate in space. Local time-stepping techniques are used to accelerate the convergence of the solution at each time step. The continuity equation is satisfied using the PISO pressure-velocity coupled algorithm. Volume of Fluid (VOF) algorithm is used to for free surface simulation. The uniform velocity locations were selected as the boundary sections. The mass flow rate was selected as the index of convergence criteria (Zeng et al., 2010).

3 CASE STUDY

Two case-studies are presented here to demonstrate the approach of applying CFD to generate data for urban hydraulic structure stage-discharge relationship study.

3.1 *Case I—CFD simulation for a spillway*

In this application, flow through urban flood control and tide intrusion prevention spillway structure was used as numerical case study. This spillway consists of a concrete ogee-shaped spillway, with two 25-foot-wide vertical lift gates installed on the crest of the spillway at elevation –0.4 ft NGVD, The historical frequency analysis shows Control Submerged (CS), Control Free (CF) and Uncontrolled Free (UF) all occurred in these structures (Ansar, M. & Chen, Z. 2009). There are only 10 field-measured flow data for CF, 5 for CS and 5 for UF. Insufficient field measurements prevent us to obtain reliable stage-discharge and well understand flow behavior. CFD is used for flow behavior study and to generate flow data as a supplement for flow rating.

3.1.1 *CFD model setup and validation*

In the model study, ANSYS-FLUENT CFD model was first established, and then this model was used to investigate flow behavior and generate flow data for flow equation study. Typical hexahedral grids of size 0.5 to 1.5 ft (total about 1, 300, 500 mesh points) were used for the computational domain. As for boundaries setup, all solid boundaries of the fluid domain were treated as no slip surfaces including: channel bottom, channel banks, and walls of the hydraulic structure. The upstream of the spillway was set as a pressure inlet boundary defined by the headwater stage. Similarly, the downstream of the structure was set as the pressure outlet boundary defined by the tailwater stage. The gates at certain operation according to the CFD planning were set as internal boundaries along with the inlet and outlet of the spillway. The free surfaces at the upstream, downstream were all set as the pressure outlet boundaries. The structural configuration, boundary conditions and one of simulation flow fields can be seen Figure 1 below:

The CFD model was then validated by comparing each pair of field-measured flow data with CFD flow data generated with identical boundary conditions (water stages and gate openings). Despite of different flow types, 20 available field measurements were all deployed for validation purpose. As shown in Table 1, the absolute relative differences between each

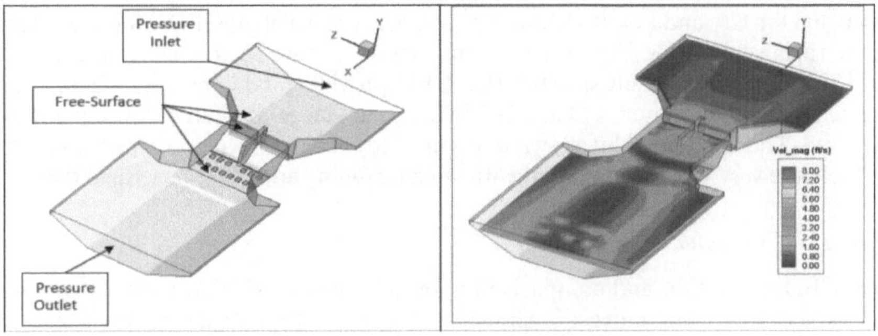

Figure 1. Spillway configuration (left) and three dimensional flow field (left).

Table 1. Field-measurements and CFD simulation results comparison.

Field SG time	Field SG instrument	Headwater stage (m)	Tailwater stage (m)	Gate opening (m)	Field measured discharge (m³/s)	CFD measured discharge (m³/s)	Relative error (%)	Flow type
09/12/96	ADCP*	2.28	0.15	0.61	39.19	39.87	1.74	CF
09/12/96	ADCP	2.18	0.10	1.22	67.85	69.06	1.79	CF
01/16/97	ADCP	2.40	0.21	0.61	39.47	41.33	4.71	CF
07/15/98	ADCP	2.58	0.01	0.14	11.04	10.97	–0.63	CF
07/15/98	ADCP	2.54	0.05	0.45	31.21	32.52	4.23	CF
06/09/99	ADCP	2.20	0.12	0.61	39.56	39.22	–	CF
06/09/99	ADCP	2.24	0.21	0.45	30.89	29.96	–3.02	CF
06/10/99	ADCP	2.43	0.11	0.31	23.36	23.25	–0.47	CF
10/18/99	ADCP	2.08	0.67	1.17	68.67	71.26	3.77	CF
09/13/01	PriceM**	2.17	0.28	1.39	79.26	81.19	2.44	CF
09/12/96	ADCP	2.31	0.22	0.24	16.59	17.53	5.62	CS
01/16/97	ADCP	2.44	0.33	0.30	19.48	20.70	6.27	CS
06/28/13	ADCP	2.42	0.60	0.15	10.76	11.24	4.46	CS
06/29/13	ADCP	2.38	0.61	0.24	16.11	16.60	3.03	CS
06/30/13	ADCP	2.35	0.52	0.37	24.64	25.07	1.76	CS
08/24/95	PriceM	2.13	0.58	2.74	115.76	109.62	–5.30	UF
10/17/95	PriceM	2.10	0.66	2.44	103.10	106.95	3.74	UF
09/12/96	ADCP	2.04	0.07	1.73	103.70	102.58	–1.08	UF
09/12/96	ADCP	2.04	0.07	1.74	95.77	102.50	7.03	UF
09/14/01	PriceM	1.89	0.36	2.44	94.69	91.56	–3.31	UF

SG—Streamgauging; ADCP—Acoustic Doppler Current Profiler; PriceM—Price Meter.

pair of CFD-generated and field-measured flow data are less than 8%, yielding an averaged value of 3.97%. This indicates that the established CFD model is really able to reflect the flow mechanism at the site circumstances. After validation, the CFD model was used for generating flow data based on recommended from CFD planning.

3.1.2 *CFD data generation and rating application*

In data generation, the hydraulic boundary conditions were planned to cover the entire ranges of hydraulic and operational conditions of CS and four data of UF flow type, and CFD simulation was implemented for each boundary condition. In quality test, the no CFD data falling out of the consistency bands, and was filtered out. All data were used for equation study. Finally, CFD data were used for flow rating together with the field measured data to for discharge-stage flow equations investigation.

Physics-based flow equations in Table 2 (Ansar & Chen 2009) were used for parameters study. After calibration of the site-specific parameters, a set of accurate flow equation can be obtained. The calibrated rating parameters are $a_1 = 1.15$ and $b_1 = 0.275$ for CS, and $a_2 = 0.847$ and $b_2 = 0.406$ for CF, and $a_3 = 0.752$ for UF, respectively. Total 40 CFD generated flow data were plotted against available filed measurement data which have been shown in Figure 2 (a), (b) and (c). This good match result suggests that CFD provide good consistency data with filed measurement. Based on these CFD and SG hybrid data, stage-discharge flow equations have been achieved. The scattering plot shown in Figure 2 (d) illustrates that obtained equations can predict discharge very well by compared with streamgauging and CFD generated data.

3.2 *Case II—CFD simulation for culvert*

A culvert CFD simulation and comparison with field-measured is given as the second case study. A typical gated box culvert is shown in Figure 3. This structure functions as flow control structures to drainage flood water from nearby community through a highway and discharge to into the canal in the other side. For real operation purpose, accuracy flow data through this structure was required, so that the real time hydro-information can be integrated into center operation system to prevent potential flooding disaster. CFD simulation was used to analyze the flow and obtain stage-discharge relationship. The simulation result was compared and validates with some available field-measured flow data.

In CFD simulation, the computational domain includes the culvert and 30 feet of upstream and downstream convey canal. The inlet and configuration of the simulation domain are

Table 2. Spillway flow equation (Ansar and Chen 209).

Flow condition	Equation	Restriction	Remarks
Controlled-Submerged (CS)	$Q = L\sqrt{gy_c^3}\ y_c = a_1 G_o \left(\frac{H-h}{G_o}\right)^{b_1}$	$\frac{h}{G_o} \geq 1.0$	Also known as submerged orifice
Controlled-Free (CF)	$Q = L\sqrt{gy_c^3}\ y_c = a_2 G_o \left(\frac{H}{G_o}\right)^{b2}$	$\frac{h}{G_o} < 1.0 \,\&\, \frac{H}{G_o} \geq \frac{1}{K}$ $K = 2/3$	Also known as free orifice
Uncontrolled-Submerged (US)	$Q = L\sqrt{gy_c^3}\ y_c = a_3 H \left(1 - \frac{h}{H}\right)^{b3}$	$\frac{h}{G_o} < 1.0, \frac{H}{G_o} < \frac{1}{K}, \& \frac{h}{H} \geq K$ $K = 2/3$	Also known as submerged weir
Uncontrolled-Free (UF)	$Q = L\sqrt{gy_c^3}\ y_c = a_3 H$	$\frac{h}{G_o} < 1.0, \frac{H}{G_o} < \frac{1}{K}, \& \frac{h}{H} < K$ $K = 2/3$	Also known as free weir

where H = Headwater above sill crest; h = Tailwater above sill crest; Q = Discharge; y_c = Critical; Depth = $(Q^2/gL^2)^{1/3}$; G_o = Gate opening; L = Sill Length; g = Acceleration due to gravity; a_i, b_i = Fitting parameters for rating i.

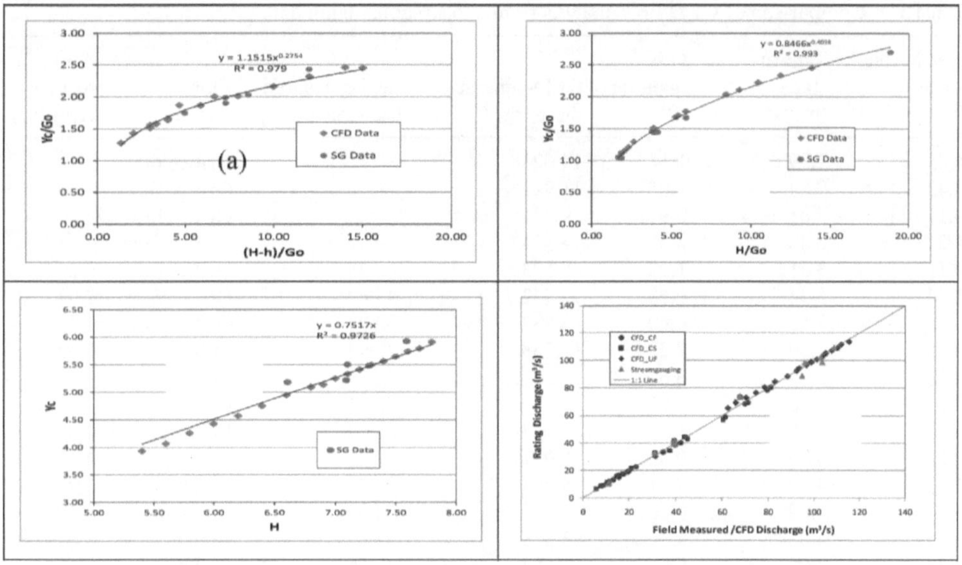

Figure 2. Comparison between the computed flow and the hybrid of the CFD-generated and field-measured flow (a. Non-dimension plot calibrated parameters for CS flow type, b. Non-dimension plot calibrated parameters for CF flow type, c. Non-dimension plot calibrated parameters for UF flow type).

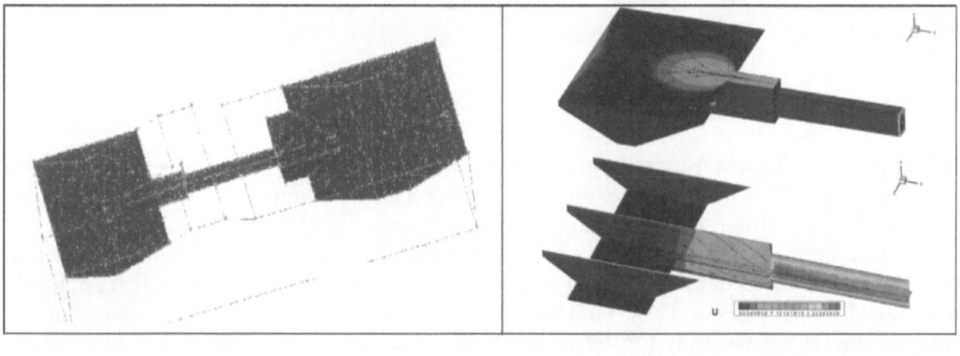

Figure 3. Culvert 3D configuration and mesh (b) CFD flow simulation flow field.

shown in Figure 3 (a). Typical hexahedral and wedge hybrid grids (about 800,000 mesh points) were used for the computational domain. In this study, typical pressure inlet and outlet, free-surface and wall boundary conditions were used in this simulation. The simulation result shows symmetry and contraction features of flow field and streamline (in Fig. 3 (b)) show that basically the CFD-generate flow field is reasonable.

In Table 3, the CFD simulated and streamgauging flow data are listed and compared. For the orifice flow regime, the Averaged Absolute Relative Error (AARE) between CFD data and field measurement data is 1.56% by using these two flow data sets together in rating calibration. For full pipe flow, ARAE is 5.45% ranging from 1.10% to 8.05%. It can be observed that CFD can provide reasonable accurate flow data which can supplement streamgauging data for culvert flow rating.

The small relative different between CFD and field measure data demonstrates that CFD can be successful applied to replica flow data. CFD is cost-effective and can simulate sufficient flow for rating and obtain accurate flow equation. As aforementioned, the accurate

Table 3. Comparison of CFD flow analysis results to streamgauging data.

Headwater stage (m, NGVD)	Tailwater stage (m, NGVD)	Gate opening (ft)	CFD-generated discharge (m^3/s)	Field-measured discharge (m^3/s)	Flow regime	Absolute relative error (%)
4.85	2.99	1.14	14.028		Orifice flow	
4.42	2.84	1.05	11.911		Orifice flow	
4.11	2.81	0.91	9.471		Orifice flow	
4.24	3.48	0.58	4.334		Orifice flow	
4.11	3.51	1.22	8.523	8.545	Orifice flow	0.26
4.24	3.54	0.60	3.942	4.062	Orifice flow	2.97
4.24	3.58	0.60	3.837	3.894	Orifice flow	1.45
4.85	4.39	2.28	11.758	12.593	Full pipe	6.62
4.18	3.69	0.60		3.334	Full pipe	
4.31	4.06	0.61		2.094	Full pipe	
4.43	4.19	1.53	6.743	6.409	Full pipe	4.95
4.87	4.43	0.92	5.071	4.845	Full pipe	4.45
4.47	4.19	0.91	4.094	3.974	Full pipe	2.94
4.24	3.62	0.60	3.691	3.850	Full pipe	4.29
4.29	4.08	0.92	3.358	3.629	Full pipe	8.05
4.27	4.05	0.91	3.657	3.616	Full pipe	1.1
4.22	3.66	0.60	3.477	3.574	Full pipe	2.8
4.29	4.01	0.76	3.153	3.321	Full pipe	5.31

flow equations are the foundation of successful decision support and operation of flood control.

4 SUMMARY

In this paper, CFD was used to study discharge capacity of prototype spillway, culvert for urban hydraulic flood and drainage control purpose. This CFD approach is based on solving RANS equations closed with the standard k-ε turbulence model. VOF method is used for free surface simulation. Field-measured flows at spillway and culvert have been used to verify the accuracy of CFD data. Overall, the relative differences are within ±10% with comparison to field measurements. These data were able to be used together with field measurement data for flow equation study for urban hydraulic structure. The flow equation then can be integrated into flood and drainage control decision support and operation system, which can improve the efficiency drainage and flood network system.

ACKNOWLEDGEMENT

This paper is a project, funded by National Social Science Fund to support major projects, Project No.: 13 & ZD162; National Social Science Fund to support projects focused on Project No.: 12AZD101.

REFERENCES

Ansar, M., and Chen, Z. 2009. Generalized Flow Rating Equations at Prototype Gated Spillways. *J. Hyd. Eng.*, Vol. 135(7), 602–608.

Bhosekar, V., Jothiprakash, V., and Deolalikar, P. 2012. Orifice Spillway Aerator: Hydraulic Design. *J. Hyd. Eng.*, Vol. 138(6), 563–572.

Gessler, D. 2005. CFD Modeling of Spillway Performance. *J. Hydr. Eng.*, pp. 1–10. doi: 10.1061/40792 (173) 398.

Johnson, M.C., and Savage, B.M. 2006. Physical and Numerical Comparison of Flow over Ogee Spillway in the Presence of Tailwater. *J. of Hydr. Engrg.*, Vol. 132(12), 1353–1357.

Savage, B.M., and Johnson, M.C. 2001. Flow over Ogee Spillway: Physical and Numerical Model Case Study. *ASCE,* 123(3), 640–649.

Weber, L., Constantinescu, S.G., Carrica, P., Politano, M., Li, S., Nakato, T., Haque. D., Turan, C., Sinha, S., Miyawki, S., Ferrati, G., and Wilson, M. 2007. *IIHR Limited Distribution Report No. 353, IIHR Hydroscience and Engineering*, The University of Iowa, Iowa, IA.

Zeng, J., Zhang L.Q., Deaton L., Damisse E., Pathak C., and González C.J. 2010 Application of Numerical Model to Generate Flow Data and Improve Flow Rating of Spillways in South Florida, *ASCE & AMWR 3rd International Perspective on Current & Future State of Water Resources & the Environment*. Chennai, India. January, 2010.

Zhao, Q., and Misra, S.K. 2004. Numerical Study of a Turbulent Hydraulic Jump, *Proceeding of 17th Engineering Mechanics Conference*, University of Delaware, NY.

Modeling and Computation in Engineering III – Zhang & Xie (Eds)
© 2014 Taylor & Francis Group, London, ISBN 978-1-138-02680-3

Application of EFDC in simulation of hydrodynamics in Poyang lake

X.H. Zhang
School of Science, Nanchang Institute of Technology, Nanchang, China

C.X. Tang, X. Xiong & W.N. Zou
Institute of Engineering Mechanics/School of Advanced Research, Nanchang University, Nanchang, China

ABSTRACT: The water volume and water quality of Poyang lake determine the water safety for the watershed of Poyang lake and the area of the middle and lower part of Changjiang river, as Poyang lake adjusts the flood of Changjiang river. The flow pattern and water surface level change influence water quality, transport of sediment, soil erosion and biodiversity etc. In this paper, EFDC (Environmental Fluid Dynamics Code) is used to simulate the hydrodynamics of Poyang lake. It is shown that EFDC can be adapted to the situation of high frequent change of water surface level and can be chosen as a useful tool for the future prediction of water resource and management of Poyang lake.

1 INTRODUCTION

Situated in the north of Jiangxi Province, Poyang lake is the largest fresh water lake in China, with five inflow rivers of Gan, Fu, Xin, Xiu and Rao, and one outlet to Changjiang river, with a watershed of 162,225 square kilometers, almost covering the whole area of Jiangxi Province. The Poyang lake resulted from the balance of the five inflow rivers and the outflow to Changjiang river.

Poyang lake is one of ten chinese ecological function areas of protection, and one of the most important ecological areas in the world by the world natural fund. It is the symbol the Jiangxi Province. The conservation and restoration are necessary for the economic development of Jiangxi Province.

Poyang lake faces the problems as flood, drought, soil erosion and decrease of biodiversity etc, with water quality worsened dramatically, and more and more eutrophication. To assess the environmental problem quantitatively, numerical simulation is necessary to understand and manage water resource, such code as POM, ECOM, EFDC, WASP and CEQUAL-W2 etc. In all simulations hydrodynamic module is the basis of the sedimentation and water quality simulation. The hydrodynamic model of Poyang lake was constructed, and the flow pattern and change of water surface level were calculated. The results shows EFDC is a convenient method to calculate the hydrodynamics of Poyang lake with a reasonable high precision. It is a good tool for the management of water resource of Poyang lake.

2 EFDC

2.1 *Background*

EFDC is a multifunctional surface water modeling system, which includes hydrodynamic, sediment-contaminant, and eutrophication components. The public domain EFDC model was originally developed at the Virginia Institute of Marine Science and is currently maintained by Tetra Tech, Inc. with support from the US EPA.

2.2 *Spatial resolution*

The EFDC model is capable of 1, 2, and 3D spatial resolution. The model employees a curvilinear-orthogonal horizontal grid and a sigma terrain following vertical grid which has the same order of accuracy both for shallow and deep water (Ji Z.G. 2012).

2.3 *Hydrodynamic capabilities*

The EFDC model's hydrodynamic component employees a semi-implicit, conservative finite volume solution scheme for the hydrostatic primitive equations with either two or three level time stepping, and with modified closed turbulent scheme (Blumberg A.F., Galoerin B., Q'Connor D.J. 1992; Galperin B., Kantha L.H., Hassid S. et al. 1988). Salinity and temperature transport are dynamically coupled which a choice of high accuracy advection schemes including MPDATA and COSMIC. Additional hydrodynamic component features include, simulation of drying and wetting (Wang, C., Sun Y.L., Zhang X.Q. 2008), representation of hydraulic control structures, vegetation resistance, wave-current boundary layers and wave induced currents. An embedded single port buoyant jet module is included for coupled near and far field mixing analysis.

2.4 *Sediment transport capabilities*

The EFDC model allows the simulation of multiple size classes of cohesive and noncohesive sediment. A sediment processes function library allows the model user to choose from a wide range of currently accepted parameterizations for settling, deposition, resuspension and bed load transport. The sediment bed is represented by multiple layers and includes a number of armoring representations for noncohesive sediment and a finite strain consolidation formulation for dynamic prediction of bed layer thickness, void ratio and pore water advection. The sediment transport component can operate in a morphological mode with full coupling with the hydrodynamic component to represent dynamic evolution of bed topography.

2.5 *Contaminant transport capabilities*

The EFDC model can represent the transport and fate of an arbitrary number of contaminants, including metals and hydrophobic organics, sorbed to any of the sediment classes and dissolved and particulate organic carbon using a three-phase equilibrium partitioning formulation. Dissolved and particulate organic carbon can be represented as independent state variables or POC can be fractionally assigned to any of the sediment classes. A contaminant processes function library allows the representation of various degradation and transformation processes.

2.6 *Eutrophication capabilities*

The EFDC model includes a variable configuration eutrophication component for simulation of aquatic carbon, nitrogen and phosphorous cycles. The full configuration of state variables is based on the CE-QUAL-ICM model including sediment diagenesis. The configuration can be readily reduced to WASP equivalent configurations. In addition to the internal eutrophication model, EFDC can create hydrodynamic transport files formatted for WASP and CE-QUAL-ICM.

2.7 *Data needed in simulation*

1. Topographical data of Poyang lake.
2. Inflow: Time series of five inflow rivers in the year 2008.
3. Meteorological condition: Speed and direction of wind, temperature of atmosphere, solar radiation, precipitation, coverage of cloud, humidity, pressure of atmosphere.

4. Outflow: Time series of water level in the year 2008 in Hukou where water of Poyang lake outflow to Changjiang river.
5. Calibration: Four sites were selected for water level calibration.

2.8 *Parameters*

Cartesian coordinate was used with a grid number of 4713 and a mesh size of 1000 m × 1000 m. Three levels in the vertical with time step of 30 s, and 0.1 m of critical depth of drying and wetting. Default value of roughness 0.02 m was adapted. After 30 days of warm-up calculation, simulation started with a period of 360 days.

3 RESULT

3.1 *Water surface level verified*

Four sites were selected for water level calibration. The figure of site Pingfeng is as follows in Figure 1.

The accuracy is higher in wet season than in dry season. The reason is the influences of topographical data on simulation accuracy is less in the high flow period, and more grids are related to the judge of drying and wetting criterion.

3.2 *Backflow*

The single outlet of Poyang lake to Changjiang controls the water level of Poyang lake (Xiong P.S. 2010). Since April every year, Poyang lake enters the dispatch period, the flow rate of five rivers entering Poyang lake starts to increase, so the water level of lake begin to rise. Before June the water level of Changjiang river is low, the water of Poyang lake can flow out freely into Changjiang river. From June to September it is the big dispatch period of Changjiang

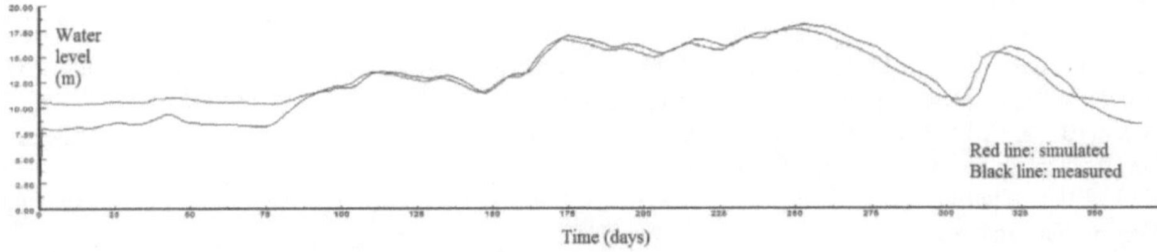

Figure 1. Calibration of water level in the site of Pinfeng.

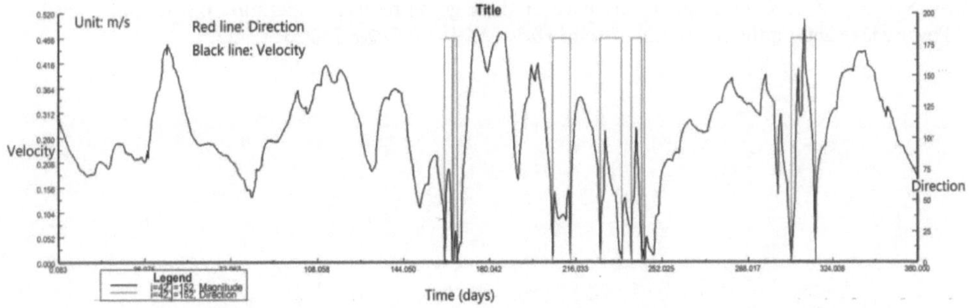

Figure 2. Velocity in the site of Hukou. Black line the value of velocity, red line indicates the direction. When water flow out, value of direction is 0 degree, and value of 180 degrees when water flow back to lake.

Table 1. Result of probability of back flow in year 2008 of the hydrologic station of Hukou.

Times	First	Second	Third	Fourth	Total days
Days of backflow measurement	209–210	230–237	244–248	314–318	20
Days of backflow simulation	206–213	226–235	239–245	306–316	32

river when the water level of river starts to rise, inhibiting the outflow of Poyang lake, even making the outflow inverse its direction back to lake. This process hinders outflow of water and sediment to Changjiang river, changing the balance of water and sediment of lake, influencing the wet ecological system of Poyang lake. For example, the increase of discharge of three gorge reservoir in May, June and October affects the Poyang lake greatly. If in the period of flood, the high water level of Changjiang river raises the water level in Poyang lake, increases the probability of flood of river meeting with flood of lake, leading a heavy burden of anti-flood (Zhao X.J., Sun Z.Y., Gao Y. 2010).

The period of backflow is from July to October, during which the water level of Poyang lake reaches its highest value. The water level starts to decrease since October.

4 CONCLUSIONS

The results of simulation are consistent well with the results measured, and the times and days of backflow in the site of Hukou are predicted. It shows EFDC can be used to simulate efficiently with reasonable results the Poyang lake of high frequent change of water level. EFDC has a flexible technology to tackle a changing boundary of water and land, a visible windows based interface, pre-processing and post-processing tools. EFDC is an efficient tool in the simulation and management of Poyang lake.

REFERENCES

Blumberg A.F., Galoerin B., Q'Connor D.J. 1992. Modeling vertical structure of open channel flows. Journal of Hydraulic Engineering 118(8): 111921134.

Galperin B., Kantha L.H., Hassid S. et al. 1988. A quasi-equilibrium turbulent energy model for geophysical flows. *Atmospheric Sci.* 45: 55262.

Ji Z.G. 2012. *Hydrodynamics and water quality: Numerical simulation of rivers, lakes and estuaries.* Beijing: Ocean press.

Wang C., Sun Y.L., Zhang X.Q. 2008. Numerical simulation of 3D tidal flow in the estuary of Jiaozhou based on EFDC model. *Journal of Ocean University of China* 38(5): 833–840.

Xiong P.S. 2010. Cause of flood and strategy of reducing damage in Poyang lake. *Yellow River of People* 32(10): 26–28.

Zhao X.J., Sun Z.Y., Gao Y. 2010. Influence of three gorge reservoir operation on the water level of Poyang lake and on the ecological system. *Forum of Three Gorge* 230(5): 19–22.

Modeling and Computation in Engineering III – Zhang & Xie (Eds)

Evaluating the impact of spatial dependence on land use change analysis in New Castle

Biao Liu
Key Laboratory of Construction Land Transformation, Ministry of Land and Resources, China
Guangdong Provincial Key Laboratory of Land Use and Consolidation, Guangzhou, China

Qian Xiang
Center for Assessment and Development of Real Estate, Shenzhen, China

Bo Huang
Department of Geography and Resource Management, The Chinese University of Hong Kong, Shatin, NT, Hong Kong

L. Zhang
The Shanghai Institute of Technical Physics of the Chinese Academy of Sciences, Shanghai, China

Yueming Hu
Key Laboratory of Construction Land Transformation, Ministry of Land and Resources, China
Guangdong Provincial Key Laboratory of Land Use and Consolidation, Guangzhou, China

ABSTRACT: Sustainability concerns have aroused great interest among policy makers in the interrelated land use change systems. Considerable attention has been devoted to study the methods in which land use change affects spatial patterns. Less specific attention has been paid to analyze the effects of spatial dependence inherent in land use change. Learning how the spatial dependence impacts the accuracy of land use change analysis is thus of our great interesting. This paper compares several widely used spatial statistic regressive methods to study the impact of spatial dependence on land use change. Specifically, logit model and the autologit model coupled with GIS are used for analyzing the impacts of spatial dependent factors such as population density, slope and surrounding land use, etc. on land use change. In addition, the impact of spatial neighbor effects with logit model for the influence of location and change is also evaluated. Case studies of land use change patterns in New Castle County, Delaware has been implemented, and the results demonstrates that spatial neighbor effects provide an illustrative description of spatial dependencies between data attributes and land use change patterns, and better support explaining the complex processes of urban development and change.

1 INTRODUCTION

Land use change analysis, which becomes a hot topic recently, plays a key role in urban planning. Great progress has been made in modeling land use change. Some statistical methods such as Markov chain analysis (Lopez et al. 2001, Weng 2002), multiple regression analysis (Theobald and Hobbs 1998), logistic regression (Wu and Yeh 1997), are employed. Afterwards, Artificial Neural Networks (ANNs) (Pajanowski et al. 2002) and Cellular Automata (CA) capable of dynamic spatial simulation and innovative bottom-up approach (Clarke and Gaydos 1998; Wu 1998; Wu 2002) are used. However, less specific attention has been paid on considering the effects of spatial dependence inherent in land use change. Actually, spatial dependence is the fact that one observation associated with a location depends on other observations at other locations. It's a very important aspect on land use change analysis.

Over the years, researchers have made considerable efforts in studying spatial neighbor effects. Anselin (1988, 1999) provides three principal methods for addressing the spatial effects that exist in land use data sets: use of spatial stochastic processes, a direct representation of correlations, and a non-parametric framework. In this study, the second method will be adopted. Some models such as autologit model are employed to demonstrate that spatial neighbor effects provide an illustrative description of spatial dependencies between data attributes and land use change patterns, and better support explaining the complex processes of urban development and change.

2 METHODOLOGY

In order to consider the spatial dependence, Conventional Logistic Regression Model and Revised Logistic Regression Model will be adopted here. A significant effect caused by spatial dependence will be shown by the comparison of estimation results of several models.

2.1 *Conventional logistic regression model*

The land use model usually involves discrete dependent variables as driving factors, thus traditional multiple regression analysis is inappropriate. Logistic regression is advocated to deal with such question. The general form of logistic regression, on which our land use change model is based, is as follows:

Model 1:

$$y_i = \log_e\left(\frac{p}{1-p}\right) = a + b_1x_{i1} + b_2x_{i2} + \cdots + b_mx_{im} \tag{1}$$

$$P(z=1) = \frac{e^y}{1+e^y} \tag{2}$$

where P stands for the probability of land use change (denoted as a binary response variable (0 or 1), value 1 means the occurrence of a change such as the transition from rural to urban, and value 0 indicates no change), $x_1, x_2, \ldots, x_m$ stands for explanatory variables and the utility function y is a linear combination function of the explanatory variables representing a linear relationship. The parameters $b_1, b_2, \ldots, b_m$ are the regression coefficients to be estimated. The goal of logistic regression is to estimate the $k+1$ unknown parameters β in equation (1). This is done with maximum likelihood estimation which entails finding the set of parameters for which the probability of the observed data is greatest.

2.2 *Revised logistic regression model*

There we adopt two ways to incorporate the spatial effects. One way is to directly add an independent variable named per_avail, which represents the percentage of urban land use in the surrounding area within 200 m radius, into Traditional Logistic Regression Model (in the following context).

Model 2:

$$y_i = \log_e\left(\frac{p}{1-p}\right) = a + b_1x_{i1} + b_2x_{i2} + \cdots + b_mx_{im} + b_{m+1}per\_avial_i \tag{3}$$

The other way which is studied here is to add WY as an independent variable into the model, which has been studied by James P. LeSage (1998). W is a known $n \times n$ spatial weight matrices. Here the weight matrices which contain functions of distances will be used. This model is called autologit model which can formally expressed as follows.

Model 3:

$$Y = \rho WY + X\beta + \varepsilon \tag{4}$$

$$y_i = a + b_1 x_{i1} + b_2 x_{i2} + \cdots + b_m x_{im} + b_{m+1} x_{i,m+1} + \sum_{j=1}^{N} \rho_{ij} y_j \tag{5}$$

N is the total number of the observations. ρ_{ij} is modeled as a negative exponential function of distance separating the two observations:

$$\rho_{ij} = b_1 \left(\frac{-D_{ij}}{b_2} \right) \tag{6}$$

where b_1 and b_2 are parameters to be estimated and D_{ij} is the distance between the two observations. As Dubin estimation method will be used here, b_1 refers to ρ in Equation (4).

3 IMPLEMENTATION

3.1 *Study area*

Three counties constitute Delaware State: they are Kent County, Sussex County and New Castle County from south to north. From Figure 1, the study area, New Castle County, covers about 1,100 square kilometers. It is the urban and manufacturing center of Delaware. About 60% of Delaware's population lives in New Castle County. Industrial and urban developments are concentrated in the northern part of the county. Changes in land use show that both the northern and southern parts are becoming progressively urbanized.

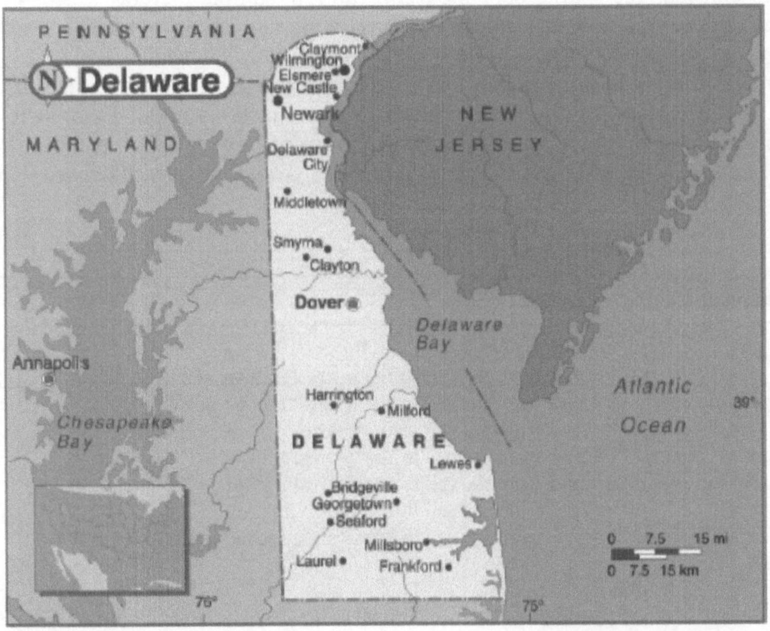

Figure 1. Map of Delaware.

3.2 *Data preparation*

The data used for the following analysis comes from three sources: 1) land use information derived from digital orthophotos, 2) Census of Population data and 3) transportation data, which has been proved to be associated with the land use change by other scholars (Cheng and Masser (2003), Landis and Zhang (2000), Verburg et al. (2001)). These three datasets are processed and combined to the same spatial coordinates.

The land use data used in this study was generated from digital orthophotos conducted by the Delaware Office of State Planning Coordination. Each of the land cells is indexed as one of the five land use classifications: agriculture, residential, commercial, industrial and barren. Residential, commercial, and industrial areas are categorized as urban regions, while the agriculture and barren type is considered to be rural areas with considerable potential for urbanization.

The Census of Population also provided key data in 1990 and 2000. The typical census block encompasses dozens of 50 m grid cells. In order to combine the datasets (across space), each block's population was assumed to increase exponentially with time, according to its own growth parameter. Block populations outside the decennial years (1990 and 2000) were interpolated/extrapolated, and these inter-decennial block group populations were distributed uniformly across all residential cells within the block.

Eleven predictor variables were compiled using ArcInfo by using a 50 m × 50 m cell size. A summary of these predictors is shown in Table 1.

The predictors can be categorized into three types: (1) site specific characteristics; (2) proximity; and (3) neighborhood characteristics. For instance, population density, which is considered to be a leading force in global land use change, is the site specific variable. The seven distance variables stand for the proximity variables. And the last three percentage variables which consider the effect of neighboring impacts within a 200 m radius could be called neighborhood characteristic variables.

Figure 2 depicts land use maps of 1984 and 1992. It can be directly found that Land use type change had happened during this period.

3.3 *Estimation results*

In order to consider the spatial dependence, a random sample method will be employed. We estimate the coefficients mentioned in the formula (1), (3) and (4). The results are in Table 2.

We can calculate the proportion of total variation explained by spatial dependence with a comparison of the fit measured by Estrella R-squared. Estrella R-squared is a new measure of fit for equations with dichotomous dependent variables. From the results, we can see that adding a simple variable per_avial does few efforts to explain the

Table 1. Summary of predictor variables for the rural-urban land conversion model.

Variable name	Description
Dens_Pop	Population density of the cell
Dist_Ind	Distance from the cell to the nearest industrial site
Dist_munc	Distance from the cell to the nearest municipality
Dist_Proad	Distance from the cell to the nearest principal roads
Dist_rail	Distance from the cell to the nearest railway
Dist_schl	Distance from the cell to the nearest school
Dist_shpc	Distance from the cell to the shopping center
Dist_sroad	Distance from the cell to the nearest second roads
Planning	Zoning of the cell
Slope	Slop of the cell
Per_avail	Percentage of urban land use in the surround area within 200 m radius

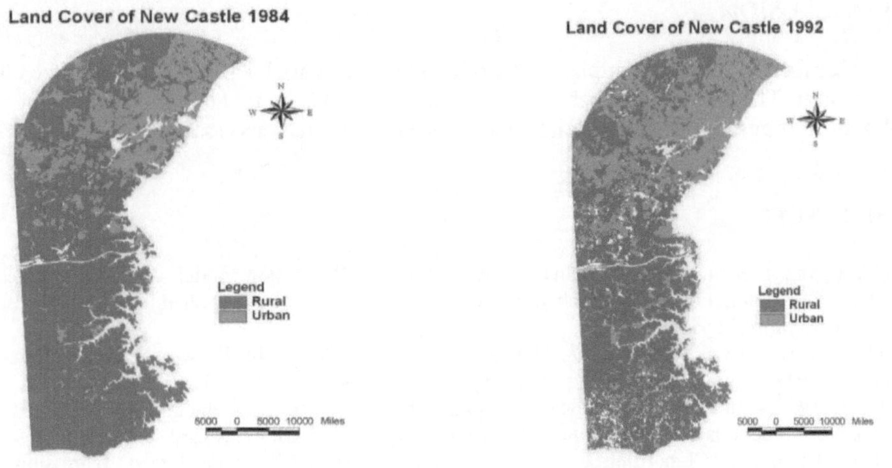

(a) Land use / cover map of 1984 (b) Land use / cover map of 1992

Figure 2. Land use maps of 1984, 1992.

Table 2. Estimation results of model 1, model 2 and model 3.

	Model 1		Model 2		Model 3	
Variable	CO	P-value	CO	P-value	CO	P-value
Intercept	1.559979	0.000000	0.868498	0.000000	–0.983291	0.000000
Dist_sroad	–0.011886	0.000000	–0.010098	0.000000	–0.008990	0.000000
Dens_Pop	0.000054	0.267855	–0.000041	0.322272	–0.000059	0.292784
Per_avail			2.476298	0.000000		
Dist_munc	0.000072	0.000000	0.000105	0.000000	0.000027	0.095207
Dist_schl	–0.000163	0.000000	–0.000156	0.000000	–0.000047	0.104219
Dist_rail	–0.000146	0.000000	–0.000146	0.000000	–0.000071	0.000645
Dist_shpc	–0.000016	0.002635	–0.000007	0.172926	0.000008	0.227028
Planning	–0.505135	0.000000	–0.000007	0.000000	–0.309437	0.000000
Slope	0.004645	0.305094	0.005135	0.262101	0.004343	0.405382
Dist_Ind	0.000146	0.000000	0.000127	0.000000	0.000044	0.045362
Dist_Proad	–0.000100	0.000000	–0.000084	0.000196	–0.000084	0.000879
ρ					4.662310	0.000000
Estrella R-squared	0.2873		0.3015		0.6276	
TPCP	82%		83%		88%	

Note: CO: Coefficients; TPCP: training percentage correctly predicted.

total variation. Maybe it is too simple to consider spatial dependence by the percentage of urban land use in the surrounding area. However, the P-value on the parameter for the spatial dependence variable per_avail is 0, indicating that this explanatory variable has a coefficient estimate that is significantly different from zero. 34.03 percent of the variation in the land use change is explained by spatial dependence, as the Estrella R-squared is roughly 62.76 percent in the model 3 that takes spatial dependence into account and 28.73 percent in the model 1 that ignores this aspect of spatial data sample. Note also that the P-value on the parameter for the spatial dependence variable ρ is 0. The TPCP also shows that the most of the variation in the land use change is explained by spatial dependence.

4 CONCLUSION

This paper focuses on some simple, easy models in which spatial dependence should be taken into account. The comparison between those models shows that when land use change was analyzed, the consideration of spatial dependence is absolutely necessary.

REFERENCES

Anselin, L. and A. Bera. 1998. "Spatial Dependence in Linear Regression Models with an Introduction to Spatial Econometrics." In A. Ullah and D. Giles (eds.), Handbook of Applied Economic Statistics. New York: Marcel Dekker, 237–289.

Anselin, L. 1988. Spatial Econometrics: Methods and Models. Dordrecht: Kluwer Academic Press.

Anselin, L. 1999. Spatial Econometrics. Working paper. Accessed July 10, 2005.

Dubin, R., 1995. Estimating logit models with spatial dependence. In L. anselin and R > J.G.M. Florax (eds), New directions in spatial econometrics. Springer_verlag, Berlin, 229–242.

Xie, C.L., Huang, B., Claramunt, C., and Chandramouli, M. 2005. Spatial logistic regression and GIS to model rural-urban land conversion. PROCESSUS Second International Colloquium on the Behavioural Foundations of Integrated Land-use and Transportation Models: Frameworks, Models and Applications, University of Toronto, June (CD-ROM).

Modeling and Computation in Engineering III – Zhang & Xie (Eds)
 ISBN 978-1-138-02680-3

Visualization of social network for scientific research evaluation

Yunsheng Du
School of Management and Economics, Beijing Institute of Technology, Beijing, China

Yuqin Liu
Academic of Printing and Packaging Industrial and Technology, Beijing Institute of Graphic Communication, Beijing, China

Pengjun Qiu
School of Management and Economics, Beijing Institute of Technology, Beijing, China

Xiaohan Shen
School of Humanities and Social Sciences, Beijing Institute of Technology, Beijing, China

ABSTRACT: In order to enrich the existing scientific research evaluation methods for scientific researchers, institutions, regions or journals that engaged in some technical field, subject or discipline, the paper reviews the shortcoming of current visualization used for scientific research evaluation. And then, the paper designs and realizes three applications of social network visualization, first improves the traditional co-author network marking signature order, then integrates correlative relationship network with dynamic threshold, and at last constructs citation network to convey the quantity, quality and influence of the object being evaluated. For each network, we propose a virtual example without real meaning to show the interpretation of the network map.

1 INTRODUCTION

Scientific research evaluation is an important part of science research management. It is not only conducive to the smooth progress of research activities, raise research level, but also conducive to promote research communication, innovation and development. How to improve the scientific research evaluation system, evaluation methods, has become an important research content of scientometrics. However, the process of scientific research evaluation is a multi-disciplinary integration process. The object being evaluated includes projects, researchers, institutions, regions, and academic journals. Evaluation methods depend on evaluation object, content, criteria. And methods commonly used include peer review, Delphi survey and other expert evaluation method, gray system, AHP, fuzzy evaluation, data envelopment analysis, comprehensive evaluation, principal component analysis, factor analysis and other statistical methods, citation analysis, word frequency analysis, and so on.

The purpose of this paper from perspective of social network visualization for scientific research evaluation is not to establish a complete and comprehensive scientific evaluation system, but try to take it as an adjunct to conventional methods, and find the neglected things before. Therefore, the study is limited to scientific research evaluation for researchers, institutions, regions and academic journals that engaged in some specific technical field, subject or discipline. The study is based on scientific literature.

2 REVIEW OF LITERATURES

With the rapid development of science and technology, science research becomes more difficult, disciplines cross and penetrates, researchers collaborate and compete. In this case, the scientific research evaluation methods need to be improved. The traditional methods cannot meet the needs of the rapid development of science and technology. For this reason, some scholars took visualization technology into scientific research evaluation. Boyack K.W. (2003) studied on analyzing and visualizing the impact of governmental funding on the amount and citation counts of research publications. Anastasios T. (2011) used multi-indicator analysis to evaluate the research activities of an institution as well as the degree of achievement for a specified research policy. He designed the system that supports academic evaluation and decision-making processes concerning research policies using visualization technology. The developed system provides user to evaluate aspects of the quality of academic research activities in the context of specific policies, criteria and make decisions on the establishment of new strategies. Robert G. (2011) used citation network visualization to rank and filter papers by network statistics, designed a tool to support user in rapidly generating readily consumable summaries of academic literature. Haiyan H. (2006) analyzed top 50 most cited authors in the international journals of science during 1995–2004, got the author co-citation map to identify the main research groups and representative scientists, revealed the highly influential authors. "Scientific evaluation and management information system" designed and developed by Junping Q. (2013) at Wuhan University in China, uses visualization technology to visualize co-author relationships, co-word relationships, co-citation relationships, and then carry out scientific research evaluation. The system provides the third software interface. "China Science and Technology Journal Citation Report" (2012) and "China Highly Cited index analysis" (2013) published by China science and technology information institute each year, draw a large number of scientific journals, researchers and institutions co-citation network to enhance the reader's understanding of the evaluation results.

As far as the current visualization technology for scientific research evaluation, the following deficiencies still exist: there is little literature related visualization and scientific research evaluation, and lacks of theoretical research. The currently most used visualization methods on the scientific research evaluation are co-author and co-citation. Co-author visualization emphasis on the overall relationship between authors ignores the individual's capability and contribution. Co-citation visualization reveals authors' research correlation, but cannot reflect the "quantity", "quality" and "influence" of the evaluated object. So the effectiveness of scientific research evaluation using co-citation needs to be optimized. Study on the visualization of scientific evaluation methods, both has significance of theoretical exploration and value to meet the practical needs.

3 SOCIAL NETWORK

A social network is a social structure made up of a set of social actors (such as individuals or organizations) and a set of the dyadic ties between these actors. The social network perspective provides a set of methods for analyzing the structure of whole social entities as well as a variety of theories explaining the patterns observed in these structures (Stanley, W., Katherine, F., 1994). The study of these structures uses social network analysis to identify local and global patterns, locate influential entities, and examine network dynamics. Social networks and the analysis of them is an inherently interdisciplinary academic field which emerged from social psychology, sociology, statistics, and graph. Social network now is employed in a number of other social and formal sciences together with visualization technology. And the paper studies the social network visualization for scientific research evaluation.

4 SOCIAL NETWORK FOR SCIENTIFIC RESEARCH EVALUATION

First, we improves the traditional co-author network with signature order, reveals the researcher's focus combined with text mining, and evaluates individual contribution to the group. Then, the paper constructs correlation network that users can dynamically adjust the threshold. Correlation network is used to investigate the relevance between researchers, institutions, regions, journals within the same time, and to evaluate their influence. The correlation network based on text mining can overcome the shortcomings of correlation network based on citation that cannot reflect the research content. Finally, citation network sort by year is built to evaluate the influence of the object being evaluated.

4.1 *A new co-author network marking signature order*

We propose a new co-author network visualization marking signature order. What we are trying to map are authors of research literature and their co-authors and their co-authors, just like a social network. Figure 1 shows a virtual example of such network without real meaning. In the network, node can be used to represent the author, institution or region, and its size is proportional to the number of papers published by them. Node label represents the name of the author, institution or region, and the number in parentheses is the quantity of papers that published by them as first author, second author, third author and subsequent. Radius of the red, green and yellow halo is proportional to the number in parentheses. The thickness of the line between two nodes indicates the number of co-authors.

Let's take look at Lin JM in the figure. It seems that he is one of the principal researchers because he published 53 papers without considering signature order. However, he is just the second or third and sequence co-author. In the group composed by Lin J.M., Wu J.H., Huang M.L. & Lan Z., the contribution of Lin M seems to be less than the others.

4.2 *A correlation network with dynamic threshold*

Correlation network visualization is used by Thomson Data Analyzer. The nodes in the network have the same meaning as above co-author network, and the lines between nodes represent a measure of similarity computed by Principal Components Analysis (PCA) according to the keywords used by the authors, institutions, regions or journals. The thickness of the line indicates the degree of similarity-a number between 0 and 1. To reduce visual clutter, only the strongest of the entire set of similarities are shown. Correlation network can be used to investigate the relevance between researchers, institutions, regions, journals within the same time, and to evaluate their influence on each other. However, when user use correlation network, he often wants to adjust similarity threshold on demand so that the network shows the lines exceeds the threshold only. But Thomson Data Analyzer cannot. Figure 2 shows a virtual example of correlation network setting the threshold to 0.5. In the figure, the nodes are same as in the Figure 1. The node label appends the most commonly used keywords associate with

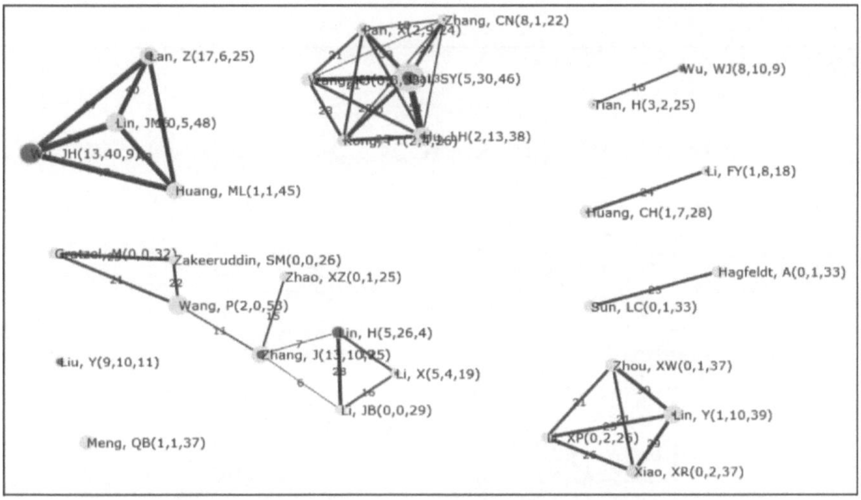

Figure 1. A virtual example of co-author network marking signature order.

the node. The correlation network changes with the threshold. This approach represents a more simple and beautiful graphic that is more intuitive and easier to understand.

Let's interpret Figure 2 from the scientific research evaluation. The co-authors have more relevance than the other authors, and the keywords difference significantly. It also notes that the nodes does not have co-authored also show correlation, as shown in the delineation of groups 2 and group 1, 3, 4, 7, they interact each other. Groups 1, 4, 7, 5, 6 have lower correlation and influence.

4.3 *A new citation network revealing research influence*

Co-citation network is used widely in bibliometrics, and often is used for scientific research evaluation. Such as "China Science and Technology Journal Citation Report" and "China Highly Cited index analysis" above, there are a lots of co-citation networks. Co-citation visualization reveals authors' research correlation, but cannot reflect the "quantity", "quality" and "influence" on each other. So the effectiveness of scientific research evaluation using co-citation needs to be optimized.

Figure 3 shows a virtual example of citation network, in which nodes represents object being evaluated, but its size is proportional to the number of papers published by them at

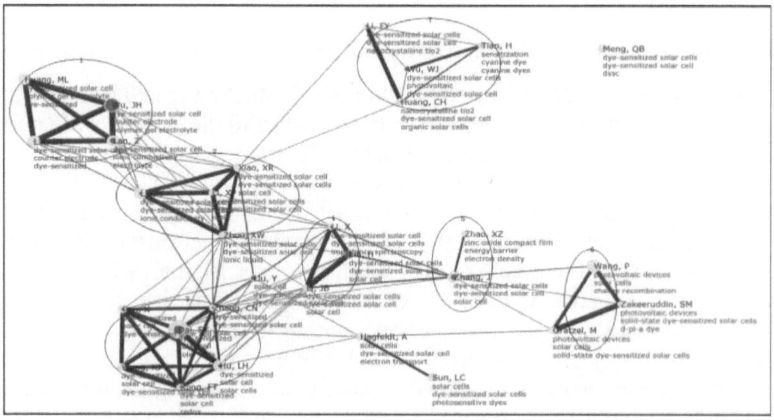

Figure 2. A virtual example of correlation network with dynamic threshold.

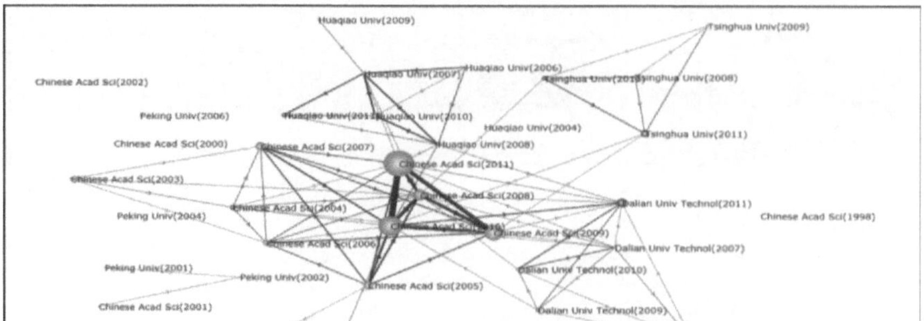

Figure 3. A virtual example of citation network revealing research influence.

specific particular year. Lines with arrow represent the citation relationship, and its width is proportional to the citation quantity. The influence between object being evaluated at some year can be reflected by the lines density and lines width, the more density and more width the more influence. For example, Chinese Academic Sci (2008) has greater influence on Chinese Academic Sci (2009, 2010, 2011) than the others. Dalian Univ Technol (2011) is more affected by Chinese Academic Sci (2010) than Dalian Univ Technol (2007, 2008, 2009, 2010).

5 CONCLUSION

As an adjunct of the existing evaluation methods, the paper build co-author relationship, subject correlation relationship, citation relationship, and use social network to visualize the relationships, to reveal the quantity, quality and influence of the object being evaluated. The effectiveness of evaluation has a direct relation with the data quality. The different spells of same authors, institutions and regions or the same spells of different authors, institutions and regions will have an impact on the evaluation results. In the practical application, like other bibliometrics evaluation method, it is necessary to clean and normalize data. In the future software design and development based on the method proposed in the paper are worthy to study.

ACKNOWLEDGEMENTS

This research is partly funded by the General Program of National Natural Science Foundation of China (Grant No. 71373019) and the National Key Technology R&D Program (Grant No. 2013BAH20F01).

REFERENCES

Anastasios, T., Sgouropoulou, C., Xydas, I. 2011. Academic Research Policy-making and Evaluation using Graph Visualization. *15th Panhellenic Conference on Informatics*, pp. 28–32. IEEE, Kastonia.

Boyack, K.W., Borner, K. Indicator-assisted Evaluation and Funding of Research:Visualizing the Influence of Grants on the Number and Citation Counts of Research Papers. *Journal of the American Society for Information Science and Technology*, vol. 54, pp. 447–461(2003).

China Science and Technology Information Institute. 2012. *China highly cited indices analysis*. Beijing Science and Technology Press, Beijing.

China Science and Technology Information Institute. 2013. *China Science and Technology Journal Citation Reports*. Beijing Science and Technology Press, Beijing.

Haiyan, H., Zeyuan, L., Yue, C. 2006. Mapping of Science Studies: Research Groups and Reprehensive Scientist. *Studies in Science and Science*, vol. 24, pp. 161–164.

Junping, Q., Hui, D. 2013. Construction of Management Information System of Scientific Evaluation. *Library and Information Science*. 1, pp. 56–62.

Robert, G., Cody, D., Ben, S. 2011. Evaluating Visual and Statistical Exploration of Scientific Literature Networks. *IEEE Symposium on Visual Languages and Human-Centric Computing*, pp. 18–22. IEEE, Pittsburgh.

Stanley, W., Katherine, F. 1994. Social Network Analysis: Methods and Applications, pp. 1–27. Cambridge University Press, Cambridge.

Modeling and Computation in Engineering III – Zhang & Xie (Eds)
 ISBN 978-1-138-02680-3

Research of vehicle routing optimization based on traveling salesman problem

Bing Li, Fan Yang & Dejing Kong
School of Management and Economics, Beijing Institute of Technology, Beijing, China

ABSTRACT: In the process of modern transport of goods, reasonable choice of vehicle routing can speed up the delivery of goods, save resources, reduce operating costs, and increase enterprise benefit. Vehicle routing planning is the most important operating decision in the modern logistics enterprise's transport. This article reviewed the literature of vehicle routing planning, commented the typical research methods, and put forward the optimizing model of vehicle routing based on the problem of traveling salesman: First of all, this article gave the initial feasible solution that can satisfy the conditions using the traveling salesman problem; then, got the most economical shipping lines by using 0–1 integer programming methods; at last, the established mathematical model would be shown by the Lingo program, which simulated the required delivery strategy and reflected the best route in the coordinate system.

1 INTRODUCTION

Modern logistics is the objective requirement of the social production and the development of national economy, and its development status play an important role in facilitating or restricting the economic development, commodity circulation and mass consumption. Today the transportation of goods is developing rapidly, taking the express industry as an example, according to *The 2012 Statistics Bulletin of The Postal Industry*, express business growth fast in 2012. National scale wholesale express service enterprises completed 5.69 billion goods all the year round, growing by 54.8%; express business income 105.53 billion Yuan, increased by an annualized 39.2%. So it is meaningful to research on vehicle routing optimization under the premise of rapid growth in demand.

2 REVIEW OF THE RESEARCH ON VEHICLE ROUTING PROBLEM

The Vehicle Routing Problem (VRP) has always been a research hot spot for the rapid development of modern logistics distribution and reasonable choice of the importance of vehicle routing. VRP was first proposed in 1959 by Dantzig and Ramser. At present, the research on VRP has produced a variety of mature algorithms: Clarke and Wright (1964) first proposed the Saving Method, which has been the most classic method solving the VRP, and many successful vehicle scheduling software were developed based on this method or its improved method. Gillet and Miller (1974) put forward the Sweep Method, which groups the demand of the nodes or arcs firstly, then solves each demand according to the TSP to design an economic route. Gendreau (1994) used the insert Method to solve the traveling salesman problem, at the same time using Greedy Method split on the route, so as to generate the initial solution. C.T. Su and H.H. Chen (1999) successfully applied self-organizing mapping network in solving planning problem of vehicle distribution area and route.

3 THE QUESTION RAISED

As is shown in Figure 1, the little red circles are the warehouse locations of a freight company, one delivery man need to send the goods from the warehouses to each destination in the city: 76 position coordinates is given, No. 1 point is the starting point, No. 2–76 points are the destinations, the scatter diagram is shown in Figure 1: Red * are No. 2–41 destinations; Blue * are the rest of the destinations.

Question 1: if any connection between 76 positions is pathway, please design the delivery line, to make the shortest delivery distance, and draw the result.

Question 2: if the pathways are shown in Figure 2, please design the delivery line, making the shortest delivery distance, and draw the result.

Question 3: on the basis of question 2, if the maximum of goods can only meet 40 destinations and the destination sorted in front must give priority to delivery. Please design the delivery line, making the shortest delivery distance, and draw the result.

4 QUESTIONS ASSUMPTIONS AND PARAMETERS SPECIFICATION

Questions assumptions: in question 1, there is a pathway between arbitrary 76 cities; all the solutions of this question are based on "each city can't be repeated"; in question 3, we can give the optimal delivery line based on "priority to send first 40th cities"; delivery man can choose vehicle routing; the cities and roads are abstracted into coordinate points and lines.

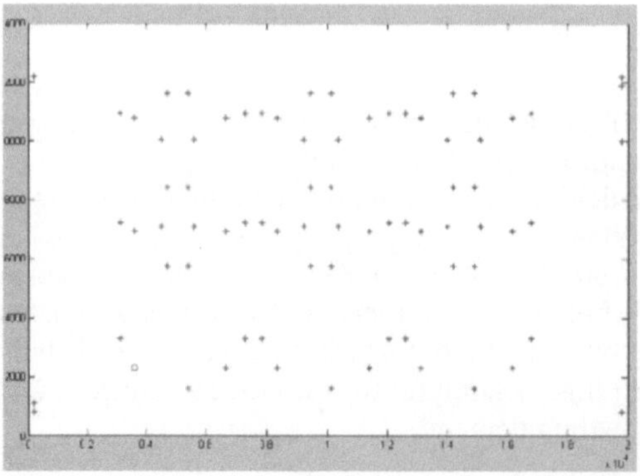

Figure 1. Delivery schematic.

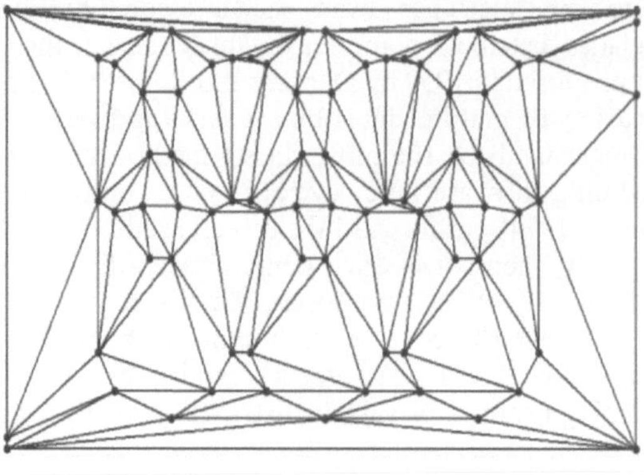

Figure 2. Pathways information.

Table 1. Parameters specification.

Symbol	Value range	Meaning
i	$i = 1...76$	City number
S_i	$i = 1...76$	City name
x_{ij}	$i = 1...76$; $j = 1...76$	0–1 variable
C_{ij}	$i = 1...76$; $j = 1...76$	Distance between city i and j
u_i	$i = 1...76$	Variable introduced
Z		Target value
$\wedge$		Set
A, B		Matrix
Y_{ij}	$i = 1...76$; $j = 1...76$	Elements in the matrix B
$D(i, j)$		Distance between city i and j
Len (i)		Sum of the distances of city i

5 MODEL COMPUTATION AND ANALYSIS

5.1 *Question analysis*

Assuming plane has n points represented the location of the n cities, and we want to look for a shortest closed path which can make each city traversed just once. This question belongs to the model of route planning and optimization, we will use the approximate model of Traveling Salesman Problem to achieve the path optimization, regarding route as travelling expenses.

1. We can use the following methods to solve question 1. For example, we revolve around city S_1-S_{10}, the destination data (coordinates) of goods and the distance data (attachments) between these points, and can obtain the minimum distance between each two points according to the distance matrix. We should consider each of the shortest distance between two points since this question is a "closed loop". So we improve the optimization method: we can conclude the shortest distance of transporting goods (exhaustive thought) by finding the minimum value of each loop. Extending the above method to 76 cities, we adopt exhaustive method to compare each path, and then conclude the shortest path required.
2. Based on the thinking method of question 1, the solution of question 2 is a more actual model of choosing cities and roads. Expressing 0–1 matrix A as the relationship between the cities, $x_{ij} = 0$ or 1, if S_i and S_j is interlinked, $x_{ij} = 1$; else, $x_{ij} = 0$. We can get the shortest path by the method of question 1 after establishing 0–1 matrix A.
3. The solution of question 3 is an extension of question 2. We will send goods of the first 40th cities (S_1-S_{41}) without repeating the same city; and return to the starting point S_1 after completing delivery; and then complete the delivery tasks of rest cities.

5.2 *Modeling*

1. We can solve shortest path problem with 0–1 programming, assuming that the starting point is 1, and the end point is n, introducing a 0–1 variable x_{ij}, if the path(i, j) is the shortest path, $x_{ij} = 1$, else $x_{ij} = 0$.

   For any point i except the starting point and end point, if $\sum_{j=0}^{n} x_{ij} = 1$, it means that there is a shortest route in the paths starting from point i, in other words, the shortest route passes the vertex. At this time all of the routes from other vertex to the vertex must have a line in the shortest possible way, so $\sum_{j=0}^{n} x_{ji} = 1$; if $\sum_{j=0}^{n} x_{ij} = 0$, it means that the shortest path do not pass vertex i, so $\sum_{j=0}^{n} x_{ij} = 0$.
   Two cases can be combined: $\sum_{j=0}^{n} x_{ij} = \sum_{j=0}^{n} x_{ji}, 1 < i < n$.

   For starting point 1, it must be satisfied with $\sum_{j=0}^{n} x_{ij} = 1$; for end point n, it must be satisfied with $\sum_{j=0}^{n} x_{ji} = 1$.

The objective function is that the sum (total) of the length of each shortest route is the shortest. The shortest path problem can be described by the following 0–1 programming:

$$\begin{cases} \min & z = \sum_{\substack{i,j=1 \\ i \neq j}}^{n} c_{ij} x_{ij} & \\ s.t. & \sum_{i=1}^{n} x_{ij} = 1, & j = 1,2,\ldots,n \\ & \sum_{j=1}^{n} x_{ij} = 1, & i = 1,2,\ldots,n \\ & u_i - u_j + n\, x_{ij} \leq n-1, & 2 \leq i \neq j \leq n \\ & x_{ij} = 0,1, & i,j = 1,2,\ldots,n \\ & u_i \geq 0, & i = 2,3,\ldots,n \end{cases} \tag{1}$$

2. Question 2 only needs to define a new 0–1 matrix B (Y_{ij} is determined by the relationship of C_i and C_j). On condition of relationship (1), we add new constraints: When $Y_{ij} = 1$ set $x_{ij} = 1$, which can make the delivery man of city S_i make a choice only from the routes in contact with it.
   The constraint:

$$x_{ij} = 1,\ Y_{ij} = 1 \quad i = 1\ldots76; j = 1\ldots76; \tag{2}$$
$$x_{ij} = 0,\ Y_{ij} = 0 \quad i = 1\ldots76; j = 1\ldots76; \tag{3}$$

3. We can find the shortest path under the condition of assuming "we will not repeat the same city", "we can reach optimization when giving priority to sending first 40 cities' goods". Sending the goods of first 40 cities in advance (S_1-S_{41}-S_1), then return to send the rest goods (S_1-S_{42}-S_{76}-S_1), we will complete the delivery in two steps, which is getting the shortest path under the condition of assuming.
4. The implementation of the exhaustive algorithm is very complex, especially using Lingo. We collect other algorithm for the implementation of this question—genetic algorithm (ga).

The description of the TSP problem is very simple, that is to find a shortest path which can traverse the n cities, or search a permutation $\pi(X) = \{V_1, V_2, \ldots, V_n\}$ of the natural subset $X = \{1, 2, \ldots, n\}$ (the elements of X stands for the number of cities), which makes $len = \sum d(V_i, V_{i+1}) + d(V_1, V_n)$ taking the minimum. The $d(V_i, V_{i+1})$ stands for the distance between city V_i and city V_{i+1}.

1. Code and creating initial population
   The code of genetic algorithm is the sequence of traversing the cities. Generate a $1 \times N$ matrix (N stands for quantity of cities) as a random path using MATLAB randperm(N). Then creating the initial population based on the $n \times N$ matrix storage group.
2. Distance matrix and the fitness function
   The locations of cities are specified before compilation, also can use the coordinates generated randomly. Distance matrix use a $N \times N$ matrix D to save, and D (i, j) stands for the distance between city i and city j. In this question, the fitness is measured by the sum of the distance, the larger the sum of the distance, the smaller the fitness, and then discusses whether it is the optimal solution.

$$D(i, j) = \text{sqrt}((x_i - x_j).\hat{}2 + (Y_i - Y_j).\hat{}2)$$

3. Selecting
   Selecting is to avoid the loss of effective gene, making high performance individuals have higher chances of survival, and then the global convergence and computational efficiency is obtained. The adaptive value function in this case is:

Fitness (i,1) = (1-((len(i,1)-minlen)/(maxlen-minlen + 0.0001))).^m;

maxlen, minlen is the maximum and the shortest path, use "fitness⟩ rand" to select solution.

4. Crossover operation is used to compose the new individual
   This instance uses the partial matching crossover strategy, and as shown in Figure 3, intersections are 2, 7, after exchanging matching section the intersections in A are 7, 6, 5, then find the value of the positions (7-3, 6-0, 5-4) in section B who is matched with A. Continue to detect conflict until there is no conflict. Do the same operations for B, and then obtain the final results.
5. Termination condition
   In order to get the better optimal solution with the increase of the number of iterations, this instance will make the optimal solution of last time join the group, to prevent the

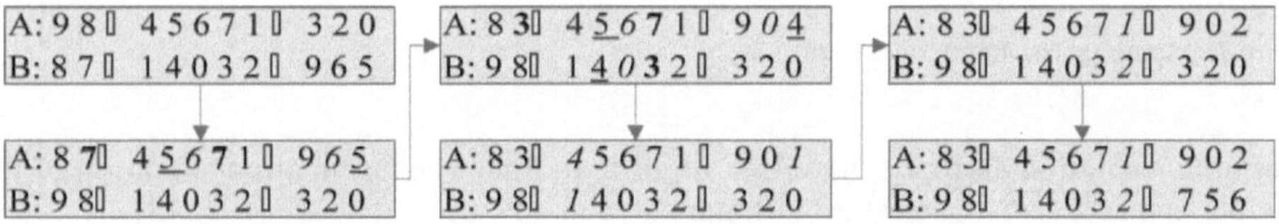

Figure 3. Path instance.

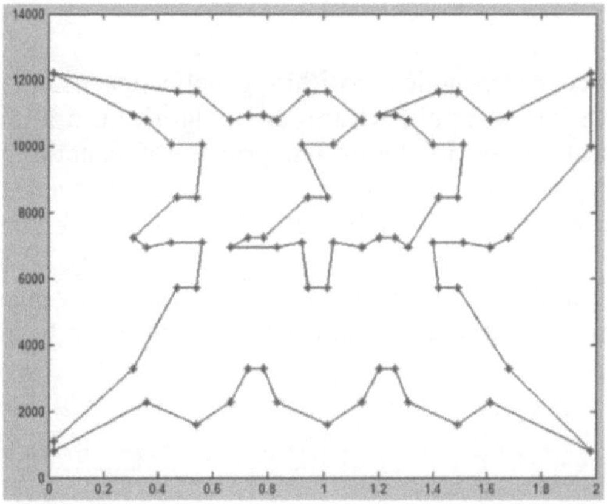

Figure 4. Optimal roadmap for question 1.

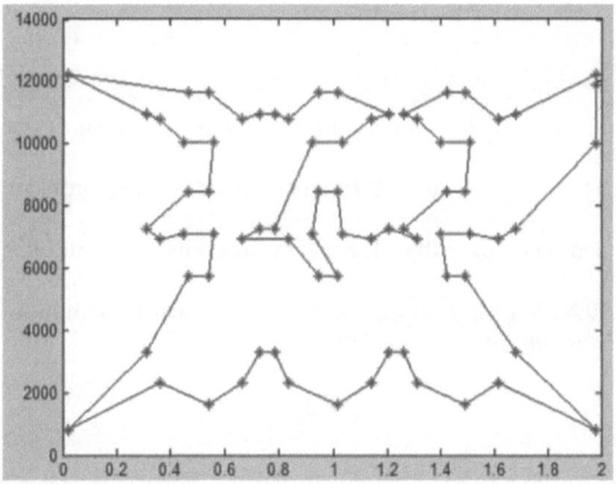

Figure 5. Optimal roadmap for question 5.

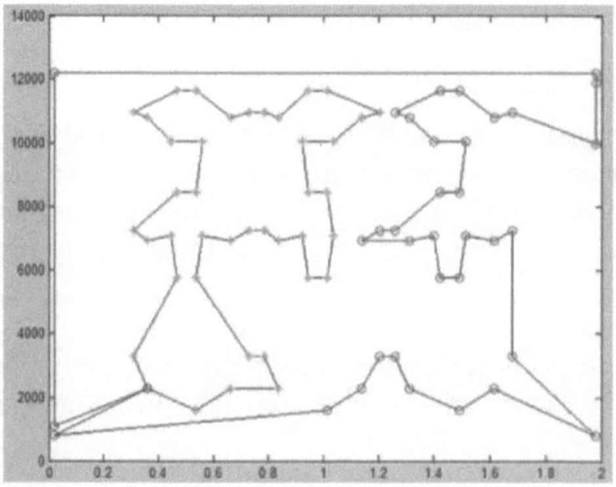

Figure 7. Optimal roadmap for question 3.

optimum solution disappear for crossover or mutation. The termination condition of this paper is that giving a biggest evolutionary step in advance.

6. The results are shown in Figures 4–7.

6 CONCLUSION

Be familiar with the process of genetic algorithm, genetic algorithm is feasible to solve TSP problem, and MATLAB has a certain advantage in algorithm optimization programming. Two principles must be followed in the design process of genetic algorithm: 1) Children have the good genetic information of parents, which is called inheritance. 2) Children have to keep the diversity of population, which is called variation. We need to plan as a whole, although these two principles often conflict in the design process. Fast computation of the algorithm is to realize, but also to achieve the optimal solution. In this paper, we believe that it is necessary to improve the standard genetic algorithm in order to improve their ability of solving.

ACKNOWLEDGEMENTS

This research is partly funded by a grant from the National High Technology Research and Development Program of China (863 Program) (No. SS2014 AA012205).

REFERENCES

Clarke. G. 1964. Scheduling of vehicles from a central depot to a number of delivery points. *Operation Research* 12(4):12–18.

Gendreau, M. 1994. A tabu search heuristic for the vehicle routing problem. *Management Science* 40:1276–1290.

Miller, B.E. 1974. A heuristic algorithm for the vehicle-dispatch problem. *Operation Research* 22:340–349.

Su C.T. & H.H. Chen. 1999. Vehicle Routing Design of Physical Distribution Center. *Journal of the Chinese Institute of Industrial Engineers* 16:410–417.

Modeling and Computation in Engineering III – Zhang & Xie (Eds)
© 2014 Taylor & Francis Group, London, ISBN 978-1-138-02680-3

Simulation research on the multipath error of BeiDou-2

Z. Lei
Shanghai Institute of Technical Physics, Chinese Academy of Sciences, Shanghai, China
Changzhou Institute of Optoelectronic Technology, Changzhou, Jiangsu, China

Yuanfei Wang
Key Laboratory of Geo-information Science, Ministry of Education, East China Normal University, Shanghai, China

Dai Ning
Shanghai Institute of Technical Physics, Chinese Academy of Sciences, Shanghai, China
Changzhou Institute of Optoelectronic Technology, Changzhou, Jiangsu, China

ABSTRACT: Multipath error tends to be short delay for BeiDou-2 satellites in the application environments. Although BeiDou-2 is currently being used for such purposes, and has been found to be accurate enough in rural and suburban environments, BeiDou-2 is not very reliable in densely urban environments, where it is often desirable to have accurate information for an automobile navigation system. This paper is to assess the levels of multipath in satellite navigation measurements for the environments using a modern receiver. Some data were collected in the open static environment, rural dynamic environment and urban environments in order to analyze and discuss the levels of multipath.

1 INTRODUCTION

Multipath error is often caused by GPS signals reflecting and diffracting off of large reflectors, such as buildings, automobiles, the ground, etc., reaching the GPS receiver, and interfering with the direct signal. Multipath depends on the relative amplitude, delay, and phase of the reflected signal, as well as the rate of change of the relative phase (Misra, P. 2001). The position solution in urban environments is degraded for several reasons (Akos, D. 2004, Kristin Larson, 2004). First of all, there is poor signal availability in cities because satellites are blocked from view by large buildings and the only satellites that the receiver should be tracking are those present within the relatively narrow strip of sky directly visible from the urban canyon. However, even with good signal availability, the position solution can also suffer because of relatively poor geometrical distribution of the satellites that can be tracked in an urban canyon, quantified by Geometric Dilution Of Precision (GDOP) values greater than three. Also, multipath will always be encountered in urban environments, where it is more likely that the received satellite navigation signal will be attenuated from the direct Line-Of-Sight (LOS) signal, making it more susceptible to interference due to multipath.

Several studies have been performed regarding the use of GPS in urban environments. They discovered that using a least squares position solution, the accuracy in an urban environment can be off by up to a kilometer, and even using a heavily filtered position solution, errors of up to 100 meters, or one city block, were present (Kristin Larson, 2004, Minami, M. 2000). However, using a high sensitivity receiver, there was 100% availability, as opposed to the 30% availability obtained with a standard sensitivity GPS receiver. For this research, the performance of standalone GPS in the position domain will be analyzed for different receivers, and signal availability will be considered as well. But in this paper, we analyze and perform the multipath error by the BeiDou-2 satellites signals of China.

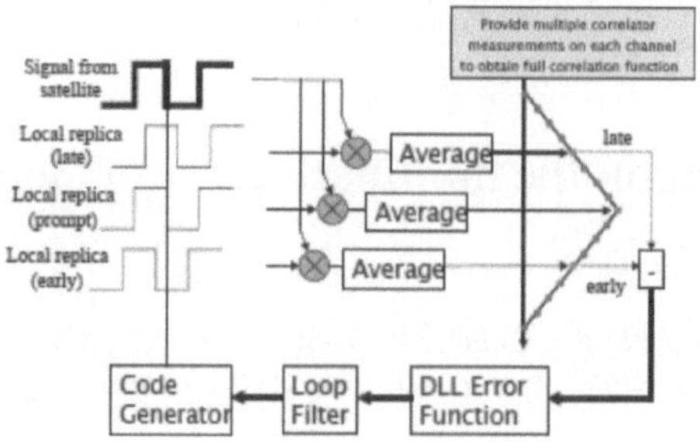

Figure 1. Correlator structure.

2 DESIGN OF SIGNAL PROCESSING

Multipath signals also distort the correlation peak in a very well understood manner. Normally only three correlators are used in the correlation peak to track the signal: early, late, and prompt (Kristin Larson, 2004). The distortions in the correlation peak due to multipath can be better observed and quantified by using several correlators along the correlation peak, as shown in Figure 1.

There are already several multipath mitigation methods designed into BeiDou-2. For example, the direct satellite navigation signal is Right Hand Circularly Polarized (RHCP), and once the signal is reflected off of an object it becomes Left Hand Circularly Polarized (LHCP). Therefore, RHCP antennas are used in BeiDou-2 to detect the LOS signal. However, a signal that has been reflected twice, four times, etc., will also be RHCP, and so a RHCP antenna will not always be effective.

Controlled antenna motion has been found to significantly improve multipath error. If an antenna is constantly rotated in a circular motion, then the fading bandwidth of the multipath is much greater than the tracking loop bandwidth, and the multipath error reduces to zero for a coherent DLL (Kristin Larson, 2004, Minami, M., 2000). Carrier smoothing of code ranges also improves multipath to some degree. Carrier smoothing techniques take advantage of the preciseness of the carrier measurements with the unambiguousness of the code measurements in order to obtain more accurate range measurements by reducing the effect of receiver noise; however, carrier smoothing tends to reduce the effect of the multipath error as well as the receiver noise. There are also several basic methods of mitigating multipath tracking errors in the DLL, the most basic of which is narrow correlator spacing. Narrow correlator spacing is very effective in improving the error due to longer delay multipath, however is still susceptible to very short delay multipath. Other ways to used to mitigate multipath are the Multipath Estimation Technology (MET), which calculates the slope on either side of the auto-correlation function in an effort to estimate the code phase delay, or the Multipath Estimating Delay Lock Loop (MEDLL), which used several correlators to estimate both code and carrier phase Error. These methods will not be used in this paper, however, because the goal is to measure multipath rather than to mitigate multipath.

3 MULTIPATH ENVIRONMENTS

3.1 *Open static environment*

In an ideal open static environment, the antenna would be located well over 300 meters away from any large objects that could act as a reflectors and with a complete view of the sky, for the entire duration of the data collection; however, this not being possible, data were instead collected on the top of the roof of the engineering building (Kristin Larson, 2004). There was a large wall relatively close the antenna, as well as various other reflectors around the rooftop. This environment, although far from ideal, was relatively open, unchanging, very convenient for collection data, and a fair representation of a typical open environment.

The data collected here were used to validate the multipath observable tool by verifying the expected results mentioned earlier.

3.2 *Rural dynamic environment*

An ideal rural dynamic environment would be completely flat for miles with an unobstructed view of the horizon in all directions for the entire duration of the data collection. Although there were some hills and buildings, this location should fairly represent typical rural dynamic environments (Kristin Larson, 2004). The rural dynamic data collected here were used to compare results between urban dynamic and rural dynamic scenarios, and to verify that higher multipath errors were observed in urban dynamic locations. Also, the effects of antenna dynamics on multipath error were evaluated by comparing the observed multipath in a rural dynamic environment to that in an open static environment.

3.3 *Urban environments*

Data were taken at a static urban location primarily to gather raw IF data to post process with multiple correlators in order to estimate the impact of urban multipath on the correlation peak. Only static scenarios were considered for this analysis because the MATLAB tracking code that adds multiple correlators to the correlation peak is not robust enough at present to re-quire the satellite once it has lost lock, which often occurs in urban dynamic scenarios (Kristin Larson, 2004, Steingass, A. 2004). Also, data collected in a somewhat static urban environment were used to evaluate the repeatability properties of multipath in an urban setting. Static urban data are also interesting because although a car driving through downtown will mostly be subjected to dynamic multipath scenarios, there will also be times when the car is stopped because of a stoplight or traffic and is momentarily in a static situation.

Because the urban environment is a very challenging multipath environment, it was desired to see if multipath repeatability could be observed in an urban environment. However, in order to observe repeatability, the antenna must be left in the exact same location for at least 48 hours, and this was a tricky feat to manage in an urban environment due to security reasons.

4 EXPERIMENTAL ANALYSE

The correlator results from a satellite at a high elevation angle in an open environment are shown in Figures 2 and 3. There should be little to no multipath error visible in these results. As seen in Figure 2, there are not any significant distortions in the correlation peak, and as seen in Figure 3, the variation in the correlators appears to be similar on each side of the correlation peak, as expected. The maximum standard deviation of measurements tends to range from about 0.01 to about 0.15.

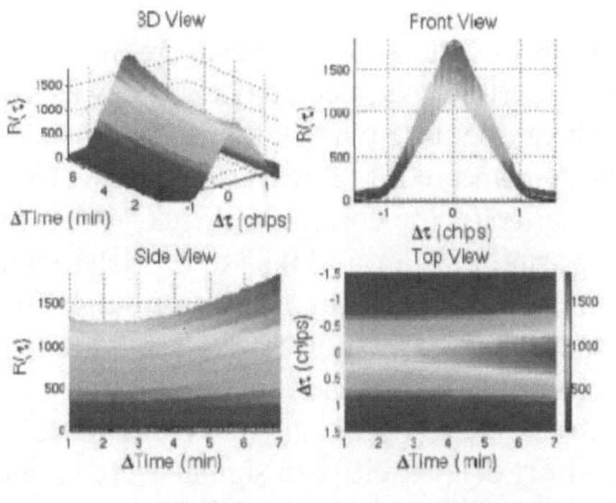

Figure 2. Correlation peak: High elevation satellite in an open environment.

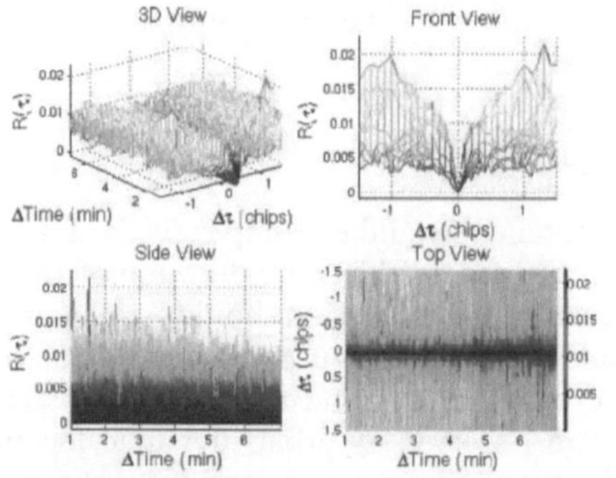

Figure 3. Correlation peak standard deviation: High elevation satellite in an open environment.

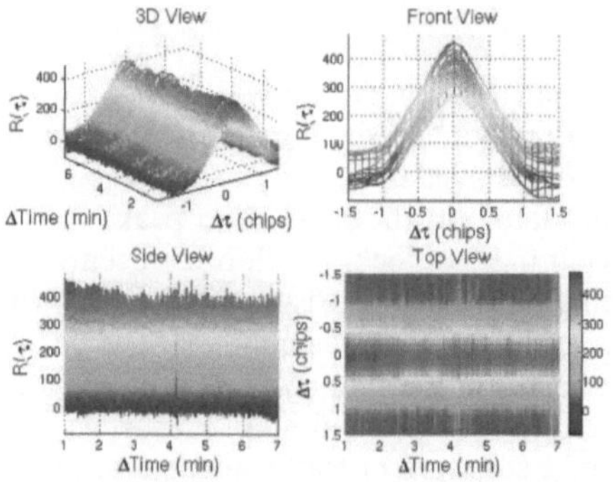

Figure 4. Correlation peak: Low elevation satellite in an open environment.

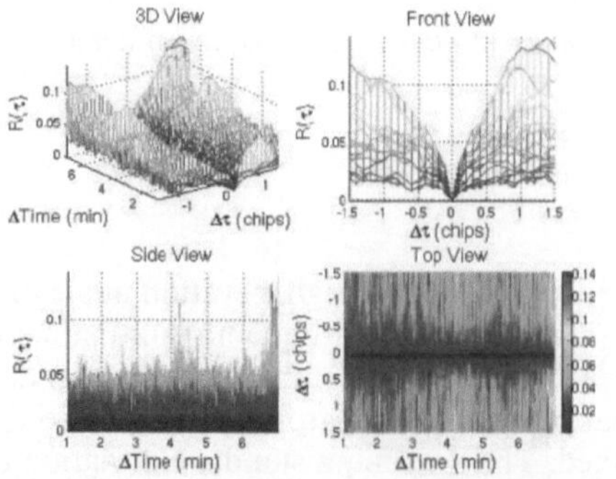

Figure 5. Correlation peak standard deviation: Low elevation satellite in an open environment.

The correlator results from a satellite at a low elevation angle in an open environment are shown in Figures 4 and 5. This satellite was selected because it was low on the southern horizon, and a wall was located north of the antenna; therefore, there should definitely be multipath error present in these results. As expected, Figures 4 and 5 show that there is more variation in the correlators on the trailing edge of the correlation peak, which indicates some longer delay multipath. Also, the level of variation is around 0.04 and occasionally fluctuates to above 0.1.

The correlator results from a satellite at a high elevation angle in an urban environment are shown Figures 6 and 7. These results clearly show strong oscillations in correlation power due to multipath; however, there does not appear to be very much difference in the levels of correlator variation on either side of the correlation peak, as shown in Figure 7, which is not indicative of multipath. However, if very short delay multipath signals were affecting the direct signal, there would not necessarily be a visible difference in correlator variation; therefore, high elevation

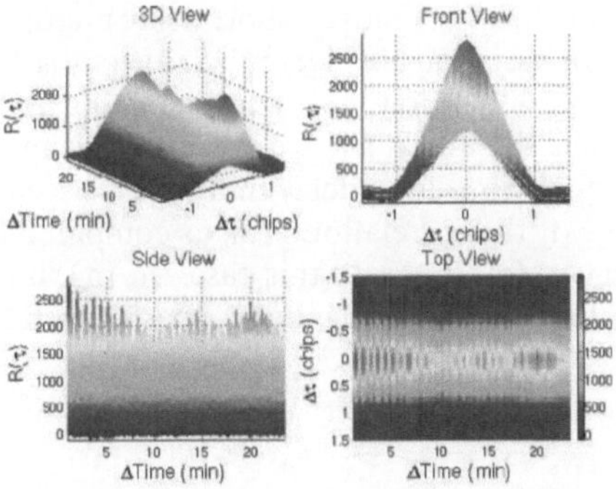

Figure 6. Correlation peak: High elevation satellite in an urban environment.

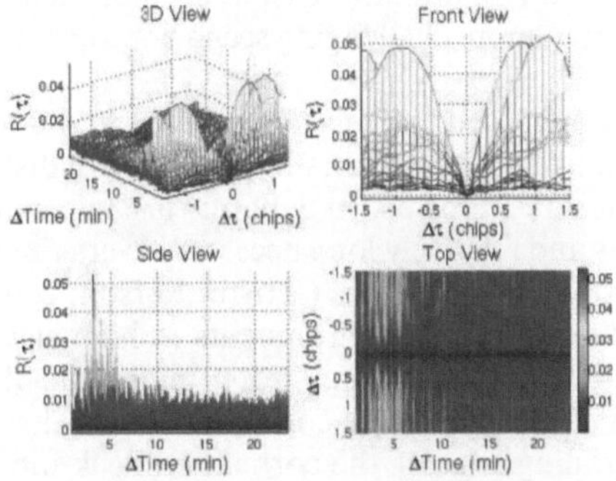

Figure 7. Correlation peak standard deviation: High elevation satellite in an urban environment.

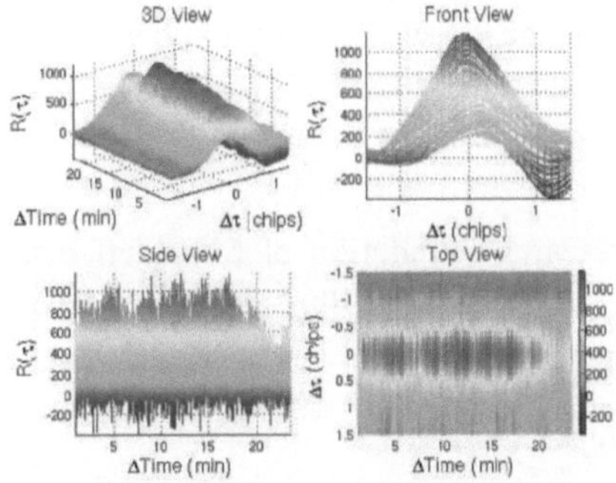

Figure 8. Correlation peak: Low elevation satellite in an urban environment.

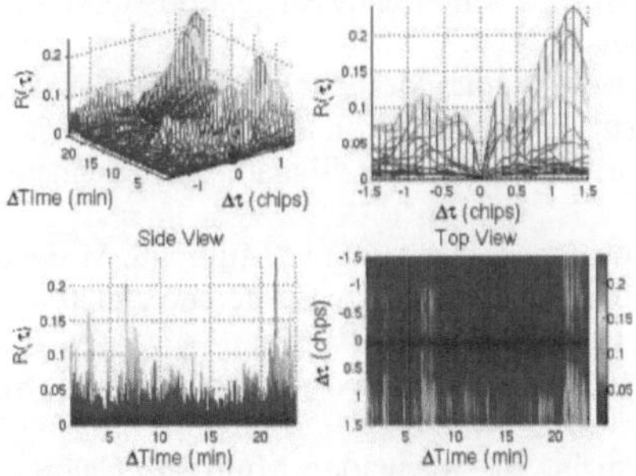

Figure 9. Correlation peak standard deviation: Low elevation satellite in an urban environment.

satellite in urban environments must see mostly short delay multipath (Kristin Larson, 2004, Lehner, A., 2005). The levels of variation are higher than seen for a high elevation satellite in an open environment, averaging slightly below 0.01, but fluctuating as high as 0.05.

The correlator results from a satellite at a low elevation angle in an urban environment are shown Figures 8 and 9. Very obvious long delay multipath is seen here because of the high variation of the trailing edge of the correlation peak as compared to the leading edge (Kristin Larson, 2004, MacGougan, G., 2001). In this case, the maximum standard deviation in measurements averages slightly less than 0.04, but ranges up to 0.2 at some points.

5 CONCLUSION

The levels of multipath were assessed in urban environments using a modern BeiDou-2 receiver. This was done by first validating the open static results from a MATLAB code that was developed to isolate the multipath observable, and then using the same multipath observable to process data from more challenging scenarios, such as rural dynamic and urban dynamic environments.

Multiple correlators were then used to analyze variations in the correlation peak caused by multipath. Although using multiple correlators to characterize distortions in the correlation peak does not have a practical real time use in dynamic urban environments because of high computational requirements and relatively long necessary averaging times, it is useful for analyzing multipath in post processing situations (Kristin Larson, 2004). Multipath error tends to be short delay for satellites in urban environments at high elevations, because although oscillations in the correlation power indicate strong multipath, the difference in standard deviation is very similar on either side of the correlation peak; therefore, if multipath reflection are not affecting the trailing edge of the correlation peak more, it must be short delay. Also, multipath from satellites at low elevation angles in urban environments tend to be long delay multipath because the trailing edge of the correlation peak has significantly more variation than the leading edge. The level of variation in both correlators and signal to noise ratios can sometimes be indicative of the presence of multipath error.

ACKNOWLEDGMENT

We thank the innovation research foundation of Shanghai Institute of Technical Physics of the Chinese Academy of Sciences, and the Changzhou City S&T Support foundation (No. CE20120018).

REFERENCES

Akos, D., Weiss J., Murphy, T., and Pullen, S., Airborne Multipath Investigation via a GPS Software Receiver. Proceedings of ION GNSS 2004, The Institute of Navigation, Long Beach, 2004.

Kristin Larson, Characterizing Multipath in Satellite Navigation Measurements for Urban Environments, thesis, B.S., University of Kansas, 2004.

Lehner, A., Steingass, A., A Novel Channel Model for Land Mobile Satellite Navigation. Proceedings of ION GNSS 2005, The Institute of Navigation, Long Beach, 2005.

Misra, P., Enge, P., Global Positioning System: Signals, Measurements, and Performance, Ganga-Jamuna Press, 2001.

Minami, M., Morikawas, H., Aoyama, T., An Adaptive Multipath Mitigation Technique for GPS Signal Reception. Proceedings of Vehicular Technology Conference, 2000.

MacGougan, G., Lachapelle, G., Nayak, R., Wang, A., Overview of GNSS Signal Degradation Phenomena. Proceedings of International Symposium on Kinematic Systems in Geodesy, Geomatics And Navigation, 2001.

Steingass, A., Lehner, A., Measuring the Navigation Multipath Channel—A Statistical Analysis. 2nd ESA Workshop on Satellite Navigation User Equipment Technologies NAVITEC, 2004.

Modeling and Computation in Engineering III – Zhang & Xie (Eds)
© 2014 Taylor & Francis Group, London, ISBN 978-1-138-02680-3

Evaluation study on listed companies of the defense industry based on improved grey relational model

Lei Zhang, Ying Huang, Yue Wang & Meng Huang
School of Management and Economics, Beijing Institute of Technology, China

ABSTRACT: The market-based reform of defense industry is necessary in reform of Chinese market economic system and it is essential to evaluate the performance of companies in the defense industry with scientific theories and methods. The paper establishes an evaluating indicator system of listed companies of the defense industry according to scientific principles and sets the index weight based on information entropy theory. Then, the paper uses the grey relational model to evaluate the performance of listed companies of the defense industry in 2012 with improved Grey Relational Analysis (GRA).

1 INTRODUCTION

The defense industry is a national strategic industry, which leads the high and new technology industry, and plays an extremely vital role in various countries' national defense development and the economic development. The defense industrial enterprises consist of military products research and production units, and they are the pillar of the defense forces. Their capacity and level influence the strength of national defense force directly. Military enterprises not only provide a variety of weapons and equipment for national security, but also make great contribution to the development of national economy. In the 17th Party Congress, we came up with an idea of getting out of a path of development of Chinese military and civilian integration. It inevitably involves the change of the organization and institution of the military enterprises. Moreover, realizing the overall efficiency is the most superior target for all the enterprises. It is an essential step for the defense industrial enterprises to carry on performance measurements on it to achieve their purpose.

A growing literature has emerged on the performance evaluation of listed companies of the defense industry in China. Bo Zhang (2006) made a study on the relationship between capital structure and company's performance by two-way empirical study and built a regression model to test the influence of the chosen variables; Min Huang (2007) used Balanced Scorecard for reference on designing evaluation index and applied the Delphi Method to calculate the weight in each index system; Jing Li (2008) constructed the synthetical appraisement index of corporate performance using factor analysis. The selected synthetical appraisement index could conquer the insufficiency of single index and simple weighted synthetical appraisement on the corporate performance. Xian-min Xie (2009) considered the character of comparative predominance of the underling enterprise in the group and advanced a method of achievement evaluation of enterprise based on the character of comparative predominance. Sheng-xiang Sun and Shu-yu Zi (2009) proposed an evaluation model based on multi-objective multi-person decision and integrated Fuzzy mathematics and genetic algorithm to solve the problem that qualitative indexes and quantitative indexes are hard to measure and the problem that decision-maker's preference is unknown effectively. Fang-yuan Li (2012) analyzed the industry characteristics and performance evaluation's current situation of China's defense enterprises with the idea of DEA, and did the efficiency analysis, scale analysis, projection analysis, sensitivity analysis and improved analysis of the target based on CCR model calculation results.

Summarizing the former research methodology of evaluation on defense industrial enterprises, we know that they have made good contribution to recent research and practice work. Owing to the special characteristics of defense industry, some fuzzy information is hidden in the actual operation process of listed companies rather than known by the public. Therefore, it is necessary to explore more scientific methods to evaluate listed companies of the defense industry effectively.

Based on that, we use the improved GRA model to analyze the operating performance of 28 China's listed defense industrial enterprises. Considering accounting standard and data availability, we select 22 indicators as criteria to assess the development level of defense industrial enterprises from the financial report data from 2011 to 2012.

The remainder of this study is organized in the following manner. Section 2 derives evaluation index system and evaluation model used in this study, while Section 3 describes the data set and discuss the results. Section 4 summarizes this study and discusses potential extensions.

2 EVALUATION INDEX AND METHODOLOGY

This section describes the evaluation index system and the methodology to evaluate the operating performance of the listed companies in defense industry.

2.1 *Evaluation index*

Different evaluation index systems produce different assessment effects. This paper builds a reasonable evaluation index system through referring to *Enterprise Performance Evaluation Standard Value in 2013* and considering the availability of financial data of listed companies in defense industry. We set the index weight based on information entropy theory, and all the observation indexes are in the following chart.

Table 1. Evaluation index of listed companies in defense industry.

Criteria layer	Weight	Observation index	Index weight
Profitability	0.3053	Profit margin on net assets	0.1446
		Return on total assets ratio	0.1182
		Net profit margin	0.2264
		Ratio of profits to cost	0.1883
		Return on invested capital	0.3225
Operation capacity	0.2681	Total assets turnover	0.0086
		Receivables turnover ratio	0.2865
		Current assets turnover	0.1646
		Inventory turnover	0.1059
		Working capital turnover rate	0.1926
		Cash recovery for all assets	0.2418
Solvency	0.1824	Debt asset ratio	0.1236
		Ratio of cash to current debts	0.2617
		Current ratio	0.1858
		Quick ratio	0.3024
		Equity ratio	0.1265
Development capacity	0.2442	Operating income growth rate	0.2089
		Capital preservation growth rate	0.2403
		Operating profit growth rate	0.1542
		Total assets growth rate	0.0642
		R&D income ratio	0.2468
		Ratio of capital accumulation	0.0856

2.2 *Improved grey relational model*

Grey Relational Analysis (GRA) is according to the grey system theory, which was pioneered in 1982 by Ju-long Deng, and the gray relational degree model, from the essential aspect, is just the comparative analysis of quantitative. Its nature is the near degrees of curve collection form included by curve ideal sequences, the closer the geometrical form, the greater the correlation degree, on the contrary, the smaller (Bing-sheng Liu etc., 2010). But the original gray relational model only presents the similarity between two curve collections. It cannot figure out the similar level of two curve collections' trend. This paper proposes an improved GRA model, which takes both curve collections' similarity and proximity into consideration, so evaluation results can be closer to the fact than other quantitative methods.

There are four steps through using this model:

Firstly, determine the index series and indicator data standardization. Suppose in j schemes, m data sequences arise as identical number of indicators are selected to evaluate the performance of defense industrial enterprises. Then there will be j data series,

$$X_{ij} = (X_{1j}, X_{2j}, \ldots, X_{mj})^T \tag{1}$$

where X_{ij} is the numerical value of the index i in the j program.

In order to choose the series and calculate the relational degree easily, we need to deal with the data standardization, to transform the raw data into comparable data series.

If the observation index is the positive indicator, the bigger the better, signed as

$$C_{ij} = \frac{X_{ij} - \min(x_{ij})}{\max(x_{ij}) - \min(x_{ij})} \tag{2}$$

If the observation index is the negative indicator, the smaller the better, signed as

$$C_{ij} = \frac{\max(x_{ij}) - X_{ij}}{\max(x_{ij}) - \min(x_{ij})} \tag{3}$$

Secondly, set weights for each observation index. There are many methods to set the index weights, and Entropy Method (EM) is applied in this paper. Define p_i is the weight of index i,

$$p_i = \frac{h_i}{\sum_{j=1}^{n} h_i} \tag{4}$$

where $\sum_{i=1}^{m} p_i = 1, p_i \geq 0$, h_i is the utility value of index i.

Thirdly, calculate the point correlation degree $r_j^{(1)}$. Based on the index series above, we construct the best reference data series, and it is consist of the best data from the indicators, signed as $C_{i*} = (C_{1*}, C_{2*}, \ldots, C_{n*})^T$. Suppose the standardized indicator data series as $C_{ij} = [C_{1j}, C_{2j}, \cdots C_{nj}]^T$, and we will get the point correlation coefficient $\xi_{ij}^{(1)}$ by the following formula:

$$\xi_{ij}^{(1)} = \frac{\min_j \min_i |C_{i*} - C_{ij}| + \rho_1 \max_j \max_i |C_{i*} - C_{ij}|}{|C_{i*} - C_{ij}| + \rho_1 \max_j \max_i |C_{i*} - C_{ij}|} \tag{5}$$

where ρ_1 is the point resolution ratio, and $\rho_1 \in (0,1)$, here $\rho_1 = 0.5$.

Based on (4) and (5) to calculate the point relational degree $r_j^{(1)}$, which is the relational degree between standardized indicator data and reference index i in the j program,

$$r_j^{(1)} = \sum_{i=1}^{n} p_i \, \xi_{ij}^{(1)} \tag{6}$$

Fourthly, calculate the slope correlation degree $r_j^{(2)}$. In fact, slope relational degree represents the similar level of two curve collections' trend during the period between year k and year $k-1$ and its calculation method is similar with the way calculating point relational degree. We will get the slope relational degree $r_j^{(2)}$ by the following formula:

$$r_j^{(2)} = \sum_{i=1}^{n} q_i \, \xi_{ij}^{(2)} \tag{7}$$

$$\xi_{ij}^{(2)} = \frac{\min_j \min_i |\Delta C_{i*} - \Delta C_{ij}| + \rho_2 \max_j \max_i |\Delta C_{i*} - \Delta C_{ij}|}{|\Delta C_{i*} - \Delta C_{ij}| + \rho_2 \max_j \max_i |\Delta C_{i*} - \Delta C_{ij}|} \tag{8}$$

where $\Delta C_{i*} = C_{i*}^k - C_{i*}^{k-1}, \Delta C_{ij} = C_{ij}^k - C_{ij}^{k-1}$, and q_1 is the weights of index i, $\sum_{i=1}^{m} q_i = 1, q_i \geq 0$, ρ_2 is the slope resolution ratio, and $\rho_2 \in (0,1)$, here $\rho_2 = 0.5$.

Fifthly, calculate the synthetical correlation degree r_j. Based on (6) and (7) to calculate the synthetical correlation degree,

$$r_j = \omega r_j^{(1)} + (1-\omega)\, r_j^{(2)}$$

where ω is the synthetical resolution ratio, here $\omega = 0.7$.

The synthetical correlation degree presents the compactedness between programs j and ideal scheme and it can help us compare the advantages and disadvantages of different programs.

3 THE EMPIRICAL ANALYSIS

This section discusses the data used to evaluate the operating performance of the listed companies in defense industry.

3.1 *Data collection*

The original data is collected from 28 defense industrial enterprises and calculated to 22 variables mentioned above from Shenzhen and Shanghai stock market ranging from 2011 to 2012.

3.2 *Improved grey relational analysis*

Based on the methodology mentioned in Section 2, we can figure out the Point Correlation Degree (PCD), Slope Correlation Degree (SCD) and Comprehensive Correlation Degree (CCD) of different sub-sectors and different property companies shown as Tables 2 and 3.

1. The development levels of different sub-sectors in defense industry were uneven, and Shipbuilding and Aerospace Equipment performed much better than Ground Equipment and Aviation Equipment as shown on Table 4. Aviation Equipment owned 10 listed companies in the defense industry, but only 1 of them broke the top 10, and 5 into the top 20. 25% of companies in Ground Equipment entered Top 10 list. And 7 companies broke

Table 2. Performance of different sub-sectors in defense industry.

Sub-sectors	Total amount	Average grade	Num. of top 5	Num. of top 10	Num. of top 20
Aviation equipment	10	0.8061	0	1	5
Aerospace equipment	6	0.8473	2	3	5
Ground equipment	4	0.8341	1	1	3
Shipbuilding	8	0.8545	2	5	7

Table 3. Performance of different property enterprises in defense industry.

	PCD		SCD		CCD	
Property*	PE	CE	PE	CE	PE	CE
Total amount	6	22	6	22	6	22
Average grade	0.8162	0.7940	0.9077	0.9132	0.8437	0.8298
Num. of top 5	2	3	1	4	1	4
Num. of top 10	3	7	1	9	3	7
Num. of top 20	5	15	3	17	5	15

*CE means Central Enterprise; PE means Private Enterprise.

the top 10, occupying 87.5% of all listed companies in Shipbuilding. Five companies of Aerospace Equipment ranked the top ten, taking up 50% of all listed companies.

2. Private Enterprises behaved well in 2012, but their development potentials were restricted by many factors. There were 6 listed companies in defense industry taking up 21.43% of all, and 3 listed companies broke the top 10 in PCD and CCD, only 1 in SCD. Compared with Central Enterprises, Private Enterprises' participation in defense industry can promote the transformation of defense industry, the increase of domestic economy, and the realization of integration of military and civilian sectors. Permitting private enterprises to enter into defense industry and participate in the R&D of military products is one of the essential policy transformations.

4 CONCLUSION

The conclusions show that the operating performance of most listed companies in defense industry in 2012 is good totally, but different sub-sectors in defense industry are uneven. Private enterprises' entry into national defense industrial market is an important trend for the development of the socialist market economy, and more effective measures should be taken to develop their future potentials. The market-oriented reform of the national defense industry should abide by common laws of the market economy as well as pay enough attention to the characteristics and regular patterns of the industry itself. With the changes of military strategy in the world, serious instability in regional development and the restart in integration process of military asset, the national defense industry will come to a brighter future.

REFERENCES

Bing-sheng L, Huiyan M, Min G. The Evaluation Study on the Competitiveness of Listed Commercial Bank Based on the Grey Relational Model [C]//E-Product E-Service and E-Entertainment (ICEEE), 2010 International Conference on. IEEE, 2010: 1–4.

Bo Z. Study on Capital Structure and Performance of China's Military Industry Listed Company [D]. Nanjing University of Aeronautics and Astronautics, 2006.

Fang-yuan L. The Construction of Defense Industry Enterprises' Performance Evaluation System and the Application Research Based on Data Envelopment Analysis [D]. Nanjing University of Science & Technology, 2012.

Jing L. Study on Relationship Between Corporate Governance and Performance of China's Listed Companies in Military Industry [D]. Harbin Engineering University, 2008.

Min H. Research on Performance Evaluation of the Military Industry Enterprise [D]. Harbin Engineering University, 2007.

Sheng-xiang S and Shu-yu Z. Study on the Military Enterprise Achievement Evaluation Based on Multi-objective Multi-person Decision Making [J]. Journal of Wuhan University of Technology (Information & Management Engineering), 2009, 05:804–807.

Xian-min X. Study of the Achievement Evaluation Method of Military Enterprise of Our Country Based on Comparative Predominance [J]. Non-Ferrous Mining and Metallurgy, 2008, 01:52–55.

Modeling and Computation in Engineering III – Zhang & Xie (Eds)
© 2014 Taylor & Francis Group, London, ISBN 978-1-138-02680-3

Comparative analysis of China's regional energy efficiency: Based on the Super-efficiency-SBM-DEA model and clustering method

Jin Tang, Hongyuan Wang & Bing Li
School of Management and Economics, Beijing Institute of Technology, Beijing, China

Xiaohan Shen
School of Humanities and Social Sciences, Beijing Institute of Technology, Beijing, China

ABSTRACT: The paper establishes the index system of total factor energy input and output and evaluates the total factor energy efficiency and energy saving and emission reduction potential in various regions of China using Super-efficient-SBM-DEA model, analyzes the regional differences based on the 30 regional data in 2012. The results show that all regions' energy efficiency is generally at a low level. More precisely, the problems for the most regions exist the phenomenon that too much energy is used and too much pollution is charged, and there is a large regional difference about energy efficiency-related indicators. The Eastern regions have the highest energy efficiency and significantly higher than that in the Midwest. In order to explore the way to improve energy efficiency in central and western regions, according to the total factor energy efficiency, energy saving and emission reduction potential, we use K-Means cluster analysis for different regions to identify the characteristics of each region and try to divide the whole country into several types. On this basis, we put forward some suggestions about energy conservation and emission reduction for each type of region.

1 INTRODUCTION

With the rapid economic development in recent years, the energy consumption in China rises dramatically. This extensive economic growth pattern of high input-low output and high pollution-low efficiency consumes massive energy resources and leads to severe pollution to the environment. Therefore, Chinese government proposed constraints on energy saving and emission reduction explicitly in national 'the 10th five year planning' and 'the 11th five year planning' and also includes it in the long term development planning of national economy (Wang Keliang, Yang Baochen and Yang Li, 2012). It is urgent for China to make effective evaluation on the energy efficiency in all regions and to improve the energy effectiveness, to implement rational emission reduction policy. This paper considers environmental pollution factors in the Decision Making Units by constructing a broader indicator system based on traditional one. Furthermore, this paper measures the Total Factor Energy Efficiency (TFEE), Energy Saving Potential (ESP) and Emission Reduction Potential (ERP) based on the Super-efficient-slacks-based measure-Data Envelop Analysis (Super-efficient-SBM-DEA) with samples from thirty regions in China in 2012. The differences and the features among the different regions are explored by K-Means clustering analysis with the aim to find ways to promote the energy use efficiency in each region, finally to propose related counter measures and suggestions for tailored emission reduction policy for each region.

2 RESEARCH METHODS

Data Envelopment Analysis (DEA) is an efficient evaluation method for multiple inputs and outputs of multiple decision-making units. Under different assumptions and constraints, DEA model has tons of forms. During the research, the data of our study needs to be considered when selecting one of the DEA models, mainly the situation of DMU, but also a combination of the purpose of efficiency evaluation. In this study, the extended model of DEA, Super-efficient-DEA, is used. Super-efficient-DEA was proposed by Banker Gifford to solve the problem that effective decision-making unit was unable to continue to sort more deeply. The basic idea is that when assessing the decision unit, excludes the decision-making unit outside of a collection unit (Cai Xiaochun, Xiao Xiaoai, 2010). In the Super-efficient-DEA model, an effective decision-making unit can increase the investment proportionately, while the efficiency value remains unchanged (Li Jinying, Cheng Yunxue, 2012). Precisely, the increasing proportion of investment is the evaluation value of Super-efficient-DEA. In addition, compared with the traditional DEA model, efficiency values of the Super-efficient-DEA model are calculated probably greater than 1. For an effective DEA decision-making unit, increasing the investment proportionately will not change the effectiveness (Hu J.L. Wang S., 2006). Namely, the decision-making unit will be at the forefront of its surface. Therefore, according to the efficiency value of the effective decision-making unit, decision-making unit can be compared and sorted (Han Yijie, Liu Xiuli, 2011). It is worth mentioning that the efficiency value of the invalid decision-making unit in Super-efficient DEA model is consistent with it in traditional DEA model (Wen Yongjun. 2009). Input-oriented Super-efficient-DEA mathematical expression models: $\min[\theta - \varepsilon(\Sigma^{m}_{i=1} s_i^- + \Sigma^{n}_{k=1} s_i^+)]$; $\Sigma^{p}_{j=1}\lambda_j x_{ij} + s_i^- = \theta x_{iq}$, $i = 1, 2, \ldots, m$; $\Sigma^{p}_{j=1}\lambda_j y_{nj} - s_n^+ = y_{nk}$, $n = 1, 2, \ldots, l$; $s_i^- >= 0$, $s_n^+ >= 0$, $\lambda_j >= 0$, $j = 1, 2, \ldots, p$. Among: λ_j represents the j-th unknown right decision unit weight; s^- and s^+ represent input and output slack variables; θ is a valid value of decision unit which is an effective input relative to output. The effectiveness of Super-efficient-DEA model includes technical efficiency and scale efficiency, according to θ, input and output slack variables to determine whether they have the validity of the decision-making unit or not (Guo Bin, Lu Yubo, 2012). $\Sigma\lambda$ is the scale value for the decision-making unit, and can be improved by the production scale of $\Sigma\lambda$.

3 EMPIRICAL ANALYSIS

3.1 *Energy indicator system and data processing*

Constructing energy input-output indicator system is the premise and foundation of total factor energy efficiency evaluation. It is commonly agreed that input factors of economic growth are energy consumption, capital and labor (Hu Caimei, 2010). According to the literature, Hu defined the total factor energy efficiency indicators such as energy consumption, capital and labor as input factors based on the total factors productivity (Hu JL. Wang S., 2006). This paper, considered the impact of environmental pollution to total factor energy efficiency by taking the pollution release (unexpected output) as one input factor, taking the economic output as the output factor. The definition of the total factor energy input-output indicator system is as following: (1) energy consumption, taking the total energy consumption in 2012 as the measurement. The data source is the China Energy Year Book 2013; the unit is 10,000 tons standard charcoal equivalent. (2) capital input, many researchers take capital stock as the indicator of capital (Wei Chu, Shen Manhong, 2007), to some large extent, there is a real utilization rate in capital, not all the capital stock is used up in the current period, so the capital stock can hardly reflect the current capital input. Furthermore, some researchers took fixed assets depreciation to measure the capital input, while the depreciation only cannot embody the part already input but not depreciated (Feng Chaoyang, 2013). This paper takes fixed asset investment of the whole society as the measurement of the capital. The data source is China Statistical Year Book 2013 and the statistical year book of related regions, the unit is 100 billion RMB. (3) Labor, the quantities of the employment in each region in 2012 are taken as the measurement, the data is from the China Statistical Year Book 2013 and the statistical year book of related regions, the unit is 10,000 persons. (4) pollutant release. Due to the externality of the production, there surely will be some impacts from the input to the environment. Some relevant studies named this kind of impact ‘unexpected output’, while it is the environmental cost input should be borne by the Decision Making Unit to obtain the economic output, so it should be taken as an input factor (Honma S., Hu J.L., 2009). This paper considers the main pollutants such as the release of sulfur dioxide, nitric oxide, smoke (dust) in the waste gas, the release of chemical oxygen demand, ammonia nitrogen and the total nitrogen, the total phosphorus in the waste water as the indicator of environmental pollution. The data source is China Energy Year Book 2013 and the unit is 10,000 tons. (5) Economic output, Gross National Product is the indicator to evaluate the economy of a country or a region (Wu Qi, Wu Chunyou., 2009). This paper takes the regional GDP as a measurement of economy output, the data source is the China Statistical Year Book 2013 and the unit is 100 billion RMB.

Before calculating the total factor energy efficiency, the dataset should be pre-processed. In the input-output indicator system constructed in this paper, there are 7 sub-indicators in pollutant release; it is complicated to take them all together into the analysis because of the information redundancy. Factor analysis is adopted to reduce the dimension of the release of sulfur dioxide, nitric oxide, smoke (dust) in the waste gas, the release of chemical oxygen demand, ammonia nitrogen and the total nitrogen, the total phosphorus, then to get the comprehensive scores of the pollutants. Because the Super-efficient DEA model requires the value of the input and the output indicators are non-negative, while the scores obtained follow standard normal distribution (the scores of about half regions will be negative) so it is necessary to map the scores to a non-negative interval. The Min-Max standardization method is implemented for the linear transformation, which is $Y = [X - \min(X)]/[\max(X) - \min(X)]$, by which the scores are mapped into [0, 1] interval. To guarantee the reliability of the research, the Pearson Correlation test is implemented in the input-output variables to test if they satisfy the extensional conditions. Extensional condition (Charnes A., Cooper W. W., Rhodes E., 1978) is a basic condition for DEA model: the output increases accordingly when the inputs of the Decision Making Unit increase. Through the 5% significant level Pearson Correlation test, there are positive correlations between the input variables and output variables, which satisfies the extensional condition and states that the DEA energy model with the sample data is reliable.

3.2 *The potential analysis on energy efficiency and energy saving emission reduction*

There is the new approach-directional technology distance function (Chambers R.G, Chung Y. 1996) proposed by Chung etc, which is the generalization of Shephard radial distance function. The new approach can process the situation of the input and output change at the same time and it also can include the environmental regulations in energy efficiency consideration (Guo Haihua, Zhang Qingyun, 2011). The SBM directional distance function proposed by Tone in 2001 can solve the problem of comparing the Decision Making Unit on the whole and considering the related slacks when the input factors increase (Zuo Zhongmei, Yang Li, 2011). This paper selects the Super-efficient-SBM-DEA model, which based on input-oriented and constant return scale. The model uses the cross sectional data of each regions in China in 2012 and takes the energy consumption, capital input, labor and pollutant release as input indicator, the gross product value as the output indicator. The Software of DEA-SOLVER Pro 5.0 is used to compute the total factor energy efficiency. On the basis of the calculation, the redundancy ratios are obtained to be the indicators of the maximum potential for energy saving and emission reduction in each region. It is can be seen from result Table 1. The regions of energy efficiency surpassing 1 are Beijing, Shanghai, Jiangsu and Guangdong on the efficiency frontier. Furthermore, the potential of energy saving and emission reduction for these four regions are all zero, which doesn't not state that they do not have any potential for energy saving and emission reduction. The result shows that these regions are DEA effective compared with other regions. There is no energy redundancy or pollutant release redundancy in these regions. Similarly, the potentials in energy saving of Zhejiang and Jiangxi are zero, which shows the non-redundancy in energy saving. From the view point of scale return, the scale return of Jiangsu, Shandong, Henan, Hubei, Guangdong and Sichuan Scale return decrease. Then increases of output in these six regions are smaller

Table 1. Total-factor energy efficiency by region (2012).

Region	TFEE	$\Sigma\lambda$	RTS	ESP	ERP
Beijing	1.209	0.886	Increasing	0.00%	0.00%
Tianjin	0.86	0.37	Increasing	13.50%	5.48%
Hebei	0.554	0.861	Increasing	51.43%	76.03%
Shanxi	0.403	0.6	Increasing	64.73%	85.02%
Neimenggu	0.594	0.294	Increasing	57.17%	71.63%
Liaoning	0.57	0.883	Increasing	41.42%	69.80%
Jilin	0.607	0.415	Increasing	29.92%	71.07%
Heilongjiang	0.479	0.678	Increasing	39.58%	84.79%
Shanghai	1.161	0.858	Increasing	0.00%	0.00%
Jiangsu	1.123	1.009	Decreasing	0.00%	0.00%
Zhejiang	0.768	0.499	Increasing	0.00%	46.70%
Anhui	0.613	0.702	Increasing	15.47%	73.71%
Fujian	0.664	0.976	Increasing	0.83%	58.59%
Jiangxi	0.617	0.588	Increasing	0.00%	75.65%
Shandong	0.681	1.391	Decreasing	29.27%	61.13%
Henan	0.548	1.467	Decreasing	29.53%	78.86%
Hubei	0.598	1.043	Decreasing	29.33%	69.97%
Hunan	0.624	0.924	Increasing	26.12%	70.62%
Guangdong	1.042	2.159	Decreasing	0.00%	0.00%
Guangxi	0.597	0.642	Increasing	19.87%	75.38%
Hainan	0.607	0.141	Increasing	4.76%	73.47%
Chongqing	0.561	0.565	Increasing	30.77%	66.75%
Sichuan	0.58	1.116	Decreasing	34.87%	71.23%
Guizhou	0.389	0.34	Increasing	60.95%	83.96%
Yunnan	0.458	0.511	Increasing	44.37%	78.07%
Shaanxi	0.574	0.716	Increasing	23.42%	74.38%
Gansu	0.41	0.28	Increasing	54.60%	83.02%
Qinghai	0.403	0.094	Increasing	69.75%	79.01%
Ningxia	0.405	0.041	Increasing	73.79%	80.14%
Xinjiang	0.373	0.372	Increasing	64.29%	88.84%

for the increase of input after the inputs increase. When the input increase will not lead to proportional increase in output, the production scale should be reduced in these regions. The rest twenty four regions are on the stage of increasing scale return, which is to say if the inputs increase at the same ratio, the output will increase more than that ratio; the production scale in these regions should be increased.

3.3 *The difference analysis on energy efficiency-related indicators*

Due to the severe unbalance regional economic development in China, there may be some regional differences existing in the energy efficiency and energy saving. Generally, China is divided into three economic parts (East-middle-west). This division is according to the economic development and geographic location after long term evolution. The energy of east, middle and west region are calculated in this paper (the average of the related indicators in each region on the basis of Table 1) to analyze the regional differences in energy saving emission reduction in China. The total factor energy efficiency score in east region (0.82) is far above the national average (0.636), but lower than the national average in the energy saving emission reduction. The energy efficiencies in middle and east region are lower than the national average, but higher than the national average in the potentials in energy saving emission reduction. It can be seen from Figure 1, the scores in total factor energy efficiency decrease from east to west and the scores in east region are far greater than the other two regions; the potential in emission reduction increases from east to west and the values of

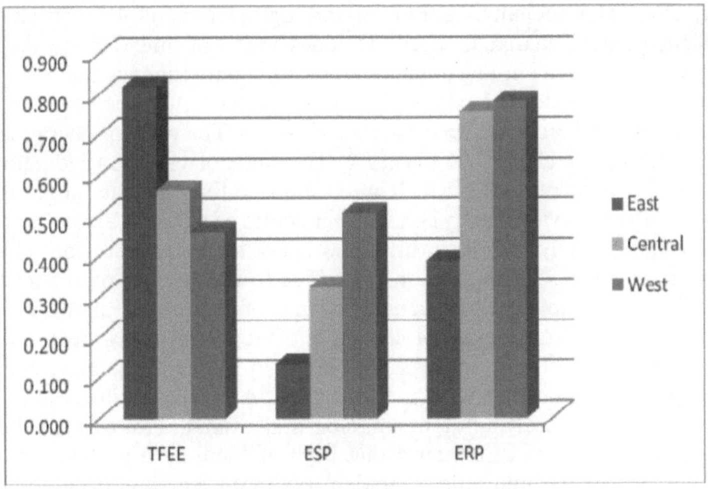

Figure 1. Relationship between energy efficiency-related indicators and three major areas in China.

middle region and west region are close and far greater than the east region. Of course, it is not appropriate to compare the total factor energy efficiency, energy saving potential and emission reduction potential by simply comparing the means, because the mean is easily influenced by the extreme value. So Least Square Deviation (LSD) is used in this paper to make multiple comparisons to explore the differences in these aspects between the regions. The result shows, on the 5% significant level, the total factor energy efficiency: east >> middle > west: the energy saving potential: west >> middle > east: the emission reduction potential: west > middle >> east. Among which, '>>' denotes significantly greater than, '>' denotes greater than (not significantly). It should be noted that the DEA model to calculate energy efficiency in this paper is, in nature, a relative efficiency analysis method. The efficiency result will change along with the sample (Wang Keliang, Yang Baochen and Yang Li. 2012). LSD test result shows that east region has the highest energy efficiency and it is significantly higher than that of middle region and west region. This conclusion is for China, if the samples fall outside the China, different results might be possible.

3.4 *K-means clustering analysis of 30 regions of China*

It is not difficult to find out that not all the regions that have high energy efficiency are in eastern part (just the energy efficiency of eastern part are universally very high) through the result of three kinds of relative energy efficiency indicators (total factor energy efficiency, energy saving potential and emission reduction potential) of using the Super-efficiency-SBM-DEA model. Similarly, not all the areas in western regions that have low energy efficiency. As a result, this article takes full advantage of the different kinds of energy efficiency targets to identify the characteristics of each region and try to divide the whole country into several types that can make the features of the area of the same kind very clear and similar. Based on the total factor energy efficiency score, energy saving potential and emission reduction potential these three variables, we proceed the K-means clustering analysis of 30 regions in China and use SPSS-Statistics 22.0 software to accomplish the clustering algorithm. It can be identified that total factor energy efficiency score, energy saving potential and emission reduction potential these three indexes have great contribution to the classification (Significance level of 5%). It means that using these three variables to proceed the classification of 30 regions of China is reasonable. Cluster 1 includes a total of five regions; they are Beijing, Tianjin, Shanghai, Jiangsu and Guangdong. Cluster 2 includes a total of four regions; they are Zhejiang, Fujian, Jiangxi and Hainan. Cluster 3 includes a total of twelve regions; they are Liaoning, Jilin, Heilongjiang, Anhui, Shandong, Henan, Hubei, Hunan, Guangxi, Chongqing, Sichuan and Shaanxi. Cluster 4 includes a total of nine regions; they are Hebei, Shanxi, Inner Mongolia, Guizhou, Yunnan, Gansu, Qinghai, Ningxia and Xinjiang.

The features of four clusters are shown in Table 2. Apparently, the score of total factor energy efficiency: cluster 1 > cluster 2 > cluster 3 > cluster 4. The value of energy saving potential: cluster 4 > cluster 3 > cluster 1 > cluster 2. The value of emission reduction potential: cluster 4 > cluster 3 > cluster 2 > cluster 1. It can be noticed that there are huge difference and different features of the relevant energy efficiency indictors of different areas. In order to make the result of the cluster analysis more intuition, we mark the areas of various clusters on the map of China. In Figure 2, "0" stands for Tibet (due to the absence of the relative data of Tibet, this article does not analyze the energy efficiency of Tibet.); "1" stands for five regions of cluster 1; "2" represent four regions of cluster 2; "3" represent twelve regions of cluster 3; "4" represent nine regions of cluster 4. As it is shown in Figure 2, the green area of cluster 1, which has high energy efficiency score is the smallest. At the same time, the yellow area of cluster 2, which has problem of pollutant emission is also small. However, the light red area of cluster 3 and deep red area of cluster 4 that both of them have problems of low energy efficiency, redundant energy input and redundant pollutant emission occupy the most area of China. Due to the huge difference and different features of the relevant energy efficiency indicators of different areas, on the one hand, the areas can make use of their own energy feature to proceed mutual cooperation (Li Jinkai, Shen Bo, Han Yafeng, Zhang Menghao. 2012). For example, they can improve the usage of energy efficiency of different kinds of provinces through complementary advantages to promote the sustainable development among the regions. On the other hand, they can formulate specific policy and take action of energy saving and emission-reduction according to their own level of energy use, efficiency of allocation of

Table 2. Clustering feature summary.

Cluster	TFEE	ESP	ERP
1	Optimal	Essentially zero potential	Essentially zero potential
2	Relatively high	Essentially zero potential	Have some potential
3	Medium	Have some potential	Have greater potential
4	Relatively low	Have greater potential	Have the greatest potential

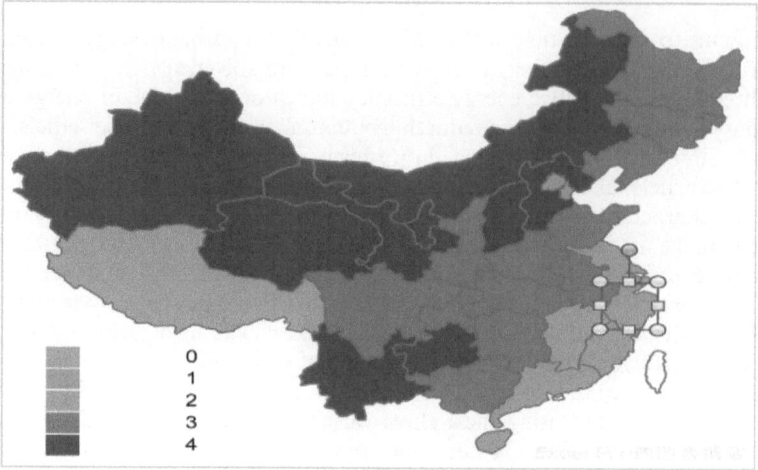

Figure 2. The administrative regions with clustering results in China.

production factors and the size of the scale to enlarge the green area of high energy efficiency and to reduce the yellow area with some issues and red area with severe problems.

4 CONCLUSION

This paper established a total factor of energy input and output index system. Based on this system, Super-efficiency-SBM-DEA model was used to evaluate the total factor energy efficiency, energy saving potential and emission reduction potential in various regions of China and analysis the regional differences. Meanwhile, we use K-Means cluster analysis for different regions based on energy efficiency-related indicators. The results of Super-efficient-SBM-DEA model evaluation showed that the average score of all regions of China's energy efficiency was 0.636, generally at a low level of energy efficiency; the average level of energy saving potential and emission reduction potential of each region were 30.32% and 61.78%. Apparently, tons of energy inputs and emissions redundancy need to be solved. Regional differences of three major economic regions of the analysis showed that the eastern region has the highest energy efficiency and significantly higher than the energy efficiency of central and western. Besides, energy efficiency is in a descending order from the eastern to the western. K-Means cluster analysis of the results showed that the energy efficiency-related indicators in different regions have large differences and different features. Each region should be based on their own situation such as the energy efficiency and the advantages and disadvantages to explore the characteristics of its own energy and energy efficiency differences in terms of policies and measures to improve energy efficiency.

ACKNOWLEDGEMENTS

This research is partly funded by the National Soft Science Research Program (Grant No. 2013GXS2B008).

REFERENCES

Cai Xiaochun, Xiao Xiaoai. Study on Energy Consumption Efficiency of China's Various Regions Based on Super-Efficiency DEA [J]. *Statistics & Information Forum,* 2010, 04:33–40.

Chambers R.G, Chung Y., Färe R. Benefit and distance functions [J]. *Journal of economic theory*, 1996, 70(2): 407–419.

Charnes A., Cooper W.W., Rhodes E. Measuring the Efficiency of Decision Making Units [J]. *European Journal of Operational Research*, 1978, 2(6):429–444.

Feng Chaoyang. Energy Efficiency Empirical Analysis of Each Region in Xinjiang Based on the "Unbalanced" Energy-saving Emission Reduction [J]. *Science Technology and Industry*, 2013, 07:69–73+78.

Guo Bin, Lu Yubo. Evaluation on Energy Saving and Emission Reduction Efficiency and Analysis on Its Influential Factor: Based on Super-efficiency-DEA and Tobit Model [J]. *Technology Economics*, 2012, 12:58–62+76.

Guo Haihua, Zhang Qingyun. The energy efficiency analysis of Chinese inter-provincial based on environmental factors [J]. *Journal of Jinan University (Philosophy and Social Sciences)*, 2011, 02:66–71+188.

Han Yijie, Liu Xiuli. Analysis of Energy Efficiency and Energy-Saving and Emission-Reduction Potential of Steel Industry in Various Regions of China Based on Super-Efficiency DEA Model [J]. *Journal of Systems Science and Mathematical Sciences*, 2011, 03:287–298.

Honma S., Hu J.L. Efficient Waste and Pollution Abatement for Regions in Japan [J]. *International Journal of Sustainable Development & World Ecology*, 2009, 16(4):270–285.

Hu Caimei. Analysis on Total Factor Energy Efficiency in Heilongjiang Province [J]. *Technology Economics*, 2010, 10:49–52.

Hu JL. Wang S. Total-factor energy efficiency of regions in China [J]. *Energy Policy*, 2006, 17:3206–3217.

Li Jinkai, Shen Bo, Han Yafeng, Zhang Menghao. The Comparative Analysis of China's Regional Energy Efficiency: Based on the DEA-Malmquist and Clustering Method [J]. *Journal of Beijing Institute of Technology (Social Sciences Edition)*, 2012, 06:1–6.
Li Jinying, Cheng Yunxue. Total-Factor Energy Efficiency Analysis of Hebei Province Based on Super-Efficiency-DEA [J]. *Industrial Engineering Journal*, 2012, 01:87–92.
Tone K. A slacks-based measure of efficiency in data envelopment analysis [J]. *European journal of operational research*, 2001, 130(3):498–509.
Wang Keliang, Yang Baochen and Yang Li. Analysis on China's Energy Efficiency and Potentials of Energy Conservation and Emissions Reduction from the perspective of Environmental Impact [J]. *Management Review*, 2012, 08:40–50.
Wei Chu, Shen Manhong. Energy Efficiency and Energy Productivity: A Comparison Based on the Panel Data by Province [J]. *The Journal of Quantitative & Technical Economics*, 2007, 09:110–121.
Wen Yongjun. Agricultural Circular Economy Development Evaluation of Each Region in Shandong Based on Super-Efficiency DEA Model [J]. *Productivity Research*, 2009, 02:21–22+49.
Wu Qi, Wu Chunyou. Research on Evaluation Model of Energy Efficiency Based on DEA [J]. *Journal of Management Sciences*, 2009, 01:103–112.
Zuo Zhongmei, Yang Li. Study on Energy Consumption Efficiency of China's Various Regions Based on SBM-DEA Model [J]. *Statistics and Decision*, 2011, 20:105–107.

Modeling and Computation in Engineering III – Zhang & Xie (Eds)
 ISBN 978-1-138-02680-3

Trends in temperature extremes during 1961–2010 in Daqing River Basin

W.J. Liu
Hebei University of Engineering, Handan, People's Republic of China
China Institute of Water Resources and Hydropower Research, Beijing, People's Republic of China

C.Z. Li, J. Liu & D.H. Yan
China Institute of Water Resources and Hydropower Research, Beijing, People's Republic of China

L.X. He
Hebei University of Engineering, Handan, People's Republic of China

ABSTRACT: Spatial and temporal characteristics of temperature have been investigated in the Daqing River Basin using daily temperature observations from 10 meteorological stations. The three indices used to assess pattern changes were average temperature, maximum temperature and minimum temperature. The yearly, quarterly, and monthly changes of these three temperature aspects were studied using a Geographic Information System (ArcGIS), mathematical statistics, and Mann-Kendall inspection. Temporal analysis showed similar trends for rising annual temperature in the Daqing River Basin. Seasonally, winter temperatures showed the most rapid warming trends and summer temperatures were the slowest. The trend ratio of minimum temperatures was the most significant in every season. By month, it showed a steady and significant increase in low temperature extremes, while the occurrence of high temperature extremes was reduced.

1 INTRODUCTION

In recent years, the issue of global warming and the environmental and socioeconomic impacts of temperature change have been of growing concern. Reports from the Intergovernmental Panel on Climate Change (IPCC) have provided scientific and technical support for these changes. The fourth assessment report from the IPCC (IPCC, 2007) pointed out that, from 1906 to 2005, global average temperatures rose 0.74 °C. In the past 100 years in China, the average surface temperature increased about 0.5 °C to 0.8 °C ("National Assessment Report on Climate Change" the writing committee, 2007). Although temperature changes in the Daqing River basin have been comparatively small, these changes directly impact the hydrological cycle and affect the supply of water. Therefore, undertaking research to better understand the nature of basin temperature changes is of great importance.

The Daqing River Basin spans across the four provinces of Hebei, Beijing, Shanxi and Tianjin. The basin is a major factor influencing climate and environment characteristics in this region, especially so in the Hebei Province and the Haihe River Basin (Wang, 2010; Wang, 2000). In recent years, climate change has had a significant effect on the water cycle of the river basin (Gong, 2012); however, there has been little research on the temperature changes in this region. In China, the study on seasonal scale and monthly scales much (Yang, 2010), Therefore, this study from three scales of annual, seasonal and monthly was analyzed on the temperature of Daqing River Basin, it can provide technical support for the study of the hydrological cycle and water resources management in basin level.

2 MATERIALS AND METHODS

We conducted an analysis using temperature data recorded at 10 meteorological stations from 1961–2010. Specifically, we studied three indexes, including average temperature, maximum temperature, and minimum temperature, on three temporal scales (year, season, and month). We processed yearly station data using sophisticated area weighting methods, such as the Thiessen Polygons method, using a Geographic Information System (GIS). Further, we performed a curve analysis trend of the interannual variability of temperature and the rate of change. We analyzed seasonal basin temperature data categorized by spring (3~5 months), summer (6~8 months), autumn (9~11 months), and winter (12~2 months) and performed a curve analysis trend of interannual variability of temperature and the rate of change. Monthly temperature data was analyzed using the threshold value according to the international general percentile threshold method (Jones, 1999; Li, 2010; Yan, 2012) and the frequency of extreme events.

3 RESULTS AND ANALYSIS

3.1 *Trends of years*

Resultly, in the past 50 years, the average annual basin temperature was 10.39 °C, with maximum and minimum temperature averages of 16.53 °C and 5.43 °C, respectively. The upper and lower range of the average temperature was 12.15 °C (1998) and 8.68 °C (1969), an increase of 0.48 °C/10a. The upper and lower range of maximum temperature was 19.80 °C (1971) and 14.62 °C (1964), an increased of 0.31 °C/10a. The upper and lower range of minimum temperature was 7.69 °C (2007) and 3.35 °C (1969), an increase of 0.71 °C /10a. The most significant basin temperature changes overall occurred in the minimum average temperatures, which were the fastest to increase. In the last fifty years, China's annual average surface temperature rose 1.1 °C, an increase of 0.22 °C (Ding, 2006). This data demonstrates that the rate of temperature increase in the Daqing River Basin is significantly higher than the national average.

3.2 *Seasonal trends*

3.2.1 *Spring*

From 1961 to 2010, the average annual basin spring temperature was 11.55 °C, with maximum and minimum temperature averages of 18.12 °C and 5.73 °C, respectively. The upper and lower range of the average spring temperature was 13.59 °C (2001) and 9.71 °C (1979), an increase of 0.49 °C/10a. The upper and lower range of maximum spring temperature was 20.54 °C (2001) and 15.42 °C (1964), an increase of 0.33 °C/10a. The upper and lower range of the minimum spring temperature was 8.19 °C (2008) and 3.42 °C (1962), an increase of 0.76 °C/10a. The most significant temperature changes overall occurred in the minimum spring basin temperatures.

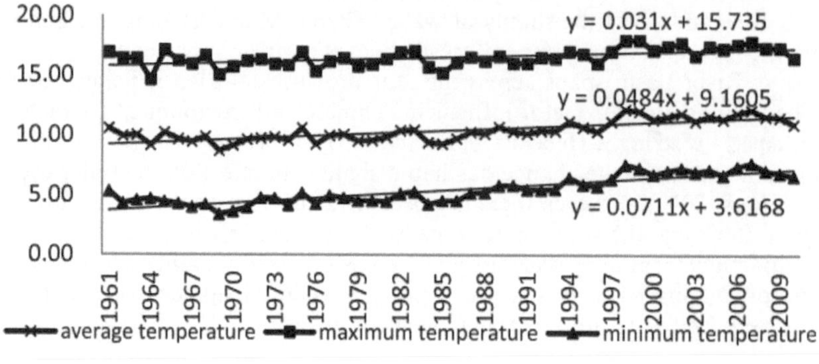

Figure 1. Temperature trends of Daqing River Basin from 1960 to 2010.

3.2.2 *Summer*

In the past 50 years, the average annual basin summer temperature was 23.42 °C, with maximum and minimum temperature averages of 29.08 °C and 18.88 °C, respectively. The upper and lower range of the average high summer temperature was 25.23 °C (2000) and 21.59 °C (1976), an increase of 0.32 °C/10a. The upper and lower range of the maximum high summer temperature was 30.92 °C (2000) and 27.04 °C (1976), an increase of 0.21 °C/10a. The upper and lower range of the average minimum summer temperature was 20.86 °C (2000) and 17.11 °C (1969), an increase of 0.50 °C/10a. The most significant temperature changes overall occurred in the minimum summer basin temperatures.

3.2.3 *Autumn*

In the past 54 years, the average annual basin autumn temperature was 10.69 °C, with maximum and minimum temperature averages of 16.99 °C and 5.88 °C, respectively. The upper and lower range of the average high autumn temperature was 13.09 °C (1998) and 8.99 °C (19.81), an increase of 0.41 °C/10a. The upper and lower range of the maximum autumn temperature was 19.24 °C (1998) and 15.31 °C (1976), an increase of 0.26 °C/10a. The upper and lower range of the minimum autumn temperature was 8.61 °C (2006) and 3.79 °C (1981), an increase of 0.61 °C/10a. The most significant temperature changes overall occurred in the minimum autumn basin temperatures.

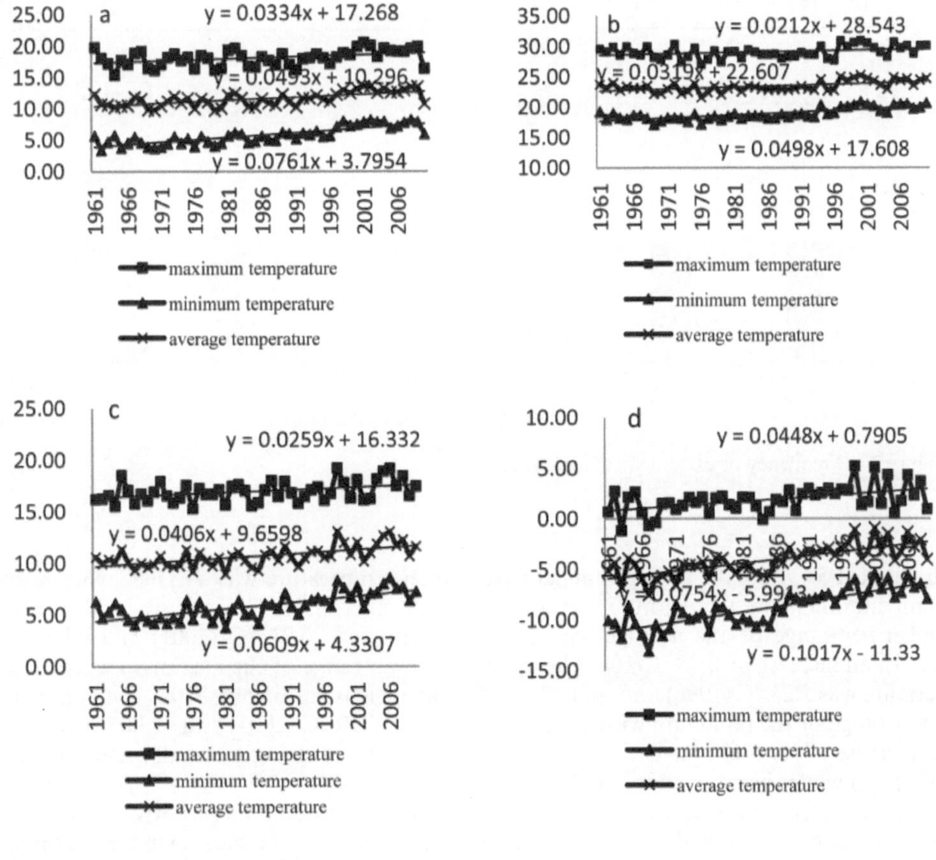

Explanation: a- Spring, b- Summer, c- Autumn, d- Winter.

Figure 2. The trends of seasonal temperature in Daqing river basin from 1961 to 2010. Explanation: a-Spring, b-Summer, c-Autumn, d-Winter.

Table 1. Frequency of extreme cold events (unit: number).

	Mount Wutai	Yu xian	Shijia zhuang	Bei jing	Ba zhou	Tian jing	Bao ding	Yao yang	Huai lai	Tang gu	The basin
1957~1965	9	6	8	4	9	5	6	9	5	5	7.16
1966~1974	5	6	8	9	7	5	9	3	7	5	6.85
1975~1983	4	9	4	10	5	5	5	4	7	6	5.47
1984~1992	2	1	2	0	1	2	1	3	0	2	1.42
1993~2001	0	0	1	0	0	4	1	1	1	2	0.72
2002~2010	0	2	1	1	1	3	2	4	3	2	1.81

Table 2. Frequency of extreme heat events (unit: number).

	Mount Wutai	Yu xian	Shijia zhuang	Bei jing	Ba zhou	Tian jing	Bao ding	Yao yang	Huai lai	Tang gu	The basin
1957~1965	0	0	1	1	1	1	1	1	0	1	0.73
1966~1974	0	0	1	0	0	1	1	2	0	1	0.66
1975~1983	0	0	0	0	0	4	0	0	0	0	0.16
1984~1992	0	1	0	1	1	2	0	0	2	0	0.48
1993~2001	10	11	11	9	11	7	8	13	12	10	10.16
2002~2010	14	12	11	13	11	9	14	8	10	11	11.18

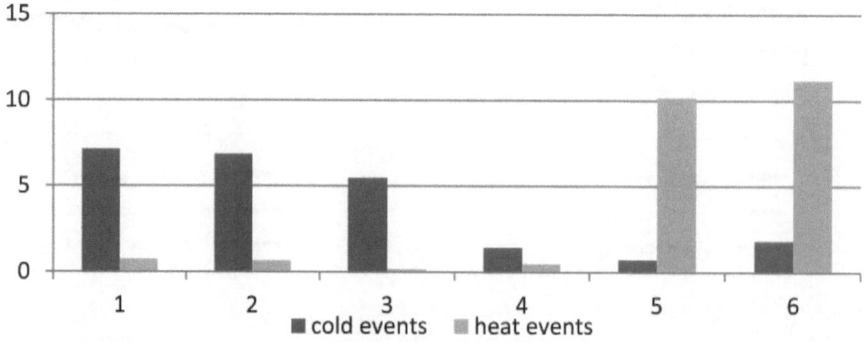

Figure 3. Frequency of extreme heat (cold) events.

3.2.4 *Winter*

In the past 54 years, the average annual basin winter temperature was –4.11 °C, with maximum and minimum temperature averages of 1.91 °C and –8.79 °C, respectively. The upper and lower range of the average winter temperature was –0.91 °C (2001) and –7.50 °C (1967), an increase of 0.75 °C/10a. The upper and lower range of the maximum winter temperature was 5.23 °C (1998) and –1.17 °C (1963), an increase of 0.45 °C /10a. The upper and lower range of the minimum winter temperature was –5.39 °C (2001) and –13.08 °C (1967), an increase of 1.02 °C/10a. The most significant temperature changes overall occurred in the minimum winter basin temperatures.

Our analysis demonstrates that there is a significant uptrend in Daqing River Basin temperature each quarter, and that this change is relatively stable. The most extreme maximum and minimum temperatures have been recorded in the twenty-first century. Throughout the year, temperature differences in the winter show more rapid changes, while differences in summertime temperatures are the slowest. Within each season, the changes in the minimum temperatures are the most significant.

In a word, the study found temperature trends basin is extreme. Therefore, we analyzed the temperature extremes.

3.3 *Month—scale trends*

An analysis of temperature data from ten stations in the Daqing River Basin showed that the extreme heat threshold is between 1.47 °C and 15.75 °C, while the extreme low temperature threshold is between –6.11 °C and 11.03 °C. The data was from 1957 to 2010.

The data in Tables 1 and 2 and Figure 3 demonstrate that, while the frequency of extreme cold events have been reduced, extreme heat events have gradually increased. In recent years, the Daqing River Basin has experienced frequent droughts and floods. Climate warming has led to more frequent occurrences of extreme temperature events that, in turn, may lead to the occurrence of extreme hydrological events.

4 CONCLUSION

1. For nearly fifty years, the Daqing River Basin temperature has been rising. Changes in the average temperature, maximum temperature, and minimum temperature were: 0.44 °C, 0.23 °C, and 0.63 °C, respectively. The most rapid increase occurred for the minimum temperatures.
2. By quarter, the most rapid changes were in the lowest temperatures, while the maximum temperatures changed more slowly. Likewise, seasonal temperature increases were more rapid in winter and slowest in the summer.
3. Reduced extreme cold events and the gradual increase in extreme heat events have significant environment impact on the Daqing River Basin.

The annual, quarterly, and monthly analyses demonstrated that the temperature of the Daqing River Basin is rising, and there is significant variation in temperature, which is consistent with the national temperature trends (Qian, 2007; Ding, 1994; Ren, 1998). The current research findings can provide important reference data for the study of the hydrological cycle of the Daqing River Basin.

ACKNOWLEDGEMENTS

This work was supported by the General Program of the National Natural Science Foundation of China (51209225), the Foundation of China Institute of Water Resources and Hydropower Research (1232), the International Science & Technology Cooperation Program of China (2013DFG70990), Open Research Fund Program of State Key Laboratory of Water Resources and Hydropower Engineering Science (2012B093).

REFERENCES

Ding YH, Ren GY, Shi GY, etc. National Assessment Report of Climate Change (I) Climate change in china and its future trend [J]. Advances in Climate Change Research, 2006, 2(1):3–8.

Ding YH, Dai XS. Temperature Variation in China during the Last 100 Years [J]. Meteorological Monthly, 1994, 12:19–26.

Gong AX, Zhang DD, Feng P. Variation trend of annual runoff coefficient of Daqing River Basin and study on its impact [J]. Water Resources and Hydropower Engineering, 2012, 43(6):1–4.

IPCC. Summary of Policymakers of Climate Change 2007: The Physical Science Basis. Contribution of Working Group I to the Fourth Assessment Report of the Intergovernmental Panel on Climate Change [M]. Cambridge, UK: Cambridge University Press, 2007.

Jones PD, Horton EB, Folland CK, et al The use of indicesto identify changes in climatic extremes [J]. Climatic Changes, 1999, 42:131–149.
Li Z, Liu WZ, Zheng FL. Analyzing Variation Trends in Extreme Precipitation Events in the Jing River Watershed during 1965–2005 [J]. Resources Science, 2010, 32(8):1527–1532.
"National Assessment Report on Climate Change" the writing committee, National Assessment Report on Climate Change [M]. Beijing: Science Press, 2007:18–40.
Qian WH, Fu JL, Zhang WW, Lin X. Changes in Mean Climate and Extreme Climate in China During the Last 40 Years [J]. Advances in Earth Science, 2007, 22(7):673–684.
Ren FM, Zhai PM. Study on changes of china's extreme temperatures During 1951~1998. Scientia Atmospherica sinica 1998.
Wang XD, Pan XB, Long BJ, Li CQ. Analysis of extreme precipitation and temperature events in Hebei province over last 50 years [J]. Chinese Journal of Agrometeorology, 2010, 31(2):170–175.
Wang LX, Zhang Y. Discussion on water environment monitoring of Hebei Province [J]. Water Resouces Protection, 2000, 2:33–35.
Yang P, Liu WD, Wang QG. The Climatic Change Trend and Seasonal Characteristics Daily Temperature Extremes in China for the Latest 40 Years [J]. Journal of Applied Meteorological Science, 2010, 21(1),29–36.
Yan XY, Zhao CY, Wang Y. Change trend of extreme temperature in Northeast China for the past 50 years. Journal of Arid Land Resources and Environment, 2012, 26(1):81–87.

Modeling and Computation in Engineering III – Zhang & Xie (Eds)
© 2014 Taylor & Francis Group, London, ISBN 978-1-138-02680-3

Research on the hydrodynamics characteristic transformation with the marine reclamation in Bohai Bay

M.C. Li
Laboratory of Environmental Protection in Water Transport Engineering, Tianjin Research Institute of Water Transport Engineering, Tianjin, China

ABSTRACT: Hydrodynamics characteristic of marine district is an important basis for evaluating the marine reclamation. In this paper, hydrodynamic model and residual current calculation equation are used to obtain the characteristic of research marine district, Tianjin marine district in Bohai Bay of Bohai Sea, China. Through the simulation and comparison of the reclamation impact, the basic hydrodynamics characteristic of research marine district could be obtained, that is very meaningful for marine reclamation management. The current field and residual current field have been changed by the marine reclamation. Multiple residual current vortexes have a disadvantage for the transformation of pollutant.

1 INTRODUCTION

The research of basic hydrodynamics characteristic of marine district is useful for managing the marine reclamation and protecting the marine environment. Many numerical models and methods are proposed for simulating and researching the marine hydrodynamics characteristic, such as the POM (Li et al. 2008), EFDC (Jiang et al. 2009) and ECOM (Wang et al. 2011), etc. In this paper, hydrodynamic model and residual current calculation equation are used to simulate and analyze the characteristic of research marine district, Tianjin marine district in Bohai Bay of Bohai Sea, China. The structure of this paper follows. In Section 2, the numerical and equation are introduced. The research marine district and the results are introduced in Section 3. Finally, the conclusions are presented in Section 4.

2 NUMERICAL MODEL

2.1 *Hydrodynamic model*

A hydrodynamic model named MIKE21 (Li et al. 2013) is employed for simulating the current field changes in this paper and triangle mesh technique is applied for close to complex coastline.

2.2 *Residual current calculation method*

The residual current is main hydrodynamics for pollutant transportation. In this paper, the residual current is calculated as follows:

$$\vec{U}_E = \frac{1}{T}\int_{t_0}^{t_0+T} \vec{U}(x_0,t)\,dt \tag{1}$$

where $\vec{U}_E$ is the residual current, $\vec{U}$ is the tidal current, T is tidal period.

3 HYDRODYNAMICS CHARACTERISTIC SIMULATION

3.1 *Research marine district*

In this paper, the hydrodynamics characteristic of Tianjin marine district in Bohai Bay is studied with, west part of Bohai Sea, China. The total area is about 15,900 km^2. It is very shallow and its averaged water depth is less than 20 m. The sea bottom is very flat and its mean slope is less than 2‰. The tide hydrodynamic is controlled by the tidal waves of the Bohai Sea, so the tidal current is complicated (Li et al. 2010).

3.2 *Coastline variation under marine reclamation*

With the development of national economy, especially in the coastal area, marine reclamation has led to coastline variation and environment transformation. Construction method of marine reclamation has caused the shoreline change sharply. Large-scale reclamation engineering of Tianjin, Caofeidian harbor and Huanghua harbor in Bohai Bay etc causes great changes of the Bohai Bay's coastline, artificial shoreline rate increase year after year.

3.3 *Results of tidal current modelling*

Through the simulation and calibration of numerical model, the tidal current field could be obtained. The results of tidal current modeling are shown in Figures 1 and 2.

3.4 *Results of residual current calculation in Bohai Bay*

Based on the tidal current, residual current could be calculated by Eq. 1. The residual current field of Bohai Bay are shown in Figures 3 and 4. Compared with the Figures 3 and 4, the difference of the influence of reclamation can be obtained.

Through the comparison of the Figures 3 and 4, the current velocity and current direction has been changed by the marine reclamation, especially in the nearshore.

Figure 1. Flood tide of Bohai Bay.

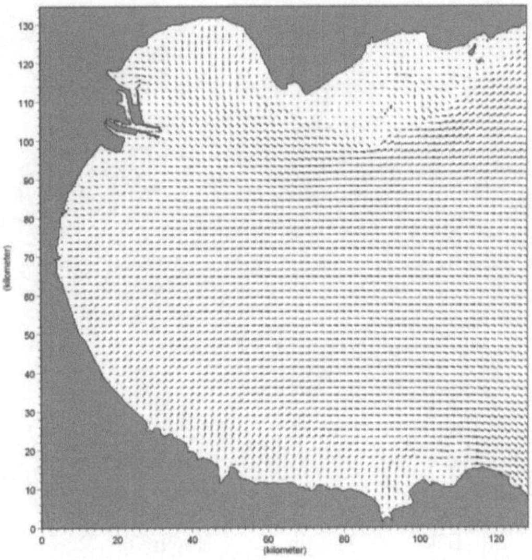

Figure 2. Ebb tide of Bohai Bay.

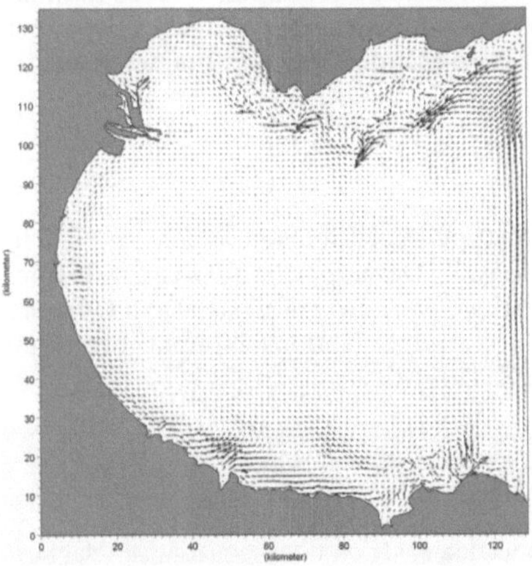

Figure 3. Initial residual current field of Bohai Bay.

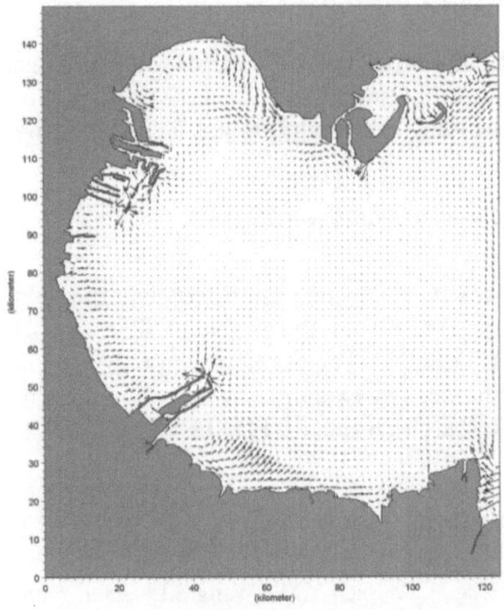

Figure 4. Residual current field of Bohai Bay after reclamation.

3.5 *Results of residual current calculation in Tianjin marine district*

The residual current field of Tianjin marine district are shown in Figures 5 and 6. Compared with the Figures 5 and 6, the difference of the influence of reclamation can be obtained.

Through the comparison of the Figures 5 and 6, the residual current has been changed by the marine reclamation, especially in the nearshore. Multiple residual current vortexes occur in the nearshore. A more scope residual current vortex flow clockwise, which has a disadvantage for the transformation of pollutant.

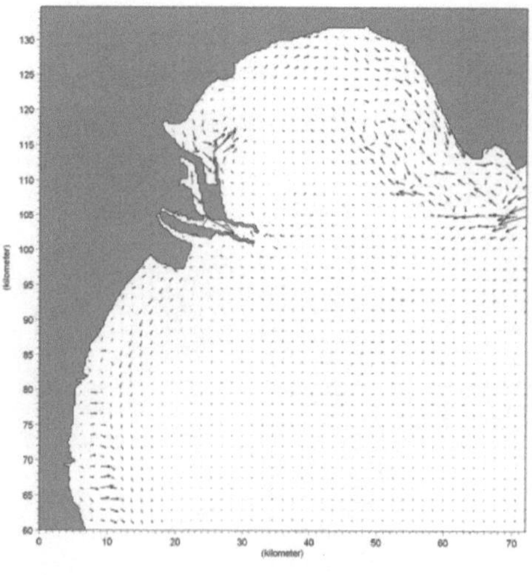

Figure 5. Residual current field of Tianjin marine district.

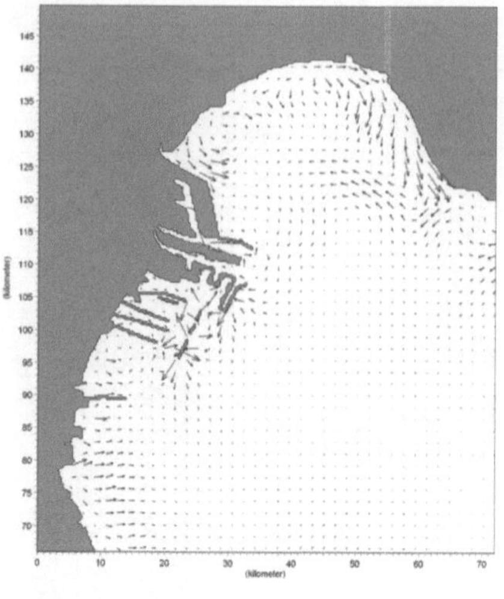

Figure 6. Residual current field of Tianjin marine district after reclamation.

4 CONCLUSION

In this paper, hydrodynamic model and residual current calculation equation are used to simulate and obtain the characteristic of research marine district. The numerical results show that the marine reclamation has a certain impact on the marine hydrodynamics characteristic.

ACKNOWLEDGEMENTS

This work was supported by the National Natural Science Foundation of China (No. 51209110), the project of Science and Technology for Development of Ocean in Tianjin (KJXH2011-17), the Research Foundation of State Key Laboratory of Coastal and Offshore Engineering Dalian University of Technology (LP1108) and the National Nonprofit Institute Research Grants of TIWTE (TKS130215 and KJFZJJ2011-01).

REFERENCES

Jiang H.Z., Shen Y.M., Wang S.D. 2009. Numerical simulation study on three-dimensional tidal flow and salinity in the Oujiang Estuary. *Chinese Journal of Hydrodynamics* 24(1):63–70.

Li M.C., Liang S.X., Sun Z.C. 2008. Study on optimization inversion method for tidal model open boundary condition. *Chinese Journal of Hydrodynamics* 23(6):646–654.

Li M.C., Liang S.X., Sun Z.C., et al. 2010. Optimal estimation for key parameters of the marine quality model using data-driven neural network. *Journal of Marine Science and Technology* 18(5):771–779.

Li M.C., Zhang G.Y., Si Q. 2013. Numerical modeling of the influence scope of marine oil spill with wind-driven. *Advanced Materials Research* 838–841:1861–1864.

Wang Y., Zhang G.Y., Li M.C. 2011. Numerical simulation of pollutant dispersion of ECOM-3D water quality model. *2011 International Conference on Remote Sensing, Environment and Transportation Engineering, Nanjing, 24–26 June 2011.*

Modeling and Computation in Engineering III – Zhang & Xie (Eds)
© 2014 Taylor & Francis Group, London, ISBN 978-1-138-02680-3

A new method to draw Rainfall-Runoff Correlation Diagram

H. Wan, J. Xia, W.H. Zhang & L.P. Zhang
State Key Laboratory of Water Resources and Hydropower Engineering Science, Wuhan University, Wuhan, Hubei Province, P.R. China

ABSTRACT: Flood forecasting has proved to be an effective and helpful non-structure method which can effectively reduce the impact of floods by providing early warnings ahead of time. Among many flood forecasting models, Rainfall-Runoff Correlation Diagram (RRCD) gives more accurate prediction by making use of large amount of history data and acting as experience expect, and thus is widely applied in operational flood forecasting. This paper proposed a new method to draw the RRCD through an equation instead of subjectivity. The method is then applied to Zhang River Reservoir with 126 flood events from historical meteorological and hydrological data from 1971 to 2000. Results show that this method gives performance as good as experience in more objective aspect and can be easily applied to consider the influence of rainfall hours cast on RRCD.

1 INTRODUCTION

Floods are one of the most frequent, destroying, long-time lasting and universal disasters. Direct loss are plunders of people's life and property. Indirect impacts are disease and confusion of the market and society. In order to minimize the damage of floods, flood forecasting from rainfall-runoff modeling is playing a vital role in flood mitigation and water management as a non-structural measure. A simple but efficient flood forecasting model, API model was first introduced by Sittner et al (1969) and gained its popularity in China soon in 1970s. After it adjusted to accommodate the situation in China, API model was developed into Rainfall-Runoff Correlation Diagram (RRCD) and was applied and verified in most part of China, especially in humid and semi-humid region. Zhang & Song (2005) revised the RRCD for Yuqiao Reservoir under the climate change and human impact. Tian (2010) compared RRCD with Unit Hydrograph and Saturation Excess Runoff Model in small to medium catchments in Dongjiang. Li (2003) proposed a new method to draw RRCD based on the theory of Saturation Excess Runoff Model and results of application in Heihe catchment showed a high accuracy.

Based on large amount of history information, RRCD acts as an expert to predict runoff with talented experience. However, the experience based work results in subjectivity with visual measurement since computer technology cannot draw a free correlation diagram. This paper proposed a new method to draw the RRCD through a simple but rational equation instead of subjectivity in previous works. Results are compared with observed 125 flood events data from Zhang River Reservoir.

2 METHOD

2.1 *Rainfall-Runoff Correlation*

Rainfall-Runoff Correlation Diagram (RRCD) the relationship between rainfall and runoff is nonlinear and was accepted and tested by many hydrologists (Minshall, 1960, Amorocho & Orlob, 1961, IGR, 1980, Nash & Brasi 1983, Singh, 1988 & Xia, 2002). When the Anticipant

Precipitation Index (API) is small, there will be less rainfall excess to generate runoff because more saturates the dry soil. Moreover, the runoff generates from the precipitation at the beginning of one flood process is small while as the soil moisture becomes wetter, the amount will be larger. The difference is mainly caused by soil moisture. While soil moisture is not easy measured, so API is often accepted as the substitution and calculated by:

$$P_{a,t+1} = K(P_{a,t} + P_t) \tag{1}$$

where $P_{a,t+1}$ and $P_{a,t}$ are Anticipant Precipitation Index (API) on the t-th and t+1-th day respectively [mm]; P_t is precipitation volume on the t-th day [mm]; K is daily depletion coefficient of soil water.

Then the RRCD can be presented as the following formulation proposed by this paper:

$$R = P^{-\beta(W_m - W_0)/P} \tag{2}$$

where W_0 is the API for one flood [mm]; W_m is the soil volume capacity [mm]; β is the coefficient; P is total precipitation volume [mm]; R is the total surface runoff depth [mm].

From equation 2, we can easily obtain when W_0 is small, i.e. the soil moisture at the beginning of the rainfall is little, the power $-\beta(W_m - W_0)/P$ will make $P^{-\beta(W_m - W_0)/P}$ small. While if W_0 is large, i.e. the soil moisture becomes larger, the power will make $P^{-\beta(W_m - W_0)/P}$ large, especially when $W_0 = W_m$, then $R = P$, i.e. after the soil is saturated all the rainfall will turn into runoff. Moreover, the denominator P in the power indicates the nonlinearity between rainfall and runoff.

2.2 *Goodness-of-fit indicators*

The absolute error (*ABE*) is used evaluate the performance of simulated runoff series compared with observed ones and is given by:

$$ABS = \sqrt{\sum_{i=1}^{n} (R(i) - \hat{R}(i))^2} \tag{3}$$

where $R(t)$ and $\hat{R}(t)$ are measured and simulated runoff depth [mm], respectively; n is data length.

According to the Standard for hydrological information and hydrological forecasting GB/T 22482-2008 (2008), the goodness-of-fit is indicated runoff permissible error, i.e. 20% of the observed runoff depth [mm], with a maximum 20 mm and minimum 3 mm. Then the qualification rate is calculated by:

$$QR = \frac{n}{m} \times 100\% \tag{4}$$

where QR is qualification rate, %; n is the amount of qualified predicted flood events; m is the total amount of flood events.

3 STUDY AREA

Zhang River, one of the east tributaries of Juzhang River which belongs to the Yangtze River, is selected as the research area in this paper. Zhang River flows from the headwater of 1400 m high to the downstream of 42 m, with a drainage area of 2980 km^2 and 202 km long main stream. It belongs to the subtropical monsoon region with scorching summer and a little warm winter. The multi-year average precipitation is approximately 993.5 mm from 1957 to 2008 with 1427.9 mm as the maximum amount in 1996 and 645.9 mm as the minimum

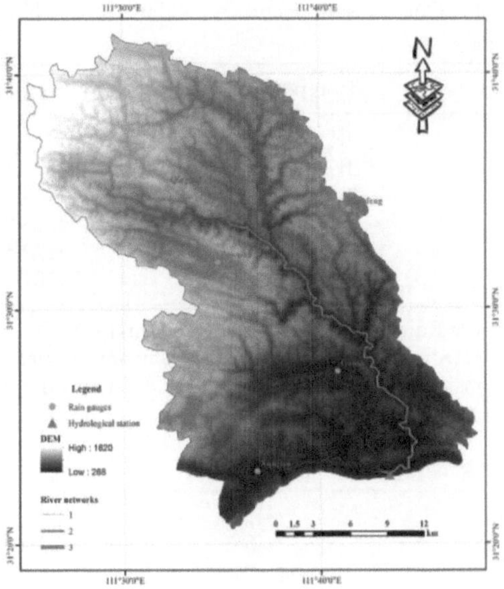

Figure 1. Location, DEM, river networks and distribution of rain gauges of Dagutai catchment above Zhang River Reservoir.

amount in 1972. There exists an extreme heterogeneity in the distribution of the precipitation through a year. 853.5 mm rainfall happens from April to October is about, accounting for 85% of annual volume. 356.6 mm occurs in July and August is about, accounting for one third of the annual amount, which is the main source contributing to large flood. The multi-year average evaporation volume is about 700~1000 mm, with monthly maximum 150 mm and minimum 16.5 mm.

Zhang River Reservoir (ZRR) was built in 1966, with catchment area about 2212 km^2 and main channel length about 171.0 km. ZRR is composed by two small ones, i.e. Guanyinshi and Jigongjian, between which there is a man-made semi-natural open channel. Water from Guanyinshi with drainage area 1957 km^2 and storage capacity about 0.8 billion m^3 flows through the open channel and concentrates in Jigongjian with drainage area 255 km^2 and storage capacity about 1.3 billion m^3. Downstream the ZRR are the Jingzhou Dyke, Jiaoliu Railway and Jingzhou City, which should be well protected from flood (Yu et al. 2004 & Luo et al. 2004). This paper takes Dagutai hydrological station which is located 75 km away from Guanyinshi Reservoir as an example with data from 1971 to 2000 to draw the rainfall-runoff correlation diagram. Detailed information, including the location, Digital Elevation Model (DEM) and river networks of Dagutai catchment can been seen in Figure 1.

4 RESULTS AND DISCUSSION

Correlation obtained in previous studies using visual measurement indicated that experiences resulted in arbitrariness and subjectivity. In our study, a simple but rational equation was presented considering the hydrological nonlinearity. Three rainfall information indicators, including Rain Hours (RH), Rain Intensity (RI) and Distance from the rainfall gravity to watershed out (Dis) were taken into consideration to improve the model accuracy. Simulated results were compared with observed runoff depth in Dagutai catchment to get the best combination. With $\beta = 1$ and $W_m = 110$ in equation 2, Table 1 listed the 6 results from 6 different combination scenarios. As expected, it can be easily seen from Table 1 that the considering W_0 and any other factors gave smaller errors than considering only W_0 with maximum 217.4 mm improved and minimum 35 mm. Considering W_0 and RH showed the

Table 1. Comparison between different combination scenarios.

Index	Components*	*ABE* (mm)
1	W_0	1453.5
2	W_0, RH	1236.1
3	W_0, RI	1418.5
4	W_0, Dis	1325.8
5	W_0, RH, Dis	1232.5
6	W_0, Dis, RH	1301.4

*RH is for Rain Hours. RI is for Rain Intensity. Dis is for Distance from rainfall gravity to catchment outlet. The order considered in this study is as the written order, from front to back.

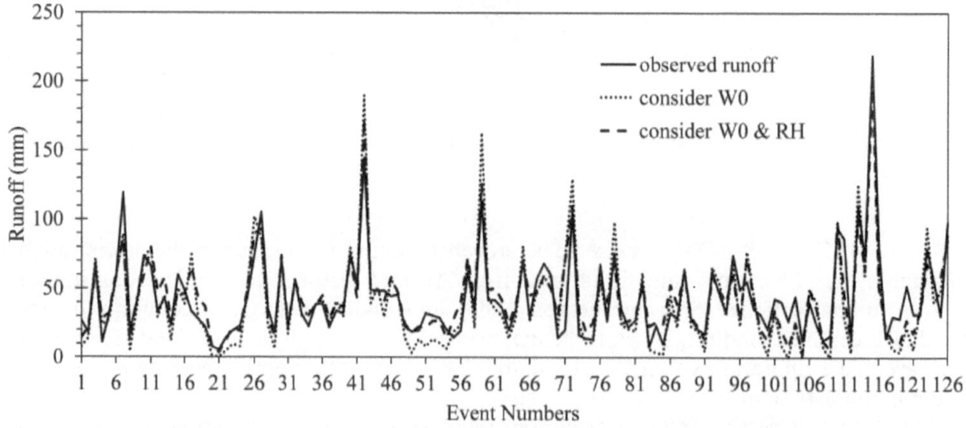

Figure 2. Performance comparison between observed runoff and simulated runoff considering only W_0 and both W_0 and Rain Hours (RH).

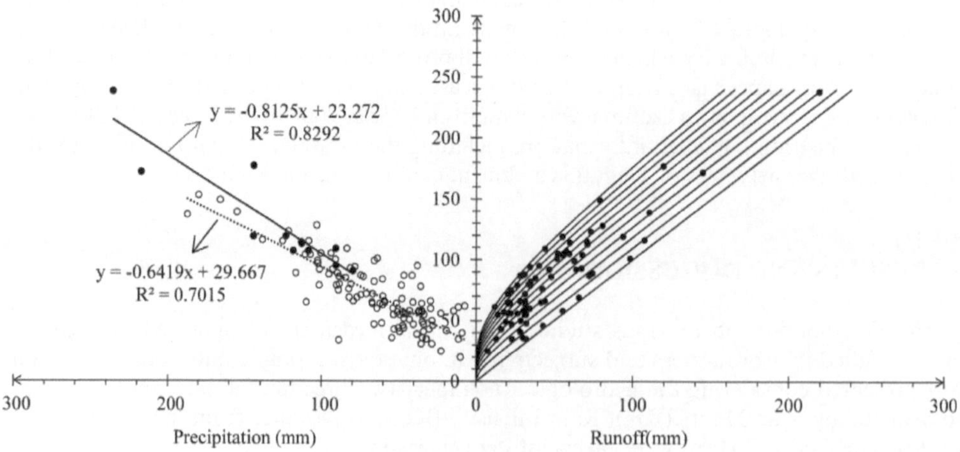

Figure 3. Rainfall-Runoff Correlation Diagram of Dagutai catchment. In the right quadrant, the 11 almost parallel lines are rainfall-runoff correlation from $W_0 = 10$ to $W_0 = 110$, from left to right respectively. In the left quadrant, hollow dots are events with rain hours less than 75 h while solid ones are that with rain hours larger than or equal to 75 h.

best combination with the ABE about 1236.1 mm, followed by considering W_0 and Dis with 1325.8 mm, then considering W_0 and RI with 1418.5 mm. This might be the caused by the closer link between flood and RH than that between RI. For higher accuracy, the combination of three components, including W_0, RH and Dis was also analyzed to demonstrate whether the order matters. As a result, an order W_0-RH-Dis was proved to the best choice with minimum *ABE* about 1232.5 mm, 68.9 mm smaller than W_0-Dis-RH. While compared with W_0-RH, W_0-RH-Dis showed a little enhancement with large amount additional work, which might not be the best choice.

Figure 2 gave vivid comparison between considering only W_0 and both W_0 and RH. Improvement can be seen clearly from the small floods. The qualification rate of RRCD considering W_0 and RH is 73.4%. And the final results of RRCD of Dagutai catchment was illustrated in Figure 3 which can be well applied in runoff generation in flood forecasting.

5 CONCLUSIONS

Prior work has demonstrated that Rainfall-Runoff Correlation Diagram (RRCD) plays a key role in operational flood forecasting in China. However, these studies obtained the correlation mainly based on visual measurement, resulting in arbitrariness and subjectivity. This paper put forward a new method to draw Rainfall-Runoff Correlation Diagram (RRCD) and applied to flood forecasting in Zhang River Reservoir. We could derive some conclusions as follows:

(1) The method was reasonable and effective not only from theoretical aspects but also from practical aspects to illustrate RRCD. With only one parameter, this approach was simple and flexible to embody the nonlinear relationship rainfall and runoff and gave accurate and satisfying prediction runoff more objectively.
(2) Fuzzy clustering analysis was helpful to divide 126 flood events into rational groups. Three additional factors besides Anticipant Precipitation Index (API), i.e. Rain Hours (RH), Rain Intensity (RI) and Distance from Rain gravity to the outlet (Dis) were compared to enhance the accuracy through different combination scenarios. A considering of API and then RH was proved to be best and easy combination to set up the operational flood forecasting model.
(3) Since hydrology especially flood forecasting is fundamentally dependent on long history series, previous Standard for hydrological information and hydrological forecasting which only consider qualification rate would not be rational enough due to poor prediction results but less risk from large amount data information. Moreover, the magnitude of data would vary much thus should be divided. A new reliability coefficient should be proposed to assess the goodness-of-fit of large information.

REFERENCES

Amorocho, J. & Orlob, G.T. 1961. *Non-linear analysis in hydrologic systems*. University of California, Sanitary Engineering Research Laboratory, Water Resources Centre. Publication 40.

Ge, S.X. 1999. *Modern Technology for Flood Forecasting*. Beijing: China Water Power Press. (in Chinese).

Institute of Geography Research (IGR). 1980. *Hydrological analysis and experiments, Special Issue of Geography*. Beijing: Science Press No.12.

Li, Y.J., Li, H.E. & Li, J.K. 2003. A method for drawing P-Pa-R correlation diagram using storage model. *Journal of Northwest Hydroelectric Power* 19(2): 5–7.

Luo, W.S., Yi, S.L., Yu, S., Li, L.H., Zhang, Q. & Song, X.Y. 2004. Research of anti-error forecasting project of Zhanghe Reservoir flood. *China Rural Water and Hydropower* 2004(12): 47–50. (in Chinese).

Minshall, N.E. 1960. Predicting storm runoff on small experimental watersheds. *Journal of the Hydraulics Division, Proceedings of the American Society of Civil Engineers* 86(HY8): 17–38.

Nash, J.E. & Brasi, B.I. 1983. A hybrid model for flow forecasting on large catchment. *Journal of Hydrology* 65: 125–137.
Singh, V.P. 1988. *Hydrological Systems: Rainfall-runoff Modeling*. Englewood Cliffs, New Jersey, USA: Prentice-Hall.
Sittner, W.T., Schauss, C.E. & Monro, J.C. 1969. Continuous hydrograph synthesis with an API-type hydrologic model. *Water Resources Research* 5(5): 1007–1022.
Standard for hydrological information and hydrological forecasting. 2008. GB/T 22482-2008.
Tian, Z.W. 2010. Comparison of different flood forecasting models in small-medium catchment in Dongjiang. *Guangdong Water Resources and Hydropower* 9: 55–57. (in Chinese).
Xia, J. 2002. *Hydrological nonlinear theories and approaches*. Wuhan: Wuhan University Press. (in Chinese).
Ye, S.Z. & Zhan, D.J. 2000. *Engineering Hydrology (the Third Edition)*. Beijing: China Water Power Press. (in Chinese).
Yu, S., Luo, W.S. S, X.Y. & Li, L.H. 2004. Research on forecasting scheme of Zhanghe Reservoir flood. *Water Resources and Power* 22(3): 33–36. (in Chinese).
Zhang, Q. & Song, Z.M. 2005. Research on the revise of rainfall-runoff correlation diagram in Yuqiao Reservoir. *Haihe Water Resources* 5: 44–45.

Modeling and Computation in Engineering III – Zhang & Xie (Eds)
© 2014 Taylor & Francis Group, London, ISBN 978-1-138-02680-3

New traits of rainfall change in the Taihu Lake Basin

P.X. Deng, Q.F. Hu, Y.T. Wang, X.Y. Chen & L.Z. Wang
State Key Laboratory of Hydrology-Water Resources and Hydraulic Engineering, Nanjing, Jiangsu, China

ABSTRACT: Change trends of various rainfall series during 1951~2011 in the Taihu Lake Basin were investigated. No statistically significant trends for the annual and flood season rainfall both for the whole basin and ninety-two gauges were tested. However, the rainfall time distribution within the flood season had significantly changed. The details are as follows: the monthly rainfall in May and September has a significant downward trend and the ratio of the rainfall in these two months to the total in the flood season declines while the ratio of the rainfall in June to August to the flood season total increases. This finding reflects a centralized tendency of precipitation temporal distribution in the flood season and has potential influence on water resources management and flood control of the study area. However, the cause remains to be further interpreted both from the aspect of global climate change and urbanization.

1 INTRODUCTION

Rainfall is the basic driving factor of the surface hydrological process. Under the background of global climate change and rapid urbanization, rainfall change and its impact on droughts and floods have raised widespread attention (Xu et al. 2009). The Taihu Lake Basin, located in the East Asian monsoon climate zone, is a typical flatland. The precipitation shows a spatial and temporal maldistribution, which results in much more floods and droughts. Since 2005, extreme hydrological events have occured frequently, such as in 2009 the serious flood caused by "Morakot" Typhoon and in 2011 the most serious drought in the last 60 years.

Many researchers have studied the precipitation changing characteristics over the Taihu Lake Basin. Wu et al. (2013) analyzed the tempo-spatial changes of the rainfall series in the past 61 years over the Taihu Lake basin and no statistically significant trends for annual and flood season rainfall were observed. Geng (2003) revealed that the annual rainfall in the Taihu Lake Basin decreased remarkably in the past ten years. Huang et al. (2009) also pointed that the rainfall in the last 50 years showed a slight increase in the Taihu Lake Basin because of the climate warming. Liu et al. (2011) showed that although the annual rainfall in the Taihu Lake Basin did not significantly increase, however, the maximum daily rainfall had a significant upward trend. His results indicated an enhancement of extreme precipitation events to some degree.

Most studies focus on the changing characteristics of rainfall amounts. However, flood and drought disasters not only correlate with the amounts, but also relate to temporal distribution. So, on the whole basin and point scale, we focus on the temporal distribution and analyze the changing trends of rainfall in the Taihu Lake basin aiming to provide the scientific basis for water resources utilization and flood control.

2 STUDY AREA AND DATA

2.1 *Study area*

The Taihu Lake Basin is located between 119°11′~121°53′E and 30°28′~32°15′N with a drainage area of 36895 km^2. The terrain is high in the west of hilly areas and low in the

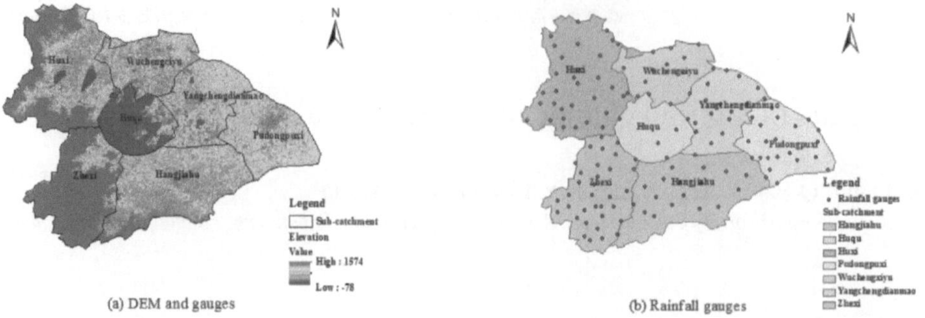

Figure 1. DEM, sub-catchment and rainfall gauges in the Taihu Lake Basin.

east plain (Fig. 1). The basin belongs to the typical subtropical monsoon climate zone with average annual precipitation about 1189 mm during the period of 1951~2011. Precipitation over the basin is seasonal uneven. Multi-year averaging rainfall in the flood season is 710 mm, amounting to about 60% of the annual total (Wang et al. 2013). According to the topography and river system, the Taihu Lake basin can be divided into 7 sub-regions, namely Huxi, Zhexi, Huqu, Hangjiahu, Wuchengxiy, Yangchengdianmao and Pudongpuxi.

2.2 *Study data*

Daily rainfall series from the whole basin and ninety-two gauges during 1951~2011 were collected, which were provided by the Taihu Lake Basin Authority. The series had undergone the quality control. Based on the daily series, the annual, flood seasonal, monthly and decadal rainfall series were obtained. Also, the ratio of monthly and decadal rainfall of the total in the flood season was calculated. Here, for description convenience, we use the symbols r_{m5}~r_{m9} to denote the ratio of monthly rainfall from May to September relative to the flood season total respectively. Besides, we use the symbols r_{x13}~r_{x27} to stand for the ratio of the rainfall in the thirteenth to twenty-seventh decade to the flood season total respectively and r_{m67} stands for the ratio of rainfall cumulation from June to July to the flood season total. Similarly, r_{m678} stands for the ratio of June to August and r_{m59} stands for the ratio of May and September.

3 RESEARCH METHOD

Based on the classic Mann-Kendall non-parametric test (MK) (Wei, 2007) and the Trend-Free Pre-Whitening Mann-Kendall non-parametric test (TFPW-MK) (Xu, 2007), changing trends of rainfall series in the Taihu basin were analyzed. These two methods have been widely adopted to test a tendency. The principle of the classic MK test is as follows:

For a time-series x_t with a sample size of n, a variable P is calculated:

$$P=\sum_{t=1}^{n} r_t,\quad r_i=\begin{cases}+1, & if\ x_i > x_j\\ 0, & if\ x_i \le x_j\end{cases}\qquad j=1,2,\ldots,i \tag{1}$$

where x_i and x_j are the sequential data values,. The mean, variance and standard normal variate are computed as respectively:

$$E=\frac{n(n-1)}{4},\quad V=\frac{n(n-1)(2n+5)}{72},\quad Z_{\alpha/2}=\frac{P-E}{V^{0.5}} \tag{2}$$

At α level of significance in two sided test for trend, the null hypothesis should be accepted if $|Z| < U_{\alpha/2}$ ($U_{\alpha/2}$ is the standard normal distribution at the $\alpha/2$ confidence level).

A positive value of Z indicates an upward trend and a negative value means a downward tendency.

Compared with the classic MK method, the TFPW-MK processes the data first to eliminate the positive relativity, and then to minimize the amplification of the significant trend. Some studies suggest that the TFPW-MK should be more effective compared with the classic MK (Bayazit et al. 2007). Pre-whitening steps of the TFPW-MK are stated as follows:

First, assuming the trend T_t in series X_t is linear, and removing the trend, we can get Y_t:

$$Y_t = X_t - T_t = X_t - \beta \cdot t \tag{3}$$

where β is the Mann-Kendall slope, X_t is the sequential data value. Subsequently, calculating the first-order auto-correlation coefficient (r_1) of Y_t, if r_1 is small, Y_t is considered to be independent. Thus, sequential data do not need pre-whitening. On the contrary, Y_t is auto-correlation. So, TFPW-MK is used to remove the auto-correlative items in the series. The independent sequence Y_t' is calculated as:

$$Y_t' = Y_t - r_1 \cdot Y_{t-1} \tag{4}$$

Then, combining the trend T_t with the independent sequence Y_t', we get a new sequence:

$$Y_t'' = Y_t' + T_t \tag{5}$$

Finally, the new sequence is adopted to test the significance in the MK method.

4 RESULTS AND DISCUSSIONS

4.1 *Variability on the whole basin scale*

4.1.1 *Inter-annual variation*

Statistics for rainfall series during 1951~2011, are given in Table 1, which includes the mean, maximum, minimum values etc. It is observed that the extreme ratios of rainfall series are 2.39~17.6 with significantly inter-annual variation. Take the annual rainfall for example, its maximum value is 1633.3 mm while the minimum value is only 682 mm. Besides, the mean value and the ratio of extreme values (E*r*) is 1189.6 mm and 2.39 respectively while the coefficient of variation (C_v) and the coefficient of skewness (C_s) is 0.16 and 0.18. The skewness of other rainfall overtakes the annual one obviously, especially in May and June when C_s reaches as high as 1.2.

Both the MK and TFPW-MK methods have been adopted to analyze the trend with a confidence level of 90% and consistent results are obtained. Only the rainfall amounts in May

Table 1. Statistics of rainfall series in the Taihu Lake Basin ($\alpha = 90\%$).

Statistics	May	June	July	August	September	Rainy season	Annual
Mean (mm)	114.3	185.9	157.6	140.7	117.2	711.0	1189.6
Maximum (mm)	308.0	609.2	355.0	319.8	298.6	1200.0	1633.3
Minimum (mm)	41.6	56.9	33.3	21.1	17.0	358.0	682.0
E*r*	7.4	10.7	10.7	15.2	17.6	3.35	2.39
C_v	0.43	0.48	0.46	0.48	0.56	0.23	0.16
C_s	1.28	1.89	0.44	0.67	0.56	0.49	0.18
Trend	↓	–	–	–	↓	–	–
Linear rates (mm/year)	–0.70	–0.17	0.62	0.72	–1.34	–0.20	–0.23
Confidence level	0.966	0.143	0.684	0.849	0.995	0.094	0.120

Note: "↓" represents downward; "↑" represents upward; "–" represents no significant trend (the same below).

and September have long-term significant downward trend with ten average linear rate –0.70 and –1.34 respectively (Fig. 2). For other rainfall series, no statistically significant trends were observed. The result reveals that rainfall in the early and late flood season declines. Hence, we should further investigate that if this condition brings about changes in the distribution of annual or seasonal precipitation.

4.1.2 *Intra-annual variation*

In the study area, rainfall in the flood season mainly caused by the Meiyu and typhoon weather system. Usually, rainfall in June and July is the richest in the flood season, accounting to 26.1% and 22.2% of the total respectively. Then it is followed by August, accounting for about 19.8%.

Further analysis of the ratio of rainfall in the months and different periods to the flood season total is taken to find whether there is a significant change for the temporal distribution. Figure 3 shows that r_{m59} shows a remarkable downward trend, whereas r_{m678} is on the contrary.

In order to obtain a better understanding of the trend, the MK and TFPW-MK were used to test the linear rates. Table 3 indicates that both the upward and downward trends of r_{m678} and r_{m59} passed the significant test with a high confidence level and, which the linear rate is up to 0.25% and –0.25% respectively. The rainfall increment in July and August contributes much to the increment of amounts in the period of June to August. Meanwhile, the increment of precipitation amounts in mid-August is the main part of the increment in August. Similarly, the reduction of precipitation in September contributed much to the decrease of the amounts in May to September. Also, the reduction in early and late September has a significant contribution to the reduction of the month.

4.2 *Variability on the point scale*

The number of rainfall gauges with long-term series is up to 92. Analyzing the temporal variability of precipitation on the point scale is intended to provide interpretations for the results at the whole basin scale.

At the confidence level of 90%, the MK and TFPW-MK methods were utilized to test the trends of the annual, rainy seasonal, monthly and decadal rainfall from the 92 gauges. The number of gauges with a significant increasing or decreasing trend is shown in Table 3.

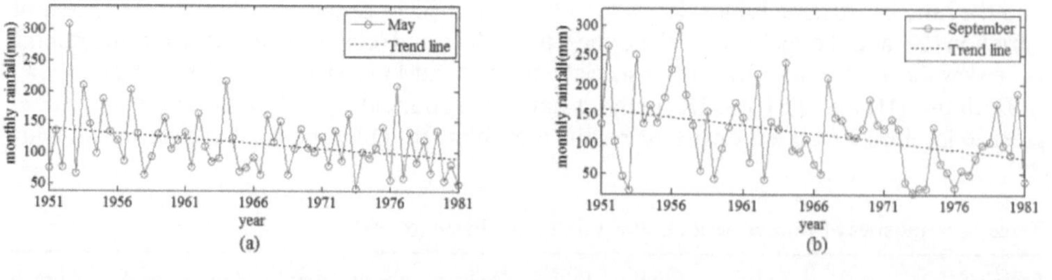

Figure 2. Annual rainfall in May and September from 1951 to 2011.

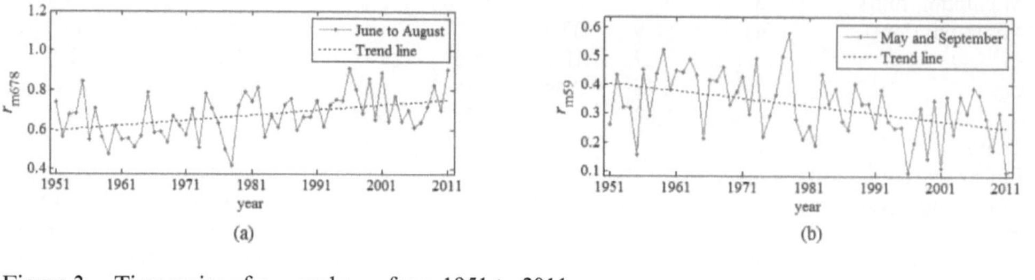

Figure 3. Time series of r_{m59} and r_{m678} from 1951 to 2011.

Table 2. Test results of rainfall proportion in the Taihu Lake Basin ($\alpha = 90\%$).

Ratio	Trend	Linear rates	Z	Confidence level	Ratio	Trend	Linear rates	Z	Confidence level
r_{m5}	↓	–0.08%	–1.96	0.950	r_{x13}	↓	–0.06%	–2.46	0.986
r_{m6}	–	0.03%	0.50	0.386	r_{x16}	–	–0.03%	–1.12	0.740
r_{m7}	–	0.10%	1.50	0.866	r_{x17}	–	–0.05%	1.12	0.740
r_{m8}	↑	0.12%	2.08	0.963	r_{x19}	–	0.04%	0.88	0.620
r_{m9}	↓	–0.18%	–2.94	0.997	r_{x23}	↑	0.05%	1.91	0.945
r_{m67}	↑	0.16%	1.91	0.945	r_{x24}	–	0.03%	1.01	0.670
r_{m678}	↑	0.25%	3.20	0.999	r_{x25}	↓	–0.08%	–2.17	0.970
r_{m59}	↓	–0.25%	–3.20	0.999	r_{x27}	↓	–0.02%	–1.67	0.906

Table 3. Number of stations with significant upward and downward trends ($\alpha = 90\%$).

Time scale	May	June	July	August	September	Rainy season	Annual
Number of gauges							
Upward	0	1	19	20	0	0	2
Downward	52	0	0	0	61	5	1

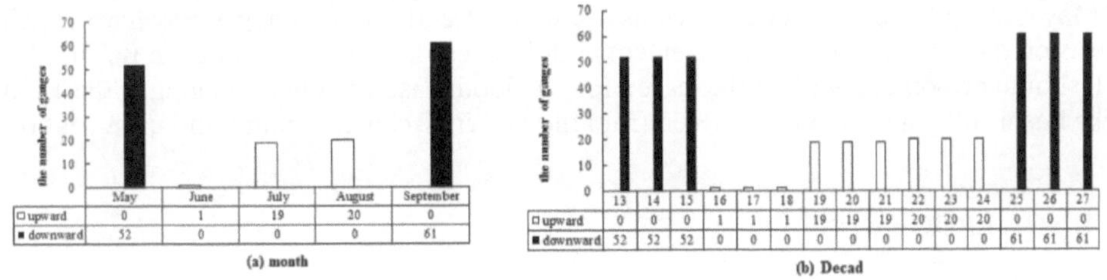

Figure 4. Number of gauges with a significant trend for monthly and decadal precipitations.

It indicates that in most of the sites no statistically significant trends for the annual and flood seasonal rainfall are tested, which resulting in the faint tendency of whole basin precipitation.

The number of sites with significant upward or downward trends for the monthly and decadal rainfall is shown in Figure 4. Rainfall at most sites shows a remarkable downward tendency in May and September. In these two months, there are respectively 52 and 61 rainfall sites showing a downward trend, and the decadal rainfall shows similar results. Nevertheless, in July and August, there are 19 and 20 gauges respectively showing a significant rising trend. As for the other months, few gauges with significant upward or downward trends are observed. As a result, the annual distribution of rainfall in the Taihu Lake Basin during the period of 1951 to 2011 changed significantly.

The test results for the ratio of monthly and decadal rainfall to the total in the flood season are shown in Figure 5. Compared with Figure 4, the number of r_{m8} with a significant upward trend increases from 20 to 28, so does the number of r_{x22}~r_{x24}. Similarly, the number of r_{m9} with significant downward trend adds from 61 to 83, so does the number of r_{x25}~r_{x27}. The sum of the r_{m7} and r_{m8} with a significant upward trend is only 50 without a dominant positions. But through test, the r_{m67} in most gauges has a significant upward trend and the number is up to 59. Meanwhile, almost all the values of r_{m678} appear significant upward trends. Accordingly, virtually all the values of r_{m59} decrease remarkably.

Therefore, at the point scale, the similar performance is found: The ratio of May and September rainfall to the total in the flood season declines, while it increases in June to August, which results in a centralized distribution tendency in the flood season.

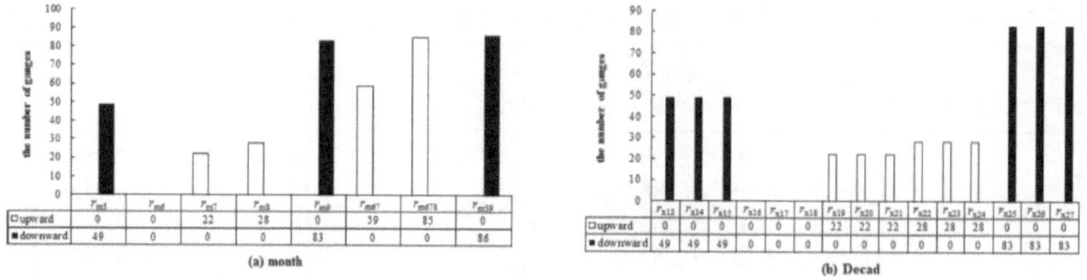

Figure 5. Number of gauges with a significant trend for proportion of monthly and decadal precipitations in flood season.

5 CONCLUSIONS

The long-term rainfall series from the whole basin and 92 rainfall gauges are used to analyze the inter-annual and intar-annual variations in the Taihu Lake Basin and new changing traits are found on the whole basin and point scale.

Most of the gauges showed that the ratio of monthly rainfall in May and September to the total in the flood season declines while it increases in June to August, which results in a significant change for the temporal distribution of annual rainfall. Meanwhile, owing to the variation characteristics of rainfall on the point scales, the ratio of rainfall between June and August shows an obvious increasing trend on the 90% confidence level, while the ratio of May and September shows an obvious decreasing trend. These changes may have certain effects on the water resources management and flood control of the Taihu Lake Basin.

It should be pointed out that the reason for the flood season rainfall changing in the Taihu Lake Basin still remains to be analyzed from the aspect of climate change and urbanization.

ACKNOWLEDGEMENT

National Nature Science Fund (51109136; 51109222).

Common Wealth Fund, Ministry of Water Resources, China (201001002:201201072-2).

REFERENCES

Bayazit M. & Önöz B. 2007. To pre-whiten or not to pre-whiten in trend analysis, *Hydrological Sciences Journal* 52(4): 611–624.

Gen, Y.Q. 2003. Analysis of spatial and temporal distribution of precipitation in the Taihu Basin, *Haihe Water Resources* (3): 33–35.

Huang, J.X. & Xu, Z.X. 2009. Spatial-Temporal characteristics of long-term trends for climate change in the Taihu Basin during 1954 to 2006. *Resources and Environment in the Yangtze Basin* 18(1):33–40.

Liu, Z.F. & Wang, X.H. & Yao, Z.J., et al. 2011. Trend and periodicity of precipitation, air temperature and runoff in the Taihu Lake Basin. *Journal of Natural Resources* 26(9): 1575–1584.

Xu, J.J. 2007. Application of a distributed hydrological model to the Yangtze River Basin. Beijing: *Tsinghua University*.

Xu, Y.P. & Ding, J.J. & Chen, Y. 2009. Impacts of urbanization on hydrology in the Yangtze River Delta, *Journal of hydraulic engineering* 4: 67–73.

Wang, Y.Y. & Han, S. & Yu, Z.Q., et al. 2013. The flood risk and flood alleviation benefit of land use management in Taihu Basin. *Journal of hydraulic engineering* 44(003): 327–335.

Wei, F.Y. 2007. Modern climate statistical diagnosis and prediction technology. Beijing: *China Meteorological Press*.

Wu, H.Y. & Wang, Y.T. & Hu, Q.F., et al. 2013. Tempo-spatial change of precipitation in Taihu Lake Basin during recent 61 years, *Hydrology* (2): 75–81.

Modeling and Computation in Engineering III – Zhang & Xie (Eds)
© 2014 Taylor & Francis Group, London, ISBN 978-1-138-02680-3

Measurement of climate complexity using permutation entropy

C.Y. Hao
College of Surveying and Land Information Engineering, Henan Polytechnic University, Henan Jiaozuo, China

ABSTRACT: Located in the southwest of Yunnan Province, Study area has a distinct regional difference in both topography and climate system. Western region belongs to the Southwestern Yunnan mountainous area with varied landforms while eastern region is a part of the Eastern Yunnan Plateau where the terrain is rather flat relatively. Different regions are influenced by different monsoon systems because the climate system is complicated and varied, which has resulted from the multiplicity of monsoon systems and the significance of great topographic effects. By a complexity parameter—permutation entropy for time series from 1971 to 2000 based on comparison of neighboring values, the tolerance to the systematic complexity of this kind of climate is simple and effectual.

1 INTRODUCTION

With the development of nonlinear science in recent years, the study on various kinds of complicated phenomenon complexity has already become the leading issue of each field no matter in social sciences or natural sciences (Hao, 2001). Although the researchers in meteorology have carried on multi-level studying to the nonlinearity and complexity in atmospheric sciences (Chou, 1997), the understanding of climate system still has important theoretical meaning and practical value because of its non-linearity, multi-level and forced dissipation (Li, 2004).

Entropy is originally a thermodynamics concept describing confused (disorder) degree of the thermodynamics system. It reflects the figure of micro-attitude or thermodynamics probability, and has direct proportion to the logarithm of thermodynamics probability. The higher thermodynamics probability indicates the more confused state, so the entropy is the measurement of the systematic complexity (Schwartz & Walczak, 2004). The conception and theory on entropy have been developed rapidly since the establishment of information theory. Among them, the permutation entropy (for short PE in the following words of this article) is a complexity parameter for time series based on comparison of neighbouring values. The advantages of this method are its simplicity, extremely fast calculation and practicality (Bandt & Pompe, 2002). By it, the disease or health degree is diagnosed from the complicated performance of heart or brain data (Pardo & Vovar, 2005).

Based on the climate physics elements for time series including daily temperature and daily precipitation from 1971 to 2000, this study attempts to measure the climate systematic complexity and makes its geography explanation in a typical region of complicated climate, the middle-south of Yunnan Province.

2 METHODS

For practical purposes, we recommend $n = 3 \ldots 7$.

Let us take a series with seven values: $x = (4, 7, 9, 10, 6, 11, 3)$.

We organize the six pairs of neighbours, according to their relative values, finding four pairs for which $x1 < xt + 1$ and two pairs for which $xt < xt - 1$. So four of six pairs of values are represented by the permutation 01 (xt > xt + 1) and two of six are represented by 10. We defined the PE of order n = 2 as a measure of the probabilities of the permutations 01 and 10. So,

$$H(2) = -(4/6)\log(4/6) - (2/6)\log(2/6) \approx 0.918 \tag{1}$$

As usual, log is with base 2, thus H is given in bit. Next, we compare three consecutive values. (4, 7, 9) and (7, 9, 10) represent the permutation 012 since they are in increasing order. (9, 10, 6) and (6, 11, 3) correspond to the permutation 201 since xt + 2 < xt < xt + 1 while (10, 6, 11) has the permutation type 102 with xt + 1 < xt < xt + 2. The PE of order n = 3 is

$$H(3) = -2(2/5)\log(2/5) - (1/5)\log(1/5) \approx 1.522 \tag{2}$$

So, the definition of PE is as followed. Consider a time {xt/t = 1, ..., T }. We study all n! permutations π of order n which are considered here as possible order types of n different numbers. For each π we determine the relative frequency

$$p(\pi) = \frac{\#\{t/t \le T - n, (x_{t+1}, \ldots, x_{t+n}) \text{ has type } \pi\}}{T - n + 1} \tag{3}$$

This estimates the frequency of π as good as possible for a finite series of values. To determine $p(\pi)$ exactly, we have to assume an infinite time series {x1,x2, ...} and take the limit for $T \to \infty$ in the above formula. This limit exists with probability 1 when the underlying stochastic process fulfils a very weak stationarity condition: for $k \leqq n$, the probability for xt < xt + k should not depend on t.

The PE of order $n \ge 2$ is defined as

$$H(n) = -\sum p(\pi)\log p(\pi) \tag{4}$$

Where the sum runs over all n! permutation π of order n. This is the information contained in comparing n consecutive values of the time series. It is clear that $0 \le H(n) \le \log n!$ where the lower bound is attained for an increasing or decreasing sequence of values, and the upper bound for a completely random system where all n! possible permutations appear with the same probability. The time series presents some sort of dynamics when $H(n) < \log n!$. Actually, in our experiments with chaotic time series, H(n) did increase at most linearly with n. So it seems useful to define the PE per symbol of order n, dividing by n – 1 since comparisons start with the second value:

$$h_n = H(n)/(n-1) \tag{5}$$

By definition of entropy, permutation entropy, as a parameter reflecting the nature of system, can measure systematic dynamic complexity. So, when applied in climate system, the larger entropy value, the higher the complexity and randomness of system is (Bandt & Pompe, 2002).

3 RESULTS

3.1 *The climatic characteristics of study area*

Located in the middle-south of Yunnan Province, the study area lies between 98°40′53″–106°11′33″E and 22°26′34″–24°27′35″N with a total area of 101 900 km2, including thirty county level administrative regions. In the west of study area, being part of the Southwestern Yunnan mountainous area, there are a series of longitudinal range-gorges including Laobie Mountain-Nandinghe River, Bangma Mountain-Lancangjiang River,

Wuliang Mountain-Babianjiang River (Amojiang River), and Ailao Mountain-Yuanjiang River (Honghe River). Among them, Ailao Mountain, with the highest peak exceed 3100 m, is a great mountain passed through from northwest to southeast in the middle part of study area, whose direction forms right angle with air current of Southwest monsoon. Comparatively, the east of study area is relatively gentle to rise and fall, belonging to the Eastern Yunnan Plateau, although it has been eroded by Yuanjiang River, Nanpanjiang River and their branches (Wang, 2002). The study area is influenced by two water vapour sources of tropical ocean (the Bay of Bengal and South Sea of China) at the same time, especially the western study area, who lies in southeast edge of Tibetan Plateau, is in the way of the southwest summer monsoon coming to China. In dry season (from November to April of next year), the study area is controlled by Plateau winter monsoon and the southern branch of westerlies, and a portion is influenced by East Asia winter monsoon, while in rainy season (from May to October), the study area was influenced by southwest summer monsoon and southeast summer monsoon. Namely, there is a obvious alternation of dry and rainy season (Qiang & Ju, 1998). In addition, the topography of study area is complicated and various, and its main longitudinal ranges and gorges are nearly right angle with main air currents. The climate of study area has remarkable particularity because of these factors mentioned above, and some synoptic climatic characteristics are exclusive for the whole country, no matter in winter or summer. The three meteorological front subjects in the world, including low latitude problem, great topographic effect problem and tropical ocean problem, appear remarkably in this region at the same time (Qin & Ju, 1997).

3.2 *Spatial pattern of PE*

By Statistica Neural Networks software that Stastisoft Company produced in the language named MATLAB 6.1 (MathWorks Inc. Copyright 1984–2000), the PEs of temperature and precipitation were calculated in Table 1, and their spatial distribution are as Figures 1 and 2.

From Figure 1 of temperature PE, we can see that the meteorological stations with larger entropies (larger than one) are distributed in the west side of Ailao Mountain while ones with smaller entropies (smaller than one) are mainly distributed in the east side of Ailao Mountain. According to the definition of PE, the complexity of temperature system in the west side of Ailao Mountain is higher than one in the east side of Ailao Mountain. There are two weakening processes of precipitation PE, including from west to east in whole study area and from middle-south to north or east in the east side of Ailao Mountain (in Fig. 2). So it concludes that there are two steam sources, and special precipitation spatial pattern of two directions. All these are tightly related with atmospheric circulation and the terrain of study area.

3.3 *Analysis*

In dry season, both the cold source over Tibetan Plateau and its southern warm seas (Zheng, 2003), including the Bay of Bengal and the Indian Ocean, make the temperature gradient point to the north. Then, this kind of thermal effect helps the presence of westerly in study area. Because of the dynamic effect of Tibetan Plateau, the westerlies form both northern branch and southern branch. As a result of these dynamic and thermodynamic effects of the Tibetan Plateau, the southern branch of westerlies trough becomes a main weather system influencing the study area in winter (Ma & Li, 2001). Because of the southern branch trough, the prevailing wind for half a year of winter mainly comes from the southwest. The southern branch of westerlies trough in winter brings a continental westerly wind, advecting from North African towards Saudi Arabia, Pakistan, India, Bangladesh, Burma, etc. This is a westerly air current with characteristic of dry-warm and extremely deep, often full of the interspaces from the ground to high air. It is a single dry-warm air mass coming from tropical continent that causes it sunny and dry, then forms the dry season.

In addition, the north cold air originated from the north often goes down in dry season, which can cause temperature drop, air pressure rise, cloudiness increase, or even frost. But the

Table 1. PEs of different meteorological stations in study area.

Stations	LONG.	Lat.	T-PEs	P-PEs
Zhenkang	24.06°N	98.96°E	1.0245	0.8701
Yongde	24.15°N	99.23°E	1.0022	0.8297
Cangyuan	23.15°N	99.26°E	1.0196	0.8402
Gengma	23.55°N	99.40°E	1.0143	0.8303
Shuangjiang	23.46°N	99.80°E	1.0166	0.8019
Lincang	23.95°N	100.21°E	1.0106	0.8129
Jinggu	23.50°N	100.70°E	1.0149	0.8344
Zhenyuan	23.88°N	100.88°E	1.0122	0.8000
Puer	23.03°N	101.28°E	1.0080	0.8194
Mojiang	23.43°N	101.71°E	1.0074	0.8050
Xinping	24.06°N	101.96°E	0.9935	0.7450
Yuanjiang	23.43°N	101.98°E	0.9847	0.7060
Lvchun	23.00°N	102.40°E	0.9792	0.8457
Honghe	23.36°N	102.43°E	0.9550	0.7268
Shiping	23.70°N	102.48°E	0.9883	0.7494
Yuanyang	23.16°N	102.75°E	0.9594	0.8759
Jianshui	23.61°N	102.83°E	0.9937	0.7315
Gejiu	23.38°N	103.15°E	0.9767	0.8151
Jinping	22.78°N	103.23°E	0.981	0.8798
Mengzi	23.38°N	103.38°E	0.9818	0.7656
Pingbian	22.98°N	103.60°E	0.9526	0.9210
Hekou	22.50°N	103.95°E	0.9578	0.8702
Qiubei	24.05°N	104.18°E	0.9891	0.8362
Wenshan	23.38°N	104.25°E	0.9706	0.7881
Yanshan	23.61°N	104.33°E	0.9675	0.8046
Maguan	23.02°N	104.41°E	0.9701	0.8900
Xichou	23.45°N	104.68°E	0.9584	0.8682
Malipo	23.13°N	104.70°E	0.9585	0.8450
Guangnan	24.06°N	105.06°E	9737	0.8137
Funing	23.65°N	105.63°E	0.9745	0.8237

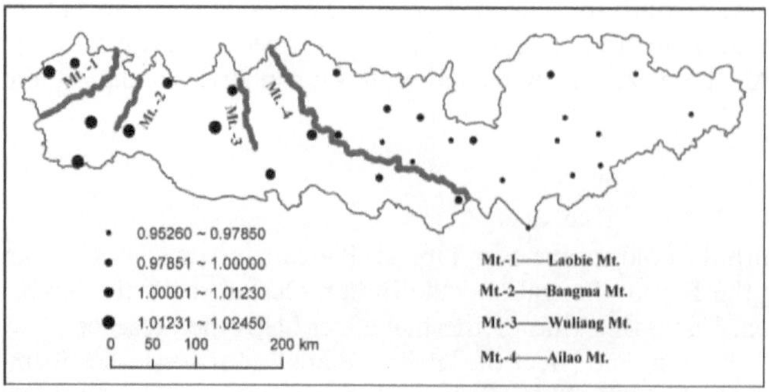

Figure 1. Spatial distribution of PE based on daily temperature series.

cold air becomes weaker and weaker influence on the weather of study area from north to south because of its long journey and crossing over the mountain after mountain, even transformation of atmosphere.

After the dry season, there are some basic characteristic of general atmospheric circulation, including the strengthening, stretching to northwest of Subtropical High and the reviving of India Low. Arrived in Indian Ocean of Northern Hemisphere, the southeast trade wind circulation rotates track to the northeast from the east of Indian Low towards India,

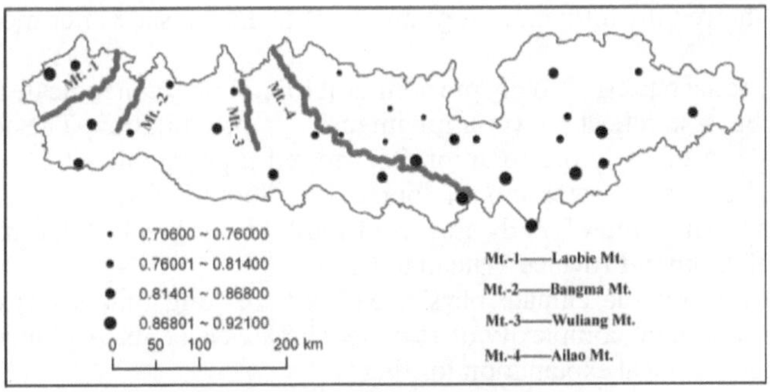

Figure 2. Spatial distribution of PE based on daily precipitation series.

Bangladesh, Burma, Yunnan Province, Tibet municipality, and Sichuan Province. This southwest air current refers to the named tropical summer monsoon or South Asia summer monsoon, and its direction is the same as the winter wind, southwester, too. But, it is characteristic of high temperature, large humidity, deep atmosphere and extremely ample vapour. Therefore, with its advance to Yunnan Province of China, the rainy season also begins and goes down to October. While sun radiates direct to south gradually, the southwest summer monsoon and its concerned tropical lows move slowly towards south. When the southern branch of westerly flow establishing, the south summer monsoon makes place to that tropical continent air mass (Chen & Li, 2000).

In conclusion, there are several different air current systems in different seasons because of special geographical location of study area. Both the Southern Asia continental air mass and Tibetan Plateau winter monsoon have influenced the study area remarkably in dry season while there is mainly south summer monsoon in rainy season. And their nearly same flowing directions have taken on a weakening trend of temperature PE from west to east. With characteristic of cold and dry, East Asia winter monsoon influences less and less from north to south in the east side of Ailao Mountain (Qin & Ju, 1997). So, there is an obvious decreasing trend from north to south. In rainy season, there are the Southwest summer monsoon and Southeast summer monsoon. But their flowing directions are different, one from the southwest and the other from the southeast, which forms two weaking processes of precipitation PE, i.e. from both southwest and southeast. In addition, owing to topographic effects, four longitudinal ranges including Laobie Mountain, Bangma Mountain, Wuliang Mountain and Ailao Mountain, have obvious barrier functions on main air currents, whose direction is nearly right angle with these ranges, and make the climatic character varied greatly too. In temperature aspect, the steady temperature in the east side of Ailao Mountain resulted from less influence by southwest wind in dry season makes the temperature PE smaller generally than in the west side of Ailao Mountain. In precipitation aspect, four longitudinal ranges formed a weakening trend of precipitation PE, namely the complexity of precipitation system is weakening from west to east. It is because of the complexity of several air current systems and great topographic effects that both daily temperature and precipitation PEs have demonstrated the unique spatial patterns in study area. In a word, PE based on the climate physics elements for long time series can reflect the systematic spatio-temporal complexity of regional climate elements, and get a good explanation for them.

4 CONCLUSIONS

Firstly, the spatial pattern of daily temperature PE can reflect the systematic complexity of regional general atmospheric circulation and the significance of great topographic effects. The weakening trend and its intensity of daily temperature PE indicated the remarkable barrier functions of four longitudinal ranges in study area. The decreasing process from north to south of PE in the east side of Ailao Mountain not only reflected the great topographic effects, but also showed the influence range and level of East Asia winter monsoon in study area.

Secondly, the spatial pattern of daily precipitation PE not only can indicate the direction of vapour flowing, but also reflect the variation intensity of precipitation. The decreasing process from west to east of daily precipitation PE showed topographic effects of longitudinal ranges and the direction of water vapour source. The South Sea vapour source affects less and less on the eastern region of study area from south to north, which has relation with the smooth terrain of Yunnan Province Plateau in China.

Thirdly, PE based on the climate physics element for long time series can reflect the systematic spatio-temporal complexity of regional climate elements. And more importantly, there is a good geographical explanation for them.

REFERENCES

Bandt, C. & Pompe, B. 2002. Permutation entropy: a natural complexity measure for time series. Physical Review Letters (88):174102.

Chen, L.J. & Li, W. 2000. On the process of summer monsoon onset over East Asia. Climatic and Environmental Research (4):345–355.

Chou, J.J. 1997. Advances of nonlinear and complexity in atmospheric sciences. Bulletin of the Chinese Academy of Sciences (5):325–329.

Hao, B.L. 2001. Characterization of complexity and the science of complexity. Physics 51(3):1–11.

Li, K. 2004. The explorations on the complexity of the behaviors of the properties and states of climate system. Scientia Meteorologica Sinica 24(1):95–103.

Ma, L. & Li, X.F. 2001. The forming and forecasting of winter snow-disaster in eastern pastoral-area of Qinghai-Xizang Plateau. Plateau Meteorology 20(3):325–331.

Pardo, E. & Tovar, F.J.R. 2005. MAXENPER: a program for maximum entropy spectral estimation with assessment of statistical significance by the permutation test. Computers & Geosciences (31):555–567.

Qiang, X.M. & Ju, J.H. 1998. A diagnostic analysis of the summer monsoon in Yunnan. Journal of Yunnan University (Natural Sciences) 20(1):75–79.

Qin, J. & Ju, J.H. 1997. Weather & climate over low latitude plateau. Beijing: China Meteorological Press.

Schwartz, T.U. & Walczak, R. 2004. Circular permutation as a tool to reduce surface entropy triggers crystallization of the signal recognition particle receptor β subunit. Protein Science (13):2814–2818.

Wang, S.Y. 2002. Geography of Yunnan. Kunming: Yunan Nationality Press.

Zheng, D. 2003. The formation, environment and development of the Tibetan Plateau. Shijiazhang: Hebei Science & Technology Press.

Modeling and Computation in Engineering III – Zhang & Xie (Eds)
© 2014 Taylor & Francis Group, London, ISBN 978-1-138-02680-3

Study on the advanced denitrification in biological aerated filter upgrading

Qi Lin & Fan Chen
Appraisal Center for Environment and Engineering, State Environment Protection Administration, Beijing, China

ABSTRACT: Some municipal wastewater treatment plant using biological aerated filter as its main process for upgrading. The pre-denitrification and post-denitrification experiment studies were carried out by a pilot equipment with 2 m^3/d designed capacity, and the parameters of each process unit in the studies were optimized. For pre-denitrification, the reflux ratio was 125%, the hydraulic retention time was 20 min, the CN filter and N filter gas-water ratio were 4:1 and 3:1 respectively. For post-denitrification, the hydraulic retention time was 45 min, the CN filter and N filter gas-water ratio were 6:1 and 5:1 respectively, and the methane dosage was 30 mg/L. Based on construction and operation cost analysis, the pre-denitrification upgrading plan was proposed.

1 INTRODUCTION

The wastewater treatment plant with national wastewater discharge standard Grade II (GB 18918-2002) of wastewater discharge applied the treatment process of high-density sedimentation sank and biological aerated filter. The function of aerated grit chamber and inclined-tube sedimentation tank which make up high-density sedimentation sank was to remove the sediment and suspended solids inorganic particles of wastewater, the biological aerated filter was to remove the organic pollutants and ammonia nitrogen (Comeau Y. 2007). According the discharge standard of pollutants for municipal wastewater treatment plant which the national standard of China issued, upgrading reconstruction is necessary for the plant to achieve the standard Grade I (GB 18918-2002) in the field of suspended solids, chemical oxygen demand, ammonia nitrogen, total nitrogen and total phosphorus (Paredes D. 2007).

Comparing the standard Grade II, the existing process of the wastewater treatment plant lacked the ability of nitrogen and phosphorus removal, and need to treat the suspended solids and organic pollutants advanced. In terms of nitrogen removal, pre-denitrification and post-denitrification were carried out (Jie Zhang. 2007).

2 MATERIALS AND METHODS

Designs on the whole process treatment for the industrial combined wastewater with the pilot of 2 t/d flow. The category of coagulant which aimed at the wastewater of the plant was determined, and the dosing position of coagulant before sedimentation sank or biological aerated filter were both researched. The final program of chemical phosphorus removal was determined by the contrastive parameters of dosage and other operational ones related. The hydrolysis acidification experiment used the perspex reactor with 3.5 m height, 0.32 m diameter and 2 t/d flow which hold anaerobic sludge. The wastewater was injected by diaphragm pump from bottom, passed through the sludge layer and ejected from water layer (Hong Yang. 2007).

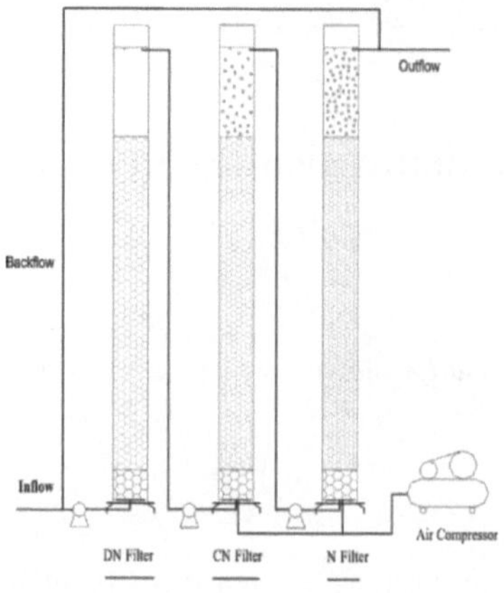

Figure 1. Pre-denitrification experiment.

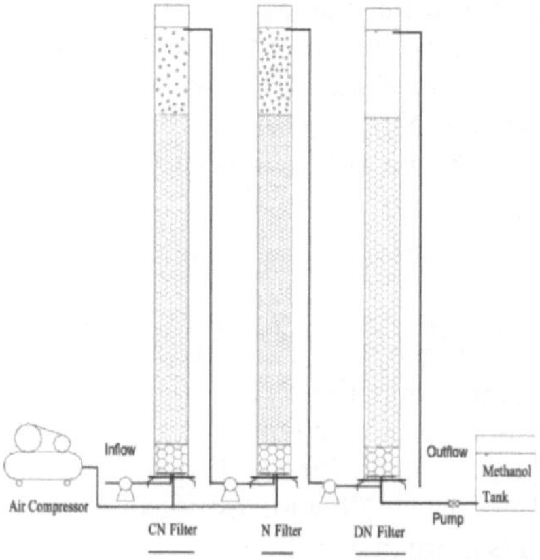

Figure 2. Post-denitrification experiment.

Denitrification combined with biological aerated filter of the wastewater treatment plant, the pre-denitrification and post-denitrification were both studied. The experiment of pre-denitrification used three-class perspex reactor with 4.3 m height, 0.32 m diameter and 2 t/d flow, including 0.3 m height supporting layer, 3.0 m height filter layer and 1.0 m height water layer (Fig. 1). The three-class reactor were Denitrification filter (DN filter), Carbon and Nitrification filter (CN filter) and Nitrification filter (N filter) respectively, the CN filter and N filter were aerated by air compressor. The wastewater was injected by diaphragm pump from bottom of filter, the filter material were lava, and the diameter of three filter were 6 mm–8 mm, 4 mm–6 mm and 3 mm–5 mm respectively. The experiment of post-denitrification used dosing methanol to denitrification, the wastewater inflew the CN filter, N filter and DN filter successively (Fig. 2).

3 RESULTS AND DISCUSSION

The main control parameters of pre-denitrification were reflux ratio, hydraulic retention time and aeration rate (Franco A. 2006). The parameters were optimized by studying on the outflow concentration of COD, TN, NH_3-N and DO.

Reflux ratio which was the most important parameter influenced the effect of total nitrogen removal directly (P.C. Sabumon. 2007). According the calculation of BOD and nitrification loading and verification of COD loading, under the condition that the single filter hydraulic retention time was 45 min and the gas-water ratio was 5:1, the outflow quality can achieve the standard Grade I. Studied on the reflux ratio in range of 50% to 250%, when it was increasing from 50% to 150%, the outflow concentration of TN was reducing, and the removal rate was increasing (Fig. 3). When the reflux ratio was lower, because of nitrate as electron acceptor lacking, the denitrification rate was hindered. When the reflux ratio was higher, the concentration of nitrate was enough to be electron acceptor, and the inflow organic was to be electron donor, the denitrification was completed without dosing external carbon source. Otherwise when the reflux ratio increasing from 150%, the outflow concentration of TN was not reducing anymore, and when the ratio increasing to 200% the removal rate was reducing. On the one hand, with the concentration of nitrate increasing, the denitrification was hindered by lacking of carbon source. On the other hand, with the reflux ratio increasing, the inflow concentration of DO which can be electron acceptor was also increasing, it can hinder the nitrate reduction and restrain the formation of nitrate reductase. For the high reflux ratio and hydraulic retention time determined the large structure capacity, considering the actual upgrading reconstruction, studying on the reflux ratio of 100%, 125%, 150%, 175% and the outflow concentration of TN variation with hydraulic retention time furthermore (Pathak. 2007).

Under the same condition of reflux ratio, the outflow concentration of TN was reducing with the hydraulic retention time increasing, the rate of microorganism removing substrate is higher (Fig. 4). Otherwise when the hydraulic retention time increasing to 20 min, the outflow concentration of TN and the removal rate was increasing slowly, and the structure capacity was also enlarging. To ensure the concentration of TN less than 15 mg/L, determined that reflux ratio was 125% and the hydraulic retention time was 20 min, at the same time the concentration of TN was 12.74 mg/L and the removal rate was 67%.

Dissolved oxygen was the important factor which maintained the metabolism of aerobic micro-organism, for the biological aerated filter, the supply of dissolved oxygen that

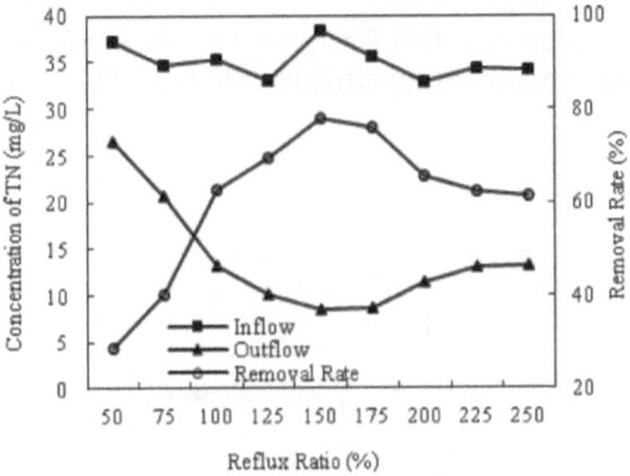

Figure 3. Inflow and outflow concentration of TN and removal rate with different reflux ratio.

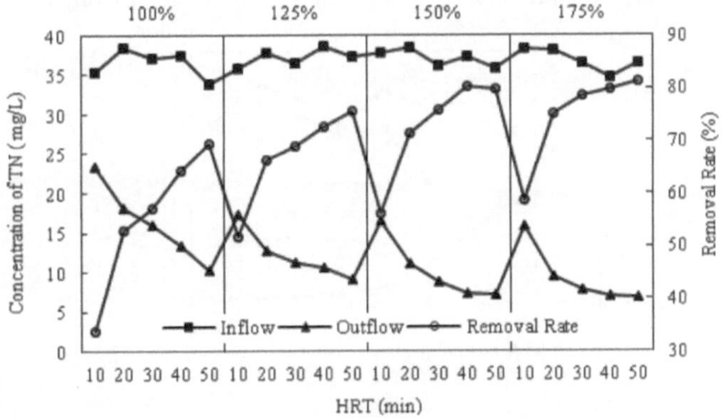

Figure 4. Inflow and outflow concentration of TN and removal rate with different reflux ratio and HRT.

the aeration rate of air compressor was the main energy consumption (Abma W.R. 2007). The lower aeration rate would reduce the ability of metabolism, the higher one would waste the power resources and cause the activity of microorganism too strong to oxygenolysis the biofilm when lack of nutrition supplying. Studied on the CN filter outflow concentration and removal rate of COD, and optimized the parameter of aeration rate. With the aeration rate increasing, the outflow concentration of COD was reducing, and the removal rate was increasing (Fig. 5). When it increased to 0.8 m^3/h, the change of two characteristics was slow, at the same time outflow concentration of DO was 2.5 mg/L and the gas-water ratio was 4:1. When the CN filter outflow which concentration of DO was higher flew into N filter, under the condition of low aeration rate, the removal rate of NH_3-N would remain higher (Fig. 6). Similar with the change of concentration of COD, the N filter outflow concentration of NH_3-N was reducing with the aeration rate increasing. To achieve the GB code first-level, determined the aeration rate of N filter was 0.6 m^3/h, at the same time outflow concentration of DO was 3.0 mg/L and the gas-water ratio was 3:1.

In the process of post-denitrification, inflow COD and NH_3-N were removed in the CN filter and N filter. For the two aerobic filters, hydraulic retention time and aeration rate were the important control parameters. Inflow COD and part of NH_3-N were removed in the CN filter, the outflow concentration of COD was reducing with the HRT increasing, when it increased to 40 min the concentration of COD reduced to 50 mg/L and changed slowly (Fig. 7). To achieve the GB code first-level, determined the HRT was 45 min. Similar with N filter, to ensure the concentration of NH_3-N less than 5 mg/L, determined that the HRT was 45 min (Fig. 8). To maintain the metabolism of aerobic microorganism of CN filter and ensure the outflow concentration of COD less than 50 mg/L, determined that the aeration rate was 0.5 m^3/h and the gas-water ratio was 6:1 (Fig. 9). To ensure the N filter outflow concentration of NH_3-N less than 5 mg/L, determined that the aeration rate was 0.4 m^3/h and the gas-water ratio was 7:1 (Fig. 10).

The outflow of N filter flew into DN filter, for the concentration of organic was low, it must dose carbon source to remove total nitrogen (Sinha B. 2007). The dosage of methanol

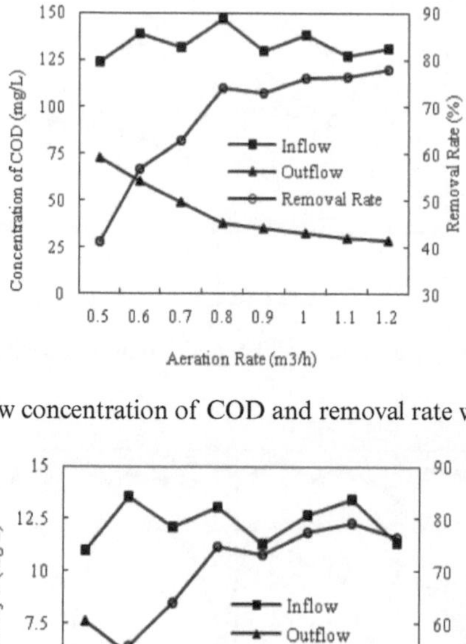

Figure 5. Inflow and outflow concentration of COD and removal rate with different aeration rate.

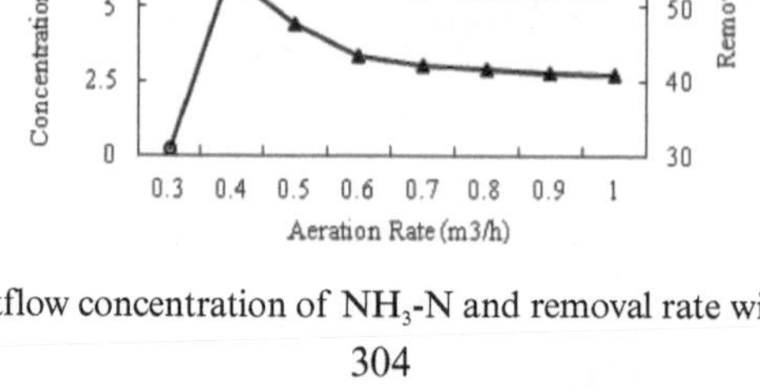

Figure 6. Inflow and outflow concentration of NH_3-N and removal rate with different aeration rate.

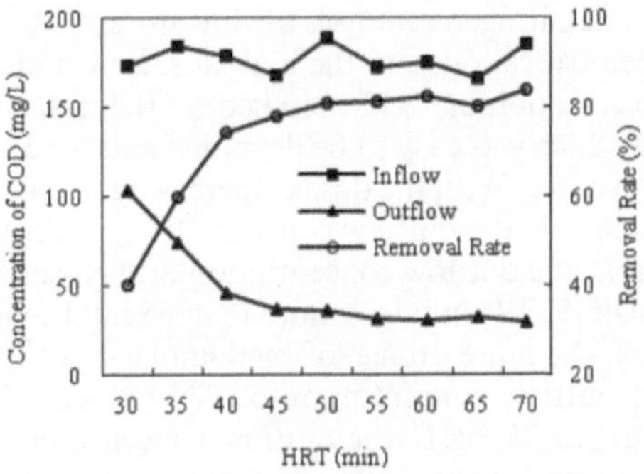

Figure 7. Inflow and outflow concentration of COD and removal rate with different HRT.

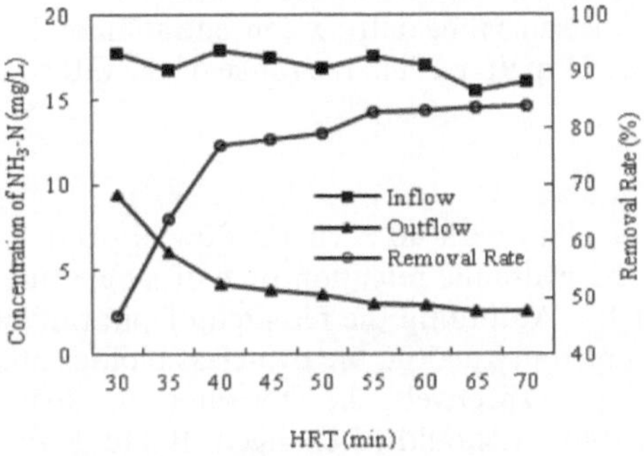

Figure 8. Inflow and outflow concentration of NH_3-N and removal rate with different HRT.

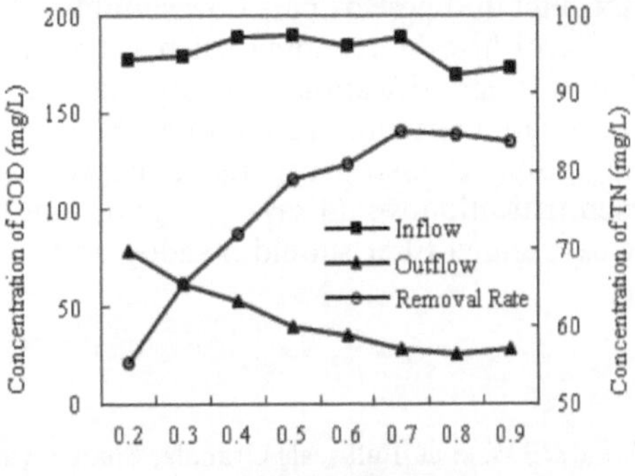

Figure 9. Inflow and outflow concentration of COD and removal rate with different aeration rate.

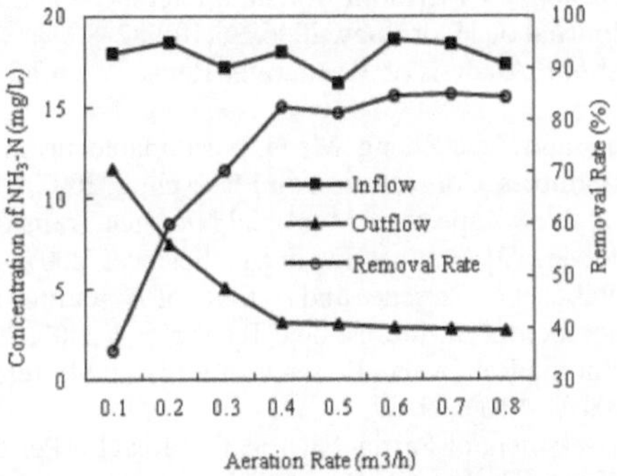

Figure 10. Inflow and outflow concentration of NH_3-N and removal rate with different aeration rate.

determined the effect of total nitrogen removal, the low dosage would cause lacking of carbon source to complete denitrification, and the high dosage would increase the economical cost and the outflow concentration of COD. Studied on the dosage of post-denitrification and selected the methanol which was easy to be degraded and used by denitrification microorganism to be the carbon source. With the dosage increasing, the inflow carbon source used by denitrification was increasing, the outflow concentration of TN was reducing. When the dosage increased to 25 mg/L, the outflow concentration of TN was less than 15 mg/L in the range of the GB code first-level. When it was more than 35 mg/L, with the concentration of nitrate and nitrite reducing, the more dosage of methanol cannot increase the reaction rate of denitrification and the outflow concentration of TN reduced slowly. When the dosage was in the range of 5 mg/L to 35 mg/L, the outflow concentration of COD also changed inconspicuously, after increasing to 40 mg/L, the methanol cannot be used by denitrification completely and caused the outflow concentration of COD increasing. Considering the influence of dosage to the outflow concentration of TN and COD, determined the dosage of methanol was 30 mg/L, at the same time outflow concentration of TN was 9.46 mg/L and the concentration of COD was 33 mg/L both in the range of the GB code first-level.

4 CONCLUSIONS

The wastewater treatment plant with 40×10^4 t/d flow applied the process of two-class biological aerated filter, the hydraulic retention time of single filter was 45 min, the gas-water ratio were 3:1 and 4:1. According the research of pilot, if post-denitrification was adopted in the upgrading recon-struction, the two-class biological aerated filter were used to be CN filter and N filter respectively, the DN filter and dosing station of methanol must be built and the air-blower should be changed. If pre-denitrification was adopted, through calculating the 125% reflux ratio and 20 min hydraulic retention time, the two-class biological aerated filter were also used to be CN filter and N filter respectively and left a bit of remainder, the DN filter and pipe of backflow should be built, the pump and air-blower were also changed. According the economic calculation, the pre-denitrification was 2 million dollars more than post-denitrification in construction cost, but it was 2.5 million dollars less than post-denitrification in annual running cost because of no dosing methanol. About 1 year later after upgrading reconstruction, the total investment of two technologies were equal, and the pre-denitrification would save a large of annual running cost. So the pre-denitrification biological aerated filter should be adopted to the nitrogen removal of upgrading reconstruction.

REFERENCES

Abma W.R., Schultz C.E., Mulder J.W. et al. Full-scale Granular Sludge Anammox Process [J]. Water Science and Technology, Biofilm Systems VI, 2007, 55(8–9): 27–33.

Comeau Y, Oldham W.K, Hall K.J. Dynamics of Carbon Reserves in Biological Dephosphatation of Wastewater, in: Biological Phosphate Removal from Wastewaters [M]. Pergamon Press, Oxford, 2007, 39–55.

Franco A. Montalto. The hydrology of Piermont Marsh, a reference for tidal marsh restoration in the Hudson river estuary [J]. Journal of Hydrology, 2006, 36(10): 82–86.

Hong Yang, Karim C. Abbaspour. Analysis of Wastewater Reuse Potential in Beijing [J]. Desalination, 2007, 212(1–3): 238–250.

Jie Zhang, Xiang-Sheng Cao and Xue-Zheng Meng. Sustainable urban sewerage system and its application in China [J]. Resources, Conservation and Recycling, 2007, 51(2): 284–293.

Paredes, D. Kuschk, Peter, et al. New Aspects of Microbial Nitrogen Transformations in the Context of Wastewater Treatment-A Review [J]. Engineering in Life Sciences, 2007, 7(1): 13–25.

Pathak, Bipin K. Kazama, Futaba, et al. Presence and Activity of Anammox and Denitrification Process in Low Ammonium-fed Bioreactors [J]. Bioresource Technology, 2007, 98(11): 2201–2206.

Sabumon P.C. Anaerobic Ammonia Removal in Presence of Organic Matter: A Novel Route [J]. Journal of Hazardous Materials, 2007, 149(1): 49–59.

Sinha B., Annachhatre A.P. Assessment of Partial Nitrification Reactor Performance through Microbial Population Shift Using Quinone Profile, FISH and SEM [J]. Bioresource Technology, 2007, 98(18): 3602–3610.

 ISBN 978-1-138-02680-3

Author index